APPROACHES TO THE
QUALITATIVE THEORY OF ORDINARY DIFFERENTIAL EQUATIONS Dynamical Systems and Nonlinear Oscillations

PEKING UNIVERSITY SERIES IN MATHEMATICS

Series Editor: Kung-Ching Chang *(Peking University, China)*

Vol. 1: An Introduction to Finsler Geometry
    by Xiaohuan Mo (Peking University, China)

Vol. 2: Numerical Methods for Exterior Problems
    by Ying Lung-An (Peking University & Xiamen University, China)

Vol. 3: Approaches to the Qualitative Theory of Ordinary Differential Equations: Dynamical Systems and Nonlinear Oscillations
    by Ding Tongren (Peking University, China)

Peking University Series in Mathematics — Vol. 3

APPROACHES TO THE
QUALITATIVE THEORY OF ORDINARY DIFFERENTIAL EQUATIONS
Dynamical Systems and Nonlinear Oscillations

Ding Tongren
Peking University, China

NEW JERSEY · LONDON · SINGAPORE · BEIJING · SHANGHAI · HONG KONG · TAIPEI · CHENNAI

Published by

World Scientific Publishing Co. Pte. Ltd.
5 Toh Tuck Link, Singapore 596224
USA office: 27 Warren Street, Suite 401-402, Hackensack, NJ 07601
UK office: 57 Shelton Street, Covent Garden, London WC2H 9HE

British Library Cataloguing-in-Publication Data
A catalogue record for this book is available from the British Library.

Peking University Series in Mathematics — Vol. 3
APPROACHES TO THE QUALITATIVE THEORY OF ORDINARY
DIFFERENTIAL EQUATIONS
Dynamical Systems and Nonlinear Oscillations

Copyright © 2007 by World Scientific Publishing Co. Pte. Ltd.

All rights reserved. This book, or parts thereof, may not be reproduced in any form or by any means, electronic or mechanical, including photocopying, recording or any information storage and retrieval system now known or to be invented, without written permission from the Publisher.

For photocopying of material in this volume, please pay a copying fee through the Copyright Clearance Center, Inc., 222 Rosewood Drive, Danvers, MA 01923, USA. In this case permission to photocopy is not required from the publisher.

ISBN-13 978-981-270-468-9
ISBN-10 981-270-468-X

Printed in Singapore.

Preface

This book is based upon the works of the author at Peking University during the last half-century. It concerns the topics of the qualitative theory of ordinary differential equations. The selection of material depends mainly upon the interests of the author. No rigorous attempt has been made to give the historic origin of the theory. The author would like to describe the following contents as the main feature of the book.

1) Chapter 1 begins with the existence and uniqueness theorem on the solution of the Cauchy problem. It provides fundamental exercises that are useful even in the advanced theory of differential equations. For example, we prove a general convergence theorem on difference methods for ordinary differential equations; that is, the numerical calculation of general difference method is convergent if and only if the solution of the corresponding Cauchy problem is unique. It means that the numerical calculation is divergent if Peano phenomenon happens to the Cauchy problem. On the other hand, it is proved that Peano's phenomena are dense among differential equations.

2) The global behavior of the solution to the Cauchy problem on infinite interval is considered in Chapter 2. According to the view-point of Poincaré, the solution on infinite interval may be continuous or discontinuous about the initial conditions, so it may be predictable or unpredictable. This leads to the Liapunov stability theory of motion.

The geometrical consideration of solutions yields the qualitative theory of differential equations, especially, with fruitful results for the autonomous differential equations (Chapter 3). On the other hand, it can be considered that the theory of dynamical systems is an abstract setting of the qualitative theory of ordinary differential equations (Chapter 5).

Finally, the theory of chaos is a natural consideration for the Liapunov unstable motions (Chapter 8). Following the current literature, we define the chaotic motion by means of the sensitive dependence on the initial conditions. Accordingly, we find a simple sufficient condition for chaotic motions, which implies that the C^1-flow on closed surface is chaotic if it has a dense orbit and finitely many (at least one) equilibrium points on the surface. The definition of chaotic motions implies obviously that all the motions in the chaotic set considered are Liapunov uniformly unstable. Therefore, it is possible that more complicated oscillations than "chaotic motions" may appear if Liapunov unstable motions coexist with Liapunov stable motions in some transitive invariant set.

3) Chapters 6 and 7 are concerned with the fixed point theorems, which are known as the main topological tools in nonlinear analysis. For example, the generalized Poincaré-Birkhoff twist theorem plays an important role for conservative systems. However, its area-preserving assumption is a severe restriction in application. We are interested in a flexible condition instead and thus obtain the bend-twist theorem for analytic maps (Chapter 7).

Using the bend-twist theorem, we prove that the dissipative super-linear Duffing equation

$$\ddot{x} + \varepsilon c\dot{x} + (ax + bx^3) = \varepsilon E \sin \omega t, \qquad (D)$$

has subharmonic motions of high order, where the generalized Poincaré-Birkhoff twist theorem does not work. In the history, the earliest strange attractor was discovered in the numerical analysis of (D) performed by the experiment in Tokyo University during the 1950's. According to Hayashi's viewpoint, the complicated strange attractor is caused by the "subharmonic motions", but his student Ueda considered the cause of "chaotic motions". Either way, the existence of subharmonic motions for the above dissipative equation (D) is now proved in this book.

4) Based on the collaboration work with F. Zanolin, we analyze the periodic Duffing equation of second order in Chapter 10. The attentive readers will find that a series of results was proved on this subject. On the other hand, Chapter 11 is devoted to the analysis of some special problems, which were solved by the author, as applications for the qualitative theory of differential equations.

The author is indeed indebted to his teaching work carried out over his years at Peking University, which has encouraged him to continue in his writing. At the same time, the preparation of this book was also facilitated by assistance from colleagues, in particular, Prof. Liu Bin and Prof. Wang Zaihong, and also Ed. Zhang Ji from the board of WSPC.

Contents

Preface   v

Chapter 1. Cauchy Problem   1
- 1.1 Fundamental Theorems   1
- 1.2 Method of Euler Polygons   15
- 1.3 Local Behavior of Integral Curves   20
- 1.4 Peano Phenomenon   24
- 1.5 Convergence Theorem on Difference Methods   33

Chapter 2. Global Behavior of Solution   47
- 2.1 Global Existence of Solution   47
- 2.2 Predictability of Solution   58
- 2.3 Liapunov Stability   61
- 2.4 Liapunov Unstability   71

Chapter 3. Autonomous Systems   73
- 3.1 Phase Portrait   73
- 3.2 Orbital Box   76
- 3.3 Types of Orbits   77
- 3.4 Singular Points   79
- 3.5 General Property of Singular Points   86
- 3.6 Closed Orbit   87
- 3.7 Invariant Torus   91
- 3.8 Limit-Point Set   95
- 3.9 Poincaré-Bendixson Theorem   97

Chapter 4. Non-Autonomous Systems — 101

- 4.1 General Systems — 101
- 4.2 Conservative Systems — 104
- 4.3 Dissipative Systems — 106
- 4.4 Planar Periodic Systems — 115
- 4.5 Invariant Continuum — 119

Chapter 5. Dynamical Systems — 123

- 5.1 The Originality — 123
- 5.2 Recurrence — 128
- 5.3 Quasi-Minimal Set — 132
- 5.4 Minimal Set — 134
- 5.5 Almost Periodic Motion — 144

Chapter 6. Fixed-Point Theorems — 155

- 6.1 Poincaré Index — 155
- 6.2 Vector Fields on Closed Surfaces — 165
- 6.3 Spatial Vector Fields — 172
- 6.4 Fixed-Point Theorems of Brouwer Type — 176

Chapter 7. Bend-Twist Theorem — 181

- 7.1 Generalized Poincaré-Birkhoff Twist Theorem — 181
- 7.2 Analytic Bend-Twist Theorem — 184
- 7.3 Analytic Poincaré-Birkhoff Twist Theorem — 189
- 7.4 Application of the Bend-Twist Theorem — 191

Chapter 8. Chaotic Motions — 199

- 8.1 Definition of Chaotic Motion — 199
- 8.2 Chaotic Quasi-Minimal Set — 201
- 8.3 Sufficient Conditions for Chaotic Sets — 202
- 8.4 Chaotic Closed Surfaces — 205
- 8.5 Applications — 209

Chapter 9. Perturbation Method — 217

- 9.1 Nonlinear Differential Equation of Second Order — 217
- 9.2 Method of Averaging — 225
- 9.3 High Frequency Forced Oscillations — 230

Chapter 10.	**Duffing Equations of Second Order**	**241**
10.1	Periodic Oscillations	241
10.2	Time-Map	250
10.3	Duffing Equation of Super-Linear Type	261
10.4	Duffing Equation of Sub-Linear Type	275
10.5	Duffing Equation of Semi-Linear Type	288
Chapter 11.	**Some Special Problems**	**313**
11.1	Reeb's Problem	313
11.2	Birkhoff's Conjecture	319
11.3	Morse's Conjecture	326
11.4	Kolmogorov's Problem	331
11.5	Brillouin Focusing System	345
11.6	A Retarded Equation	355
11.7	Periodic Lotka-Volterra System	365
Bibliography		**377**

Chapter 1
Cauchy Problem

1.1 Fundamental Theorems

1.1.1 Statement of Cauchy Problem

We study the system of ordinary differential equation

$$\frac{dx}{dt} = f(t,x), \qquad (t,x) \in D \qquad (1.1)$$

where D is a region in the space $\mathbb{R} \times \mathbb{R}^n$, and $f(t,x)$ is an n-vector[1] valued function defined on D. The variable x stands for an *unknown function* $x = x(t)$ valued in $\mathbb{R}^n$, and the integer $n\, (\geq 1)$ is called the order of the system (1.1). The region D is, in general, assumed to be open in the space $\mathbb{R} \times \mathbb{R}^n$.

If $x = \varphi(t)$ is a differentiable function in the interval $J \subset \mathbb{R}$, satisfying

$$\frac{d\varphi(t)}{dt} = f(t, \varphi(t)) \qquad (t \in J),$$

then $x = \varphi(t)$ is called a *solution* of (1.1) on J. In geometry, a solution $x = \varphi(t)$ of (1.1) represents a curve Γ in D, called an *integral curve* of (1.1).

Initial-Value Problem Given a condition

$$x(t_0) = x_0, \qquad (t_0, x_0) \in D, \qquad (1.2)$$

find a solution $x = x(t)$ of (1.1) on some interval J, such that it satisfies the *initial condition* (1.2) (see [21]).

---
[1] In this book, a vector will be always understood as a column-vector even if it is sometimes written in a row-form for simplicity.

The above initial-value problem is sometimes denoted by (1.1)+(1.2), or briefly by

$$(E): \quad \frac{dx}{dt} = f(t, x), \quad x(t_0) = x_0,$$

where (t_0, x_0) is called an *initial point* in D. In literature, an initial-value problem is also called a *Cauchy problem*, for a memory of the contribution of Cauchy to the theory of differential equations.

In geometry, a solution $x = \varphi(t)$ of (E) stands for an integral curve of (1.1) passing through the initial point (t_0, x_0) in D.

On the other hand, it is easy to show that $x = x(t)$ is a solution of (E) on J if and only if $x = x(t)$ satisfies the *integral equation*

$$x = x_0 + \int_{t_0}^{t} f(s, x) \, ds, \quad \forall \, t \in J. \tag{1.3}$$

The initial-value problem (E) is mainly concerned with the *existence* and *uniqueness* of solution, and as well as the *dependence* of solution on initial condition (or parameter). We will consider these problems in the fundamental theorems mentioned below (see [78], [21], [56] and etc.).

1.1.2 Auxiliary Lemmas

Now, we introduce the following lemmas, which are the basic techniques in the analysis of differential equations.

Arzelà-Ascoli Lemma *Let* X *be a compact region in the space* $\mathbb{R}^p$ *(*$p \geq 1$*). If*

$$f_n : \quad X \quad \to \quad \mathbb{R}^q \quad (n \in \mathbb{Z}^+)$$

is a sequence of uniformly bounded[2] *and equi-continuous*[3] *mappings, then there is at least a uniformly convergent subsequence of* $\{f_n\}$.

---

[2] That is, there is a constant $B_0 > 0$, such that

$$|f_n(z)| < B_0 \quad (\forall \, z \in X), \quad \text{for all } n \in \mathbb{Z}^+.$$

[3] That is, given any $\varepsilon > 0$, there is a $\delta = \delta(\varepsilon) > 0$, such that whenever $|u - v| < \delta$ $(u, v \in X)$, we have

$$|f_n(u) - f_n(v)| < \varepsilon, \quad \text{for all } n \in \mathbb{Z}^+.$$

Proof. The proof is well-known in literature (see [21]). □

Gronwall Lemma *If $u(t)$ is a continuous function, satisfying the integral inequality*

$$0 \leq u(t) \leq C + K \int_{t_0}^{t} u(t)\, dt \qquad (t_0 \leq t \leq t_1), \tag{1.4}$$

where $C \geq 0$ and $K > 0$ are constants, then

$$u(t) \leq C\, e^{K(t-t_0)}, \qquad t_0 \leq t \leq t_1. \tag{1.5}$$

Proof. We apply a useful technique in the analysis of inequality. Letting

$$w(t) = \int_{t_0}^{t} u(t)\, dt \qquad (t_0 \leq t \leq t_1),$$

and using (1.4), we obtain

$$\frac{dw(t)}{dt} - Kw(t) \leq C, \qquad w(t_0) = 0.$$

It follows that

$$\frac{d}{dt}\left[w(t) e^{-Kt}\right] \leq C e^{-Kt}.$$

Then, integrating from t_0 to $t(> t_0)$, we have

$$w(t) \leq \frac{C}{K}\left[e^{K(t-t_0)} - 1\right],$$

which together with (1.4) yields the desired inequality (1.5). □

1.1.3 Peano Theorem

Cauchy was the first person in history to prove the existence and uniqueness of solution of the initial-value problem (E), during the period of 1830's, under the assumption that $f(t,x)$ and $f'_x(t,x)$ are continuous in $(t,x) \in Q$, where Q is a local neighborhood at the initial point (t_0, x_0). In 1886, Lipschitz improved the Cauchy theorem by using a simplified condition (i.e., the Lipschitz condition defined below) to replace the continuity of $f'_x(t,x)$. In 1890's, Peano proved the local existence of solution of (E) merely under the continuity condition of $f(t,x)$ on Q.

Now, we begin to consider the most elementary theorem; namely, the Peano Theorem. Without loss of generality, assume $f(t,x)$ is a continuous function defined in the rectangular region

$$Q := \{(t,x) \in D : \ |t - t_0| \leq a, \ |x - x_0| \leq b\},$$

centered at the point (t_0, x_0), with some constants $a > 0$ and $b > 0$. Then fix the constants

$$M > \sup_{(t,x) \in Q} |f(t,x)| \quad \text{and} \quad h = \min\left\{a, \frac{b}{M}\right\}.$$

Now, we are ready to prove the *Peano Theorem*.

Theorem 1.1 *If the function $f(t,x)$ is continuous in $(t,x) \in Q$, then the Cauchy problem (E) has at least a solution $x = \varphi(t)$ on the interval $I = [t_0 - h, t_0 + h]$.*

Proof. We apply the method of Tonelli approximation (1910's) to the initial-value problem (E) as follows.

For a given integer $m \geq 1$, define

$$\phi_m(t) = \begin{cases} x_0, & \text{as } t \in [t_0 - h/m, t_0 + h/m], \\ x_0 + \int_{t_0}^{t - \frac{h}{m}} f(t, \phi_m(t))\,dt, & \text{as } t \in [t_0 + h/m, t_0 + h], \\ x_0 + \int_{t_0}^{t + \frac{h}{m}} f(t, \phi_m(t))\,dt, & \text{as } t \in [t_0 - h, t_0 - h/m]. \end{cases} \quad (1.6)$$

It can be seen from (1.6) that for $t \geq t_0$, the function $\phi_m(t)$ is first defined in the sub-interval $[t_0, t_0 + h/m]$; and next it is extended to the sub-interval $[t_0 + h/m, t_0 + 2h/m]$ via the integral; and in a similar manner, to the third sub-interval $[t_0 + 2h/m, t_0 + 3h/m]$; and finally to the last sub-interval $[t_0 + (m-1)h/m, t_0 + h]$. Therefore, $\phi_m(t)$ is defined on the right sub-interval $[t_0, t_0 + h]$. Similarly, the function $\phi_m(t)$ is defined on the left sub-interval $[t_0 - h, t_0]$.

The above-defined sequence $\{\phi_m(t)\}$ is called the *Tonelli sequence* on I. It can be verified by using (1.6) that the Tonelli sequence satisfies

$$|\phi_m(t) - x_0| \leq Mh \qquad (t \in I), \qquad (1.7)$$

and

$$|\phi_m(t_2) - \phi_m(t_1)| \leq M|t_2 - t_1| \qquad (t_1, t_2 \in I), \qquad (1.8)$$

for $m \geq 1$. The property (1.7) states that the Tonelli sequence $\{\phi_m(t)\}$ is uniformly bounded on I, while the property (1.8) says that the Tonelli sequence $\{\phi_m(t)\}$ is equi-continuous on I.

Then, using the Arzelà-Ascoli lemma, we conclude that the sequence $\{\phi_m(t)\}$ has at least a uniformly convergent sub-sequence $\{\phi_{m_i}(t)\}$. Then,

$$\phi(t) = \lim_{m_i \to \infty} \phi_{m_i}(t) \qquad (t \in I)$$

is a continuous function. Let $m = m_i$ in (1.6). It follows from the uniform convergence of $\{\phi_{m_i}(t)\}$ that

$$\phi(t) = x_0 + \int_{t_0}^{t} f(t, \phi(t))\, dt \qquad (t \in I).$$

This proves that $x = \phi(t)$ is a solution of (E) on I.

The proof of the Peano theorem is thus completed. $\square$

As an application of Peano theorem, we prove the following famous result in analysis.

Implicit-Function Theorem *Given the equation*

$$g(t, x) = 0, \tag{1.9}$$

where $g(t, x)$ is a C^1 - differentiable (n-vector valued) function in $(t, x) \in Q$, if it satisfies the condition

$$g(t_0, x_0) = 0 \quad \text{and} \quad \det[g'_x(t_0, x_0)] \neq 0, \tag{1.10}$$

then the equation (1.9) defines a unique differentiable implicit-function $x = \varphi(t)$ on $t_0 - \alpha \leq t \leq t_0 + \alpha$, satisfying the initial condition $x(t_0) = x_0$, where $\alpha > 0$ is some small constant.

Proof. Let us consider the auxiliary Cauchy problem

$$(E^*): \qquad \frac{dx}{dt} = -\{g'_x(t, x)\}^{-1} g'_t(t, x), \qquad x(t_0) = x_0.$$

The assumption on $g(t, x)$ implies that the function

$$f(t, x) := -\{g'_x(t, x)\}^{-1} g'_t(t, x)$$

is continuous in the neighborhood of (t_0, x_0). Hence, the Peano theorem asserts that the Cauchy problem (E^*) has at least a (differentiable) solution

$x = \varphi(t)$ on some local interval $|t - t_0| \leq \alpha$. Then we have

$$\frac{dg(t, \varphi(t))}{dt} = g'_t(t, \varphi(t)) + g'_x(t, \varphi(t))\frac{d\varphi(t)}{dt}$$

$$= g'_t(t, \varphi(t)) + g'_x(t, \varphi(t))(-\{g'_x(t, x)\}^{-1}g'_t(t, x)) = 0,$$

which yields

$$g(t, \varphi(t)) = C \qquad (|t - t_0| \leq \alpha),$$

where C is an arbitrary constant. Then, using the initial condition given in (1.10), we obtain

$$C = g(t_0, \varphi(t_0)) = g(t_0, x_0) = 0.$$

Therefore, $x = \varphi(t)$ is a solution of the equation (1.28), satisfying the initial condition $x(t_0) = x_0$. This proves the existence of implicit-function.

For the uniqueness part, assume $x = \varphi_1(t)$ and $x = \varphi_2(t)$ are any implicit-functions of (1.9) in the interval $|t - t_0| \leq \alpha$, satisfying the initial condition $\varphi_1(t_0) = x_0 = \varphi_2(t_0)$.

Let $g_i(t, x)$ be the i-th component of $g(t, x)$, and let

$$h_i(\lambda) = g_i(t, \varphi_1(t) + \lambda(\varphi_2(t) - \varphi_1(t))).$$

It follows from the mean value theorem that

$$h_i(1) - h_i(0) = h'_i(\lambda_i), \qquad 0 < \lambda_i < 1,$$

which yields

$$g_i(t, \varphi_2(t)) - g_i(t, \varphi_1(t)) = \frac{\partial g_i}{\partial x}(t, \tilde{\xi}_i)(\varphi_2(t) - \varphi_1(t)),$$

where $\tilde{\xi}_i$ is a vector given by

$$\tilde{\xi}_i = \varphi_1(t) + \lambda_i(\varphi_2(t) - \varphi_1(t))).$$

Hence, we have

$$0 = g(t, \varphi_1(t)) - g(t, \varphi_2(t)) = \mathbb{G}(\varphi_1(t) - \varphi_2(t)), \qquad (1.11)$$

where $\mathbb{G}$ is an $n \times n$-matrix given by

$$\begin{pmatrix} \dfrac{\partial g_1}{\partial x_1}(t,\tilde{\xi}_1) & \cdots & \dfrac{\partial g_1}{\partial x_n}(t,\tilde{\xi}_1) \\ \vdots & \vdots & \vdots \\ \dfrac{\partial g_n}{\partial x_1}(t,\tilde{\xi}_n) & \cdots & \dfrac{\partial g_n}{\partial x_n}(t,\tilde{\xi}_n) \end{pmatrix}, \quad \text{for } |t-t_0| \leq \alpha.$$

It is noticed that $(t,\tilde{\xi}_i)$ tends to (t_0, x_0) whenever α approaches to 0. Hence, the matrix $\mathbb{G}$ is thus sufficiently near to the non-singular matrix $g'_x(t_0, x_0)$. It follows that $\mathbb{G}$ is invertible if α is small enough. Using (1.11) leads to

$$\varphi_1(t) = \varphi_2(t) \quad (|t-t_0| \leq \alpha).$$

We have thus proved the uniqueness of the implicit function.
The proof of the implicit-function theorem is thus completed. □

1.1.4 Cauchy-Lipschitz Theorem

Let us introduce the Lipschitz condition. Assume the function $f(t,x)$ is defined on Q. If there is a constant $L > 0$, such that for any points (t,u) and (t,v) in Q we have

$$|f(t,u) - f(t,v)| \leq L|u-v|, \tag{1.12}$$

then we say that $f(t,x)$ satisfies the *Lipschitz Condition* on Q (with respect to x), and L is called the *Lipschitz constant*. It is obvious that the Lipschitz condition is merely the simple property of differentiability condition, but a Lipschitzian function may be not differentiable.

In literature, the following existence and uniqueness theorem of solution is called *Cauchy-Lipschitz theorem*.

Theorem 1.2 *If $f(t,x)$ is continuous in $(t,x) \in Q$ and satisfies the Lipschitz condition (1.12), then the initial-value problem (E) has one and only one solution $x = \varphi(t)$ on I.*

Proof. Since $f(t,x)$ is continuous in $(t,x) \in Q$, then Peano theorem guarantees the existence of solution of (E) on I. Hence, it suffices to prove the uniqueness of solution of (E).

Assume $x = \phi(t)$ and $x = \psi(t)$ are solutions of (E) on the interval $|t - t_0| \leq \alpha$ for some constant $\alpha > 0$. Then we have

$$\phi(t) = x_0 + \int_{t_0}^{t} f(t, \phi(t))\, dt \qquad (|t - t_0| \leq \alpha),$$

and

$$\psi(t) = x_0 + \int_{t_0}^{t} f(t, \psi(t))\, dt \qquad (|t - t_0| \leq \alpha).$$

It follows from the Lipschitz condition that

$$|\psi(t) - \phi(t)| \leq \left| \int_{t_0}^{t} |f(t, \psi(t)) - f(t, \phi(t))|\, dt \right|$$

$$\leq \left| \int_{t_0}^{t} L|\psi(t) - \phi(t)|\, dt \right|, \qquad |t - t_0| \leq \alpha.$$

If $t \geq t_0$, we have

$$|\psi(t) - \phi(t)| \leq \int_{t_0}^{t} L|\psi(t) - \phi(t)|\, dt \qquad (t_0 \leq t \leq t_0 + \alpha).$$

It follows from the Gronwall lemma (with $u(t) = |\psi(t) - \phi(t)|$, $C = 0$ and $K = L$) that

$$|\psi(t) - \phi(t)| \leq 0 \qquad (t_0 \leq t \leq t_0 + \alpha).$$

Hence

$$\psi(t) = \phi(t) \qquad (t_0 \leq t \leq t_0 + \alpha).$$

If $t \leq t_0$, in a similar manner, we can prove

$$\psi(t) = \phi(t) \qquad (t_0 - \alpha \leq t \leq t_0).$$

The uniqueness of solution of (E) is thus proved. So, the proof of Theorem 1.2 is complete. $\square$

It follows from the Cauchy-Lipschitz theorem that if the initial condition is given and the differential equation is sufficiently regular, the solution of the corresponding Cauchy problem is predicable. In this sense, the Cauchy problem is *well-posed* or *deterministic*.

1.1.5 Dependence of Solution on Parameter

1) Now, we consider the dependence of solution on initial condition.

As Poincaré pointed out, even if it were the case that the natural law of differential equations had no longer any secret for us, we would still only know the initial data approximately by measuring. Suppose the initial condition is measured approximately by

$$x(\tau) = \xi. \tag{1.2*}$$

Naturally, there is an error between the approximate condition (1.2*) and the given condition (1.2). Hence, we have to consider the error between the solutions of the corresponding Cauchy problems

$$(E): \quad \frac{dx}{dt} = f(t,x), \quad x(\tau) = \xi,$$

and

$$(E_0): \quad \frac{dx}{dt} = f(t,x), \quad x(t_0) = x_0,$$

where (E) is the *perturbation* of the Cauchy problem (E_0). For convenience, let $p = (\tau, \xi)$ and $p_0 = (t_0, x_0)$ be the corresponding conditions. Hence, the problem concerns the *dependence* of solution of (E) on the initial data p.

Let

$$G = \{ (\tau, \xi) \in Q : \quad |\tau - t_0| \leq h/4, \quad |\xi - x_0| \leq b/2 \}.$$

It follows from Peano theorem that for each initial condition $p \in G$, the Cauchy problem (E) has at least a solution, existing on the interval $I_0 = [t_0 - h/4, t_0 + h/4]$. It is noticed thar the solution of (E) may be unique or not. Anyway, let $x = \phi(t,p)$ be a solution of (E).

Lemma 1.1 *If the solution of (E_0) is unique on the interval I_0, then for any given constant $\varepsilon > 0$, there is a constant $\delta > 0$, such that*

$$|\phi(t,p) - \phi(t,p_0)| < \varepsilon, \qquad \text{for } t \in I_0,$$

whenever $|p - p_0| < \delta$.

Proof. Assume the contrary. Then there exists $\varepsilon_0 > 0$, such that for any small constant $\delta_i > 0$, there is a point $(t_i, p_i) \in I_0 \times B_{\delta_i}(p_0)$, satisfying

$$|\phi(t_i, p_i) - \phi(t_i, p_0)| \geq \varepsilon_0. \tag{1.13}$$

Let $\delta_i \to 0$ when $i \to \infty$. It follows that $p_i \to p_0$ as $i \to \infty$, and $\{t_i\}$ has at least a cluster point $\hat{t} \in I_0$. Without loss of generality, assume $t_i \to \hat{t}$ as $i \to \infty$.

On the other hand, it can be easily proved that $\{\phi(t, p_i)\}$ is a uniformly bounded and equi-continuous sequence on the interval I_0.

Using the Arzelà-Ascoli lemma, we conclude that there is a uniformly convergent subsequence $\phi(t, p_{i_k})$. Moreover, we can easily prove that

$$\psi(t) = \lim_{i_k \to \infty} \phi(t, p_{i_k}) \qquad (t \in I_0)$$

is a solution of (E_0). Then, the uniqueness of solution of (E_0) implies the identity: $\psi(t) = \phi(t, p_0)$, for all $t \in I_0$.

On the other hand, using (1.13) for its limit as $i \to \infty$, we obtain

$$|\psi(\hat{t}) - \phi(\hat{t}, p_0)| \geq \varepsilon_0,$$

which is in conflict with the identity: $\psi(t) = \phi(t, p_0)$.

We have thus proved Lemma 1.1 by contradiction. $\square$

The following theorem is a consequence of Lemma 1.1. It states the continuous dependence of the solution $x = \phi(t, p)$ of (E) on the initial condition $p \in G$.

Theorem 1.3 *Assume $f(t, x)$ is continuous in $(t, x) \in D$. If the solution $x = \varphi(t, p)$ of (E) is unique for each initial point $p \in G$, then $x = \varphi(t, p)$ is continuous in $(t, p) \in I_0 \times G$.*

2) Next we consider the dependence of solution on parameter.

It follows from the substitutions $t = s + \tau$ and $x = z + \xi$ that the initial-value problem (E) is transformed into the following one

$$(E_\mu): \qquad \frac{dz}{ds} = g(s, z; \mu), \qquad z(0) = 0,$$

where $\mu = (\tau, \xi)$ is a parameter and $g(s, z; \mu) = f(s+\tau, z+\xi)$ is a continuous function of $(s, z; \mu)$.

Sometimes, the natural law of differential equation may indeed depend on certain parameter.

It leads us to consider a general form of Cauchy problem as follows

$$(E_\lambda): \qquad \frac{dx}{dt} = f(t, x; \lambda), \qquad x(t_0) = x_0,$$

where λ is a parameter-vector in a compact region $\Lambda \subset \mathbb{R}^s$ and $f(t, x; \lambda)$ is a continuous function of $(t, x; \lambda) \in Q \times \Lambda$, which is valued in $\mathbb{R}^n$.

Similar to Cauchy problem (E), fix the constants

$$h^* = \min\left\{a, \frac{b}{M^*}\right\} \quad \text{and} \quad M^* > \max_{(t,x;\lambda)\in Q\times \Lambda} |f(t,x;\lambda)|,$$

with the interval $I^* = [t_0 - h^*, t_0 + h^*]$.

Proposition 1.1 *If (E_λ) has a unique solution on I^* for any $\lambda \in \Lambda$, then its solution $x = \varphi(t, \lambda)$ is continuous in $(t, \lambda) \in I^* \times \Lambda$ (i.e., the solution of (E_λ) is continuous for the parameter $\lambda \in \Lambda$).*

Proof. Given $(\tau, \sigma) \in I^* \times \Lambda$, assume $(t_k, \lambda_k) \in I^* \times \Lambda$ is any sequence tending to (τ, σ) as $k \to \infty$.

It suffices to prove that for any given $\varepsilon > 0$, we have

$$|\varphi(t_k, \lambda_k) - \varphi(\tau, \sigma)| < \varepsilon, \quad \text{as } k \to \infty.$$

Assume the contrary. Then there are a constant $\varepsilon_0 > 0$ and a sequence (t_k, λ_k) tending to (τ, σ), such that

$$|\varphi(t_k, \lambda_k) - \varphi(\tau, \sigma)| \geq \varepsilon_0, \quad \text{as } k \to \infty. \tag{1.14}$$

Since $x = \varphi(t, \lambda_k)$ is the solution of (E_{λ_k}), we have

$$\varphi(t, \lambda_k) = x_0 + \int_{t_0}^{t} f(t, \varphi(t, \lambda_k); \lambda_k))\, dt, \quad t \in I^*, \tag{1.15}$$

which leads to the following conclusions:

1) $|\varphi(t, \lambda_k)| \leq |x_0| + M^* h^*$, for $t \in I^*$
   (it means that $\varphi(t, \lambda_k)$ is uniformly bounded in I^*);
2) $|\varphi(t, \lambda_k) - \varphi(s, \lambda_k)| \leq M^*|t - s|$, for $t, s \in I^*$
   (it implies that $\varphi(t, \lambda_k)$ is equi-continuous in I^*).

It follows from the Arzelà-Ascoli lemma that there is a uniformly convergent subsequence of $\varphi(t, \lambda_k)$ on the interval I^*. Without loss of generality, assume $\varphi(t, \lambda_k)$ is uniformly convergent on I^*, and let

$$\lim_{k\to\infty} \varphi(t, \lambda_k) = \psi(t), \quad t \in I^*,$$

which together with (1.15) yields

$$\psi(t) = x_0 + \int_{t_0}^{t} f(t, \psi(t); \sigma)\, dt, \quad t \in I^*.$$

Then the uniqueness of solution implies
$$\psi(t) = \varphi(t, \sigma), \qquad t \in I^*. \tag{1.16}$$

On the other hand, it follows from (1.14) that
$$|\psi(\tau) - \varphi(\tau, \sigma)| \geq \varepsilon_0 > 0,$$

which is in conflict with (1.16).

Proposition 1.1 is thus proved by contradiction. $\square$

As remarked above, since (E) can be reduced to (E_μ), Theorem 1.3 is a consequence of Proposition 1.1. Hence, (E) is well-posed whenever $f(t, x)$ is continuous in $(t, x) \in Q$ and the solution of (E) is unique.

Note that the following result is an important consequence of Lipschitz condition.

Corollary 1.1 *If $f(t, x; \lambda)$ is continuous in $(t, x; \lambda) \in Q \times \Lambda$ and satisfies the Lipschitz condition with respect to x, then the solution $x = \varphi(t, \lambda)$ of (E_λ) is continuous for the parameter $\lambda \in \Lambda$.*

3) Finally, we consider the differentiability of solution for parameter.

Theorem 1.4 *If $f(t, x; \lambda)$ is continuous in $(t, x; \lambda) \in Q \times \Lambda$ and C^1-differentiable with respect to (x, λ), then the solution $x = \varphi(t, \lambda)$ of (E_λ) is existent on I^* and C^1- differentiable for the parameter $\lambda \in \Lambda$.*

Proof. Let $\{\phi_m(t, \lambda)\}$ be the Tonelli sequence of (E_λ), defined by

$$\phi_m(t, \lambda) = x_0, \qquad \text{as } t \in [t_0 - h^*/m, t_0 + h^*/m];$$

$$\phi_m(t, \lambda) = x_0 + \int_{t_0}^{t-h^*/m} f(s, \phi_m(s, \lambda); \lambda) \, ds,$$

$$\text{as } t \in [t_0 + h^*/m, t_0 + h^*];$$

$$\phi_m(t, \lambda) = x_0 + \int_{t_0}^{t+h^*/m} f(s, \phi_m(s, \lambda); \lambda) \, ds,$$

$$\text{as } t \in [t_0 - h^*, t_0 - h^*/m],$$

$(m = 1, 2, \cdots)$.

Since $f(t, x; \lambda)$ is C^1 differentiable with respect to (x, λ), we have

$$\frac{\partial \phi_m(t, \lambda)}{\partial \lambda} = 0, \qquad \text{as } t \in [t_0 - h^*/m, t_0 + h^*/m];$$

$$\frac{\partial \phi_m(t, \lambda)}{\partial \lambda} = \int_{t_0}^{t-h^*/m} \left[\frac{\partial f}{\partial x} \frac{\partial \phi_m(s, \lambda)}{\partial \lambda} + \frac{\partial f}{\partial \lambda} \right] ds,$$

$$\text{as } t \in [t_0 + h^*/m, t_0 + h^*];$$

$$\frac{\partial \phi_m(t, \lambda)}{\partial \lambda} = \int_{t_0}^{t+h^*/m} \left[\frac{\partial f}{\partial x} \frac{\partial \phi_m(s, \lambda)}{\partial \lambda} + \frac{\partial f}{\partial \lambda} \right] ds,$$

$$\text{as } t \in [t_0 - h^*, t_0 - h^*/m],$$

where

$$\frac{\partial f}{\partial x} = f'_x(s, \phi_m(s, \lambda); \lambda) \qquad \text{and} \qquad \frac{\partial f}{\partial \lambda} = f'_\lambda(s, \phi_m(s, \lambda); \lambda),$$

$(m = 1, 2, \cdots)$.

It follows from the Gronwall lemma that

(1) The sequence $\dfrac{\partial \phi_m(t, \lambda)}{\partial \lambda}$ is uniformly bounded in I^*;

(2) The sequence $\dfrac{\partial \phi_m(t, \lambda)}{\partial \lambda}$ is equi-continuous in I^*.

Using Arzelà-Ascoli lemma, assume without loss of generality that $\partial \phi_m(t, \lambda)/\partial \lambda$ is uniformly convergent on I^*. Let

$$\lim_{m \to \infty} \frac{\partial \phi_m(t, \lambda)}{\partial \lambda} = \zeta(t, \lambda).$$

It follows that

$$\zeta(t, \lambda) = \int_{t_0}^t [f'_x(t, \varphi(t, \lambda); \lambda)\zeta(t, \lambda) + f'_\lambda(t, \varphi(t, \lambda); \lambda)] \, dt.$$

Therefore, $z = \zeta(t, \lambda)$ is the solution of the Cauchy problem

$$\frac{dz}{dt} = f'_x(t, \varphi(t, \lambda); \lambda)\zeta(t, \lambda)z + f'_\lambda(t, \varphi(t, \lambda); \lambda), \qquad z(t_0) = 0,$$

which together with Proposition 1.1 implies that $z = \zeta(t, \lambda)$ is continuous in (t, λ). On the other hand, as shown above, $\phi_m(t, \lambda)$ uniformly converges to

$\varphi(t,\lambda)$ and $\frac{\partial \phi_m(t,\lambda)}{\partial \lambda}$ uniformly converges to $\zeta(t,\lambda)$ when $m \to \infty$. Hence, we have the partial derivative

$$\varphi'_\lambda(t,\lambda) = \zeta(t,\lambda),$$

which is continuous in $(t,\lambda) \in I^* \times \Lambda$.

On the other hand, it follows from $\varphi'_t(t,\lambda) = f(t,\varphi(t,\lambda);\lambda)$ that $\varphi(t,\lambda)$ is C^1 differentiable in $(t,\lambda) \in I^* \times \Lambda$.

The proof of Theorem 1.4 is thus complete. □

Theorem 1.5 *If $f(t,x)$ is continuous in $(t,x) \in Q$ and C^1 differentiable with respect to x, then the solution $x = \varphi(t,\tau,\xi)$ of the Cauchy problem (E) is C^1 differentiable with respect to the initial condition $(\tau,\xi) \in G$.*

Proof. As mentioned in the Cauchy problem (E_μ), this theorem seems to be a special case of Theorem 1.4. However, we have to make a remark as follows.

Indeed, the differentiability of $x = \varphi(t,\tau,\xi)$ with respect to ξ can be proved through Theorem 1.4 by using a substitution $x = z + \xi$, where the corresponding partial derivative $\partial \phi_m/\partial \xi$ is continuous.

However, it can be seen that the differentiability of $x = \varphi(t,\tau,\xi)$ with respect to τ cannot be proved in this way, since the continuity of the partial derivative $\partial \phi_m/\partial \tau$ depends on the continuity of $f'_t(t,x)$, which is in question since we do not assume even the existence of $f'_t(t,x)$. Therefore, we have to change the method of proof.

For this aim, let us modify the Tonelli sequence $\{\phi_m\}$ of (E) as follows:

$$\phi_m(t,\tau,\xi) = \xi + \int_\tau^t f(t,\xi)\,dt, \qquad \text{as } t \in [\tau - h/4m, \tau + h/4m];$$

$$\phi_m(t,\tau,\xi) = \xi + \int_\tau^{\tau+\frac{h}{4m}} f(t,\xi)\,dt + \int_\tau^{t-\frac{h}{4m}} f\left(t+\frac{h}{4m}, \phi_m(t,\tau,\xi)\right) dt,$$

$$\text{as } t \in [\tau + h/4m, \tau + h/4];$$

$$\phi_m(t,\tau,\xi) = \xi + \int_\tau^{\tau-\frac{h}{4m}} f(t,\xi)\,dt + \int_\tau^{t+\frac{h}{4m}} f\left(t-\frac{h}{4m}, \phi_m(t,\tau,\xi)\right) dt,$$

$$\text{as } t \in [\tau - h/4, \tau - h/4m],$$

$(m = 1, 2, \cdots)$.

We see that $\{\partial \phi_m/\partial \tau\}$ is continuous. Therefore, this sequence $\{\phi_m\}$ is available to prove Theorem 1.5 in a similar manner as before. □

1.1.6 Carathéodory Theorem

In 1920's, Carathéodory generalized the initial-value problem

$$(E_0): \qquad \frac{dx}{dt} = f(t,x), \qquad x(t_0) = x_0$$

to the following sense (see [21]).

If $x = \varphi(t)$ is a function absolutely continuous in the interval J, such that it satisfies the differential equation

$$\varphi'(t) = f(t, \varphi(t)) \qquad (\text{for almost all } t \in J),$$

and the initial condition $\varphi(t_0) = x_0$, then $x = \varphi(t)$ is called a *generalized solution* of (E_0) on J.

The following theorem is called the *existence theorem of Carathéodory*, which is widely applied in the modern theory of differential equations.

Theorem 1.6 *Let the function $f(t,x)$ be defined on Q. Assume $f(t,x)$ is measurable in t for each fixed x, and continuous in x for each fixed t. If there exists a Lebesgue-integrable function $m(t)$ on the interval $|t - t_0| \leq a$, such that*

$$|f(t,x)| \leq m(t) \qquad ((t,x) \in Q), \tag{1.17}$$

then there exists a generalized solution $x = \varphi(t)$ of (E_0) on a local interval $|t - t_0| \leq \alpha$ (for some constant $\alpha > 0$).

Proof. The proof of this Theorem is similar to that of Theorem 1.1 if we use the Lebesgue's integral instead of the Riemann integral therein. □

1.2 Method of Euler Polygons

Roughly speaking, there are three elementary methods in the theory of differential equations for approximating the solution of the initial-value problem (E_0). When the function $f(t,x)$ is lipschitzian for the variable x, the Picard's successive approximation is the most concise method of approximation. But if $f(t,x)$ does not satisfy the Lipschitz condition for x, the *Müller's example* expresses that the Picard method is not useful (see [21]). However, both the methods of Euler's polygonal approximation and Tonelli's approximation can still work in that case. By the way, we point out that the Tonelli method is a unified method for proving all the fundamental theorems as we did above.

In the next subsection, we will simplify the proof due to Gardner [68] for the existence of primitive function in calculus through the method of Euler's polygonal approximation without use of integration.[4]

1.2.1 *Existence of Solution off Integration*

In particular, the Euler polygonal approximation in the planar case can be used to prove the Peano theorem without the helps of the Arzelà-Ascoli lemma and even the idea of integration (see [68]).

Now, we introduce the proof of Gardner with modifications.

1) Consider the Cauchy problem

$$(E_0): \quad \frac{dx}{dt} = f(t,x), \quad x(t_0) = x_0,$$

of first order differential equation, where $f(t,x)$ is a continuous function on the planar rectangular region Q, centered at (t_0, x_0). Let M and h be the positive constants and $I = [t_0 - h, t_0 + h]$ be the interval defined as above. For simplicity, we are confined in the case of right-hand interval $[t_0, t_0 + h]$. The case of left-hand interval $[t_0 - h, t_0]$ can be considered in a similar way.

Given an integer $m \geq 1$, divide the interval $[t_0, t_0 + h]$ by the points

$$t_0 < t_1 < \cdots < t_{m-1} < t_m (= t_0 + h), \qquad (1.18)$$

with

$$\delta_m = \max_{1 \leq i \leq m} \{(t_i - t_{i-1})\} \to 0, \quad \text{as } m \to \infty.$$

Let L_m be the Euler polygon of (E_0) with the cusps:

$$P_i = (t_i, x_i) \qquad (i = 0, 1, \cdots, m),$$

where

$$x_{i+1} = x_i + f(t_i, x_i)(t_{i+1} - t_i), \qquad i = 0, 1, \cdots, m-1.$$

---

[4]It is already known in analysis that the existence of the primitive function without use of integration can be proved by the Weierstrass's polynomial approximation theorem.

Hence, the Euler polygon L_m is given by the formula

$$x = L_m(t) = x_i + f(t_i, x_i)(t - t_i) \quad (t \in [t_i, t_{i+1}]), \qquad 0 \leq i \leq m-1.$$

It follows from $|f(t,x)| \leq M$ that when $t_0 \leq t \leq t_0 + h$, we have

$$|L_m(t) - x_0| \leq \sum_{k=0}^{i-1} |f(t_k, x_k)|(t_{k+1} - t_k) \leq M.$$

It follows that the sequence of Euler's polygons $\{L_m(t)\}$ is uniformly bounded on $t_0 \leq t \leq t_0 + h$.

Denote by $\underline{s_m}$ and $\overline{s_m}$ the lower-limit and the upper-limit of a sequence s_m, respectively. It is well-known in textbook that

$$\underline{L_m(x) - L_m(y)} \leq \overline{L_m(x)} - \overline{L_m(y)} \leq \overline{L_m(x) - L_m(y)}. \tag{1.19}$$

Let $\tau, s \in [t_0, t_0+h]$ with $s \neq \tau$. It follows from (1.18) that $\tau \in [t_p, t_{p+1})$ and $s \in [t_q, t_{q+1})$ for some integers p, q satisfying $0 \leq p, q \leq m-1$. Without loss of generality, assume $p \leq q$.

2) Then we will prove the inequality:

$$\min_{p \leq i \leq q} f(t_i, x_i) \leq \frac{L_n(s) - L_n(\tau)}{s - \tau} \leq \min_{p \leq i \leq q} f(t_i, x_i). \tag{1.20}$$

In fact, let $A = (t, L_m(t))$ and $B = (s, L_m(s))$. Consider the auxiliary points

$$F_a = (t, M_a(s-t)) \quad \text{and} \quad F_b = (t, M_b(s-t)),$$

where

$$M_a = \max_{p \leq i \leq q} f(t_i, x_i) \quad \text{and} \quad M_b = \min_{p \leq i \leq q} f(t_i, x_i).$$

Then we have a triangle Δ having the vertices A, F_a and F_b.

It is noticed that the slopes of the sides $\overline{AF_a}$ and $\overline{AF_b}$ are equal to M_a and M_b, respectively, $(M_b \leq M_a)$.

Now, let us consider the sub-polygon $L_m[A, B]$ on L_m with the cusps

$$A, \quad E_{p+1}, \quad \cdots, \quad E_q, \quad B.$$

Since the line-segments

$$\overline{AE_{p+1}}, \quad \overline{E_{p+1}E_{p+2}}, \quad \cdots, \quad \overline{E_q B}, \tag{1.21}$$

have the slopes
$$f(t_p, x_p), \quad f(t_{p+1}, x_{p+1}), \quad \cdots, \quad f(t_q, x_q),$$
respectively, we can prove by induction that each line-segment in (1.21) lies in the triangle Δ. Hence, we have
$$L_m[A, B] = \overline{AE_{p+1}} \cup \overline{E_{p+1}E_{p+2}} \cup \cdots \cup \overline{E_q B} \subset \Delta,$$
which implies that the line-segment $\overline{AB}$ is contained in Δ. It follows that the slope of $\overline{AB}$ is bounded between the slope M_a of the side $\overline{AF_a}$ and the slope M_b of the side $\overline{AF_b}$. We have thus proved the inequality (1.20).

3) Finally, define the function
$$Y(t) := \limsup_{m \to \infty} L_m(t), \qquad t_0 \leq t \leq t_0 + h.$$

Let $\tau, \tau + \sigma \in [t_0, t_0 + h]$ ($\sigma \neq 0$), and let $\xi = Y(\tau)$. Then we have
$$\frac{Y(\tau + \sigma) - Y(\tau)}{\sigma} = \frac{1}{\sigma}\left[\limsup_{m \to \infty} L_m(\tau + \sigma) - \limsup_{m \to \infty} L_m(\tau)\right].$$

It follows from (1.19) that
$$\liminf_{m \to \infty} \frac{L_m(\tau + \sigma) - L_m(\tau)}{\sigma} \leq \frac{Y(\tau + \sigma) - Y(\tau)}{\sigma}$$
$$\leq \limsup_{m \to \infty} \frac{L_m(\tau + \sigma) - L_m(\tau)}{\sigma},$$
which together with (1.20) yields
$$\min_{(t,x) \in Q_\sigma} \{f(t, x)\} \leq \frac{Y(\tau + \sigma) - Y(\tau)}{\sigma} \leq \max_{(t,x) \in Q_\sigma} \{f(t, x)\},$$
where
$$Q_\sigma = \{(t, x) \mid |t - \tau| \leq |\sigma|, \ |x - \xi| \leq M|\sigma|\}.$$
Then, letting $\sigma \to 0$, we get
$$Y'(\tau) = f(\tau, \xi) \qquad (t_0 \leq \tau \leq \tau + h).$$
It follows from $\xi = Y(\tau)$ and $Y(t_0) = x_0$ that $x = Y(t)$ is a solution of the Cauchy problem (E_0) on the interval $[t_0, t_0 + h]$.

Therefore, we have proved the existence of solution of the planar Cauchy problem (E_0) without using the Arzelà-Ascoli lemma and even the idea of integration.

We can prove in a similar manner that
$$x = Z(t) = \liminf_{m \to \infty} L_m(t) \qquad (t_0 \leq t \leq t_0 + h)$$
is a solution of the planar Cauchy problem (E_0), with the property that
$$Z(t) \leq Y(t), \qquad t_0 \leq t \leq t_0 + h.$$

Remark Since the solution $y = F(x)$ of the Cauchy problem
$$\frac{dy}{dx} = f(x), \quad y(0) = 0$$
is a primitive function of $f(x)$ (i.e., $F'(x) = f(x)$), we have proved, as a corollary of the above result, the existence of primitive functions for the continuous functions without using the idea of integration.

1.2.2 Maximal Solution and Minimal Solution

Consider a planar Cauchy problem
$$(E_0): \qquad \frac{dx}{dt} = f(t, x), \quad x(t_0) = x_0,$$
where $f(t, x)$ is a continuous function in the region
$$Q: \qquad |t - t_0| \leq a, \quad |x - x_0| \leq b.$$

Now, let k be a positive integer. Assume $x = \phi_k(t)$ is a solution of the Cauchy problem
$$(E_{1/k}): \qquad \frac{dx}{dt} = f(t, x) + \frac{1}{k}, \quad x(t_0) = x_0,$$
and $x = \psi_k(t)$ is a solution of the Cauchy problem
$$(E_{-1/k}): \qquad \frac{dx}{dt} = f(t, x) - \frac{1}{k}, \quad x(t_0) = x_0.$$

Lemma 1.2 *If $x = \chi(t)$ is a solution of the Cauchy problem (E_0), then*
$$\psi_k(t) \leq \chi(t) \leq \phi_k(t), \qquad t_0 \leq t \leq t_0 + h. \tag{1.22}$$

The proof of (1.22) is trivial, and is thus omitted.

Moreover, it can be shown that the sequences $\{\phi_k(t)\}$ and $\{\psi_k(t)\}$ are uniformly bounded and equi-continuous in the interval $[t_0, t_0+h]$. It follows from the Arzelà-Ascoli lemma that they have uniformly convergent subsequences, respectively. Without loss of generality, assume

$$\Phi(t) = \lim_{k \to \infty} \phi_k(t) \quad \text{and} \quad \Psi(t) = \lim_{k \to \infty} \psi_k(t),$$

for $t \in [t_0, t_0 + h]$. It follows from $(E_{1/k})$ and $(E_{-1/k})$ that $x = \Phi(t)$ and $x = \Psi(t)$ are solutions of the Cauchy problem (E_0), and (1.22) implies

$$\Psi(t) \leq \chi(t) \leq \Phi(t), \qquad t_0 \leq t \leq t_0 + h, \tag{1.23}$$

where $x = \chi(t)$ is any solution of (E_0).

In this sense, $x = \Phi(t)$ is called the *maximal solution* and $x = \Psi(t)$ is called the *minimal solution* for the planar Cauchy problem (E_0).

Hence, when $f(t, x)$ is continuous in $Q \subset \mathbb{R}^2$, there exist a maximal solution and a minimal solution of (E_0) on $[t_0, t_0+h]$. The maximal solution equals to the minimal solution if and only if the solution of (E_0) is unique on $[t_0, t_0 + h]$.

A similar conclusion can be derived on the left-hand interval $[t_0 - h, t_0]$.

1.3 Local Behavior of Integral Curves

1.3.1 *Integral Box*

Let $f(t, x)$ be a continuous n-vector field in the region $D \subset \mathbb{R} \times \mathbb{R}^n$. It follows from the Peano theorem that there is at least a solution $x = \varphi(t, \tau, \xi)$ of (1.1) passing through each initial point $(\tau, \xi) \in D$.

Assume the solution $x = \varphi(t, \tau, \xi)$ passing through each point $(\tau, \xi) \in D$ is unique. Then it can be shown that the local behavior of the solutions $\{\varphi(t, \tau, \xi)\}$ is quite simple in geometry.

In fact, it follows from Theorem 1.3 that the solution $x = \varphi(t, \tau, \xi)$ of (E) is continuous in $(t, \tau, \xi) \in I_0 \times G$. In particular, for the fixed $\tau = t_0$, the solution $x = \psi(t, \xi) = \varphi(t, t_0, \xi)$ represents the integral curve Γ_ξ of (1.1) passing through the initial point (t_0, ξ). It follows that the solution $x = \psi(t, \xi)$ is continuous in the box

$$B_0 = \{(t, \xi) \in \mathbb{R} \times \mathbb{R}^n : |t - t_0| \leq h/4, \ |\xi - x_0| \leq b/2\},$$

with the property that

$$\psi(t,\xi_1) \neq \psi(t,\xi_2) \ (or \ \Gamma_{\xi_1} \cap \Gamma_{\xi_2} = \emptyset) \iff \xi_1 \neq \xi_2.$$

Let

$$\Upsilon_0 = \{(t,x) \in Q : \ x = \psi(t,\xi), \ |t-t_0| \leq h/4 \ (|\xi-x_0| \leq b/2)\}.$$

Then

$$\mathscr{T}: \quad (t,\xi) \quad \mapsto \quad (t,\psi(t,\xi)) \tag{1.24}$$

is a topological transformation from B_0 onto Υ_0, with the property that the straight line-segment

$$L_\xi = \{(t,\xi) \in B_0 : \ |t-t_0| \leq h/4\}$$

is transformed onto the integral curve

$$\Gamma_\xi = \{x = \psi(t,\xi) : \ |t-t_0| \leq h/4\}$$

(i.e., $\mathscr{T}(L_\xi) = \Gamma_\xi$). It follows that the family of integral curves $\{\Gamma_\xi\}$ in Υ_0 is topologically equivalent to that of parallel lines $\{L_\xi\}$ in B_0, such that $L_\xi = \mathscr{T}^{-1}(\Gamma_\xi)$ for any $\Gamma_\xi \in \Upsilon_0$.

The set Υ_0 of integral curves is called an *integral box* of the differential equation (1.1) across the initial point (t_0, x_0) (see the following Fig. 1).[5]

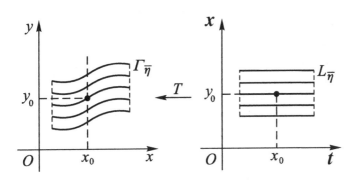

(Fig. 1)

We have thus proved the following result:

---
[5] Where the (t,x)–space is replaced by (x,y)–space.

Theorem 1.7 *If the solution $x = \varphi(t, \tau, \xi)$ of differential equation (1.1) satisfying the initial condition $x(\tau) = \xi$ is unique for each $(\tau, \xi) \in G$, then there is an integral box Υ_0 of (1.1) across the initial point (t_0, x_0).*

1.3.2 Peano Broom

If the solution of the Cauchy problem (E) is not unique, there are at least two integral curves of (E). Then

$$\Omega(E) = \{\,\Gamma : \quad \Gamma \text{ is an integral curve of } (E) \text{ in } Q\,\}$$

is called the *Peano broom* of (E). Sometimes, for convenience, a Peano broom is also considered as the point-set occupied by the integral curves therein.

Example 1.1 Consider the planar Cauchy problem

$$(E_1): \quad \frac{dy}{dx} = \sqrt{|y|} + b, \quad y(x_0) = 0, \tag{1.25}$$

where b is a constant. It can be shown that

(1) If $b \neq 0$, then the solution of (E_1) is unique. Hence, (E_1) has no Peano broom;

(2) If $b = 0$, then the solution of (E_1) is not unique. It is not hard to find the Peano broom

$$\Omega(E_1) = \begin{cases} 0 \leq y \leq \dfrac{1}{4}(x - x_0)^2, & 0 \leq x - x_0 < \infty; \\ -\dfrac{1}{4}(x - x_0)^2 \leq y \leq 0, & -\infty < x - x_0 \leq 0. \end{cases}$$

Hence, there is a Peano broom at each point x_0 on the line $y = 0$, which is a singular solution of the differential equation $dy/dx = \sqrt{|y|}$.

In fact, Peano broom exists at each point of a singular solution. We have trivial examples of differential equations, which have Peano brooms across all points of the singular solutions. But it is not trivial to find a differential equation having Peano brooms across all points in a region.

Example 1.2 In the paper [84], Lavrentief gave such an example

$$\frac{dy}{dx} = g^*(x, y), \tag{1.26}$$

where $g^*(x,y)$ is a continuous function defined on a closed square

$$Q = \{(x,y) \in \mathbb{R}^2 : |x| \leq 1, |y| \leq 1\},$$

such that the differential equation (1.26) has a Peano broom across each point in the square Q. It is said in literature (see [109] and [102]) that the differential equation (1.26) admits *Lavrentief Phenomenon* on Q.

Example 1.3 Let

$$t = t, \quad x = r\cos\theta, \quad y = r\sin\theta$$

be the cylindrical coordinates in the space (t,x,y). Consider a spatial Cauchy problem

$$(E_2): \begin{cases} \dfrac{dr}{dt} = 0, \quad \dfrac{d\theta}{dt} = \sqrt{|\sin\theta|}; \\ r(0) = 1, \quad \theta(0) = 0. \end{cases}$$

Since the integral

$$\int_0^c \frac{d\theta}{\sqrt{|\sin\theta|}} \quad (\text{for } c \neq 0)$$

is convergent, it follows from the Osgood's criterion (see [21]) that the solution of (E_2) is not unique. Hence, the Cauchy problem (E_2) has a Peano broom $\Omega(E_2)$ on the cylindrical surface Z_1 $(r = 1)$. The intersection of $\Omega(E_2)$ and the plane T_c $(t = c > 0)$ is an arc A_c on the circle $S_c = Z_1 \cap T_c$. If the condition

$$c < \int_0^{2\pi} \frac{d\theta}{\sqrt{|\sin\theta|}}$$

is satisfied, then the arc A_c is a proper sub-arc of S_c; if

$$c \geq \int_0^{2\pi} \frac{d\theta}{\sqrt{|\sin\theta|}},$$

then the arc A_c covers the circle S_c. In any case, the Peano broom $\Omega(E_2)$ is a two-dimensional region on the cylindrical surface Z_1.

This example shows that the Peano broom of differential equations in three-dimensional space $\mathbb{R}^3$ may contain no three-dimensional interior point. However, it can be seen from the Lemma 1.3 below that the Peano broom of differential equations in two-dimensional plane $\mathbb{R}^2$ contains at least a two-dimensional interior point. Generally speaking, the structure of Peano brooms in higher dimensional space is rather complicated.

1.4 Peano Phenomenon

Consider the differential equation

$$\frac{dx}{dt} = f(t, x), \tag{1.27}$$

where the vector field $f(t,x)$ is continuous in the region $D \subset \mathbb{R} \times \mathbb{R}^n$. Since we consider the local behavior of the solutions of (1.27), there is no loss of generality to assume that the region D is a strip

$$S = \{(t, x) : \ |t| \leq 1, \ x \in \mathbb{R}^n\},$$

centered at the origin $\mathfrak{o}$, and the vector field $f(t, x)$ has the property that

$$f(t, x) = 0, \quad (t, x) \in S \quad (\text{for } |x| \gg 1). \tag{1.28}$$

It can be seen that the solution $x = \varphi(t)$ of (1.27) passing through each point $p = (\tau, \xi) \in S$ (i.e., $\varphi(\tau) = \xi$) exists on the interval $I = [-1, 1]$.

It follows from Theorem 1.7 that the differential equation (1.27) has an integral box Υ_p across the point $p \in S$ if the solution $x = \varphi(t)$ of (1.27), which passes through each initial point $z \in S$ near the point p, is unique.

If the solution $x = \varphi(t)$ of (1.27) passing through the point $p \in S$ is not unique, then there is a Peano broom

$$\Omega_p = \{(t, x) \in D : \ x = \varphi(t) \text{ is a solution of (1.27) passing through } p\}$$

across the point $p \in S$. In this case, the solution of (1.27) passing through the initial point p is not predicable. Really, *Peano phenomenon* happens to the Cauchy problem at p.

1.4.1 Density Theorem on Peano Phenomena

Now, consider the Cauchy problem

$$(E): \quad \frac{dx}{dt} = f(t, x), \quad x(0) = 0,$$

where the continuous vector field $f(t, x)$ satisfies the property (1.28). It is obvious that the solution of (E) exists on the interval $I = [-1, 1]$.

The Cauchy problem (E) is called *regular* if it has just one solution on I, and *singular* if it has at least two distinct solutions on I.

Denote by $\mathcal{F}$ the family of continuous functions $\{f(t,x)\}$ having the property (1.28). Define the norm
$$\|f\| = \sup_{(t,x)\in S} |f(t,x)|.$$

It is clear that $\mathcal{F}$ is a linear functional space with the norm $\|\cdot\|$.

Consider the partition
$$\mathcal{F} = \mathcal{F}_U \cup \mathcal{F}_P,$$
where $\mathcal{F}_U$ and $\mathcal{F}_P$ are defined, respectively, by

1) $f \in \mathcal{F}_U$ if and only if (E) is regular;
2) $f \in \mathcal{F}_P$ if and only if (E) is singular.

If (E) is singular, then the Peano phenomenon happens to the Cauchy problem (E). In this case, the solution of (E) is not predicable, and there is the *Peano broom*
$$\Omega(E) = \{(t,x) \in S : \quad x = \phi(t) \ (t \in I) \text{ is a solution of } (E)\}$$
at the initial point o which describes the chaotic behavior of the Peano phenomenon.

There arises a natural problem:

How often does Peano phenomenon happen to (E)? or *What is the probability of Peano phenomena happening to (E)?*

In what follows, we will prove the following result, which estimates, in certain sense, the "lower bound" of the probability of Peano phenomena happening to differential equations.

Theorem 1.8 $\mathcal{F}_P$ *is dense in* $\mathcal{F}$.

Proof. Let $f_0 \in \mathcal{F}$. Given $\varepsilon > 0$, using the approximation theorem, we have a C^∞-differential function $\hat{f}$ in $\mathcal{F}$, such that
$$\|\hat{f} - f_0\| < \frac{\varepsilon}{2}. \tag{1.29}$$

It is clear that the Cauchy problem
$$(\hat{E}): \quad \frac{dx}{dt} = \hat{f}(t,x), \quad x(0) = \xi,$$

has a unique solution $x = \Phi(t,\xi)$. It is obvious that $\Phi(t,\xi)$ is C^∞-differentiable in $(t,\xi) \in S$ and satisfies

$$\Phi(t,\xi) \equiv \xi, \qquad \text{when } |\xi| \gg 1. \tag{1.30}$$

Denote by Γ_ξ the integral curve of $(\hat{E})$; that is,

$$\Gamma_\xi = \{(t,x) \in S : \quad x = \Phi(t,\xi), \ t \in I\}.$$

Consider the integral box

$$\Upsilon = \{\Gamma_\xi : \quad \xi \in \mathbb{R}^n\},$$

which agrees with the strip S as a point set.

Then we have a topological mapping

$$\mathscr{T}_0 : \quad (t,\xi) \mapsto (t, \Phi(t,\xi))$$

from S onto Υ, such that

$$\mathscr{T}_0(L_\xi) = \Gamma_\xi,$$

where

$$L_\xi = \{(t,x) \in B : \quad x = \xi, \ t \in I\}$$

is a horizontal straight line-segment in S.

It follows from the property of the mapping $\mathscr{T}_0$ that the family of integral curves $\{\Gamma_\xi\}$ in Υ is topologically equivalent to the family of parallel line-segments L_ξ in S.

Furthermore, $\mathscr{T}_0$ has a continuous tangent map

$$\mathscr{T}_0' : \quad S \times (\mathbb{R} \times \mathbb{R}^n) \quad \to \quad \Upsilon \times (\mathbb{R} \times \mathbb{R}^n).$$

If the roles of the regions S and Υ are ignored for brevity, $\mathscr{T}_0'$ is a linear map

$$\mathscr{T}_0' : \quad \mathbb{R} \times \mathbb{R}^n \quad \to \quad \mathbb{R} \times \mathbb{R}^n,$$

defined by

$$\mathscr{T}_0' = \begin{pmatrix} 1 & 0_n \\ \hat{f}(t, \Phi(t,\xi)) & \Phi_\xi'(t,\xi) \end{pmatrix}$$

where 0_n denotes the n-row null vector, and $\Phi'_\xi(t,\xi)$ is the Jacobian matrix. It follows from (1.30) that $\Phi'_\xi(t,\xi)$ is the unit matrix for $(t,\xi) \in S$ when $|\xi| \gg 1$, and is thus bounded in S. Let

$$\|\Phi'_\xi\| < B_0, \tag{1.31}$$

where B_0 is a positive constant. It follows that $\mathscr{T}'_0$ is a linear continuous, invertible and bounded map, such that

$$\mathscr{T}'_0 \begin{pmatrix} 1 \\ \hat{0} \end{pmatrix} = \begin{pmatrix} 1 \\ \hat{f} \end{pmatrix} \quad \text{or} \quad (\mathscr{T}'_0)^{-1} \begin{pmatrix} 1 \\ \hat{f} \end{pmatrix} = \begin{pmatrix} 1 \\ \hat{0} \end{pmatrix},$$

where $\hat{0}$ is the n-column null vector.

Finally, let $\Lambda(u)$ be a continuous function in $u \in \mathbb{R}$, satisfying

$$\Lambda(u) = \begin{cases} \sigma\sqrt{|u|}, & |u| \leq 1; \\ 0, & |u| > 2, \end{cases}$$

with some small parameter $\sigma > 0$. It is clear that the Cauchy problem

$$\frac{du}{dt} = \Lambda(u), \quad u(0) = 0 \quad (u \in \mathbb{R})$$

is singular.

Then consider correspondingly the Cauchy problem

$$(E^*): \quad \frac{dz}{dt} = g(z), \quad z(0) = 0 \quad ((t,z) \in S),$$

where

$$g(z) = \begin{pmatrix} \Lambda(z_1) \\ \vdots \\ \Lambda(z_n) \end{pmatrix}, \quad z = \begin{pmatrix} z_1 \\ \vdots \\ z_n \end{pmatrix} \in \mathbb{R}^n.$$

Take a sufficiently small constant $\sigma > 0$, such that

$$\|g\| < \frac{\varepsilon}{2B_0}. \tag{1.32}$$

It can be seen that (E^*) has a Peano broom $\Omega(E^*)$ in S. Let

$$\begin{pmatrix} 1 \\ \check{f} \end{pmatrix} = \mathscr{T}'_0 \begin{pmatrix} 1 \\ g \end{pmatrix}.$$

Then the Cauchy problem

$$(\check{E}): \quad \frac{dx}{dt} = \check{f}(t,x), \quad x(0) = 0,$$

has the Peano broom $\Omega(\check{E}) = \mathscr{T}_0(\Omega(E^*))$. It implies

$$\check{f} \in \mathcal{F}_P. \tag{1.33}$$

On the other hand, we have

$$\begin{pmatrix} 1 \\ \check{f} \end{pmatrix} - \begin{pmatrix} 1 \\ \hat{f} \end{pmatrix} = \mathscr{T}_0' \begin{pmatrix} 1 \\ g \end{pmatrix} - \mathscr{T}_0' \begin{pmatrix} 1 \\ \hat{0} \end{pmatrix} = \mathscr{T}_0' \begin{pmatrix} 0 \\ g \end{pmatrix},$$

which yields

$$\check{f} - \hat{f} = \Phi_\xi' g.$$

It follows from (1.31) and (1.32) that

$$\|\check{f} - \hat{f}\| = \|\Phi_\xi' g\| \le B_0 \|g\| < \frac{1}{2}\varepsilon.$$

Finally, using (1.29), we have

$$\|\check{f} - f_0\| < \varepsilon,$$

which together with (1.33) implies that $\mathcal{F}_P$ is dense in $\mathcal{F}$.

The proof of Theorem 1.8 is thus completed. □

Theorem 1.8 shows that the Peano phenomena are dense among the differential equations of order n for any integer $n \ge 1$.

1.4.2 *Scarcity Theorem on Planar Peano Phenomena*

1) Now, consider a planar Cauchy problem

$$(E_0): \quad \frac{dx}{dt} = f(t,x), \quad x(t_0) = x_0,$$

where $f(t,x)$ is a continuous and bounded function defined on the planar strip

$$S_2 = \{(t,x) \in \mathbb{R}^2 : \ |t| \le 1, \ x \in \mathbb{R}^1\}.$$

Denote by $\mathfrak{F}$ the family of continuous and bounded functions on S_2, and let
$$\|f\| = \sup_{(t,x)\in S_2} |f(t,x)|,$$
be the norm of a function $f \in \mathfrak{F}$. It can be verified that $\mathfrak{F}$ is a (complete) Banach space.

It is noted that for $f \in \mathfrak{F}$, each solution of the differential equation $dx/dt = f(t,x)$ exists on the interval $I = [-1, 1]$.

Assume the solution of (E_0) is not unique. Then (E_0) has a Peano broom $\Omega(E_0)$ across the initial point (t_0, x_0).

On the other hand, when the solution of (E_0) is not unique, there exist a unique maximal solution $\Phi(t)$ of (E_0) and a unique minimal solution $\Psi(t)$ of (E_0) on the interval I. Let
$$\Omega^* = \{\, (t,x) \in S : \quad \Psi(t) \leq x \leq \Phi(t), \quad t \in I \,\}.$$

Lemma 1.3 *If the solution of (E_0) is not unique, then the Peano broom*
$$\Omega(E_0) = \Omega^*.$$

Proof. Let
$$\Gamma: \qquad x = \phi(t) \quad (t \in I),$$
be an integral curve of (E_0). It is follows from
$$\Psi(t) \leq \phi(t) \leq \Phi(t) \quad (t \in I)$$
that $\Gamma \subset \Omega^*$.

On the other hand, given a point
$$(\tau, \xi) \in \Omega^*,$$
the differential equation $dx/dt = f(t,x)$ has at least a solution $x = u(t)$ passing through the initial point (τ, ξ) (i.e., $u(\tau) = \xi$). Since $f(t,x)$ is bounded on S, $x = u(t)$ exists on the interval I.

If $x = u(t)$ is bounded by the minimal solution $\Psi(t)$ and the maximal solution $\Phi(t)$ on the interval I, then we have
$$\Psi(t) \leq u(t) \leq \Phi(t), \qquad t \in I,$$
which implies $\Psi(t_0) = u(t_0) = \Phi(t_0) = x_0$. It follows that $x = u(t)$ is a solution of (E_0), which passes through the point (τ, ξ).

If $x = u(t)$ is not bounded by $\Psi(t)$ and $\Phi(t)$, then there exist a constant $\alpha \in [-1, \tau)$ and a constant $\beta \in (\tau, 1]$, such that

$$\Psi(t) \leq u(t) \leq \Phi(t), \qquad t \in (\alpha, \beta),$$

satisfying

$$u(\alpha) = \Phi(\alpha) \ or \ \Psi(\alpha) \qquad and \qquad u(\beta) = \Phi(\beta) \ or \ \Psi(\beta).$$

For definiteness, assume

$$u(\alpha) = \Phi(\alpha) \qquad and \qquad u(\beta) = \Psi(\beta).$$

Then, let

$$w(t) = \begin{cases} \Phi(t), & as \ t \in [-1, \alpha], \\ u(t), & as \ t \in (\alpha, \beta), \\ \Psi(t), & as \ t \in [\beta, 1]. \end{cases}$$

It can be easily seen that $x = w(t)$ is a solution of (E_0), which passes through the point (τ, ξ).

Therefore, we have proved that the set Ω^* is covered by the integral curves of (E_0).

Lemma 1.3 is thus proved. □

Corollary 1.2 *Since $\Phi(t) \not\equiv \Psi(t)$ on I, the planar Peano broom $\Omega(E_0)$ contains at least an interior point (and thus at least a rational interior point $p^* = (t^*, x^*)$, where t^* and x^* are rational numbers).*

Remark 1.1 *It can be seen from the Example 1.3 that a spatial Peano broom may contains no spatial interior rational point.*

2) Now, let

$$\mathfrak{F} = \mathfrak{F}_U \cup \mathfrak{F}_P$$

be the partition defined by the planar Cauchy problem (E_0) in a similar manner as the partition of the space $\mathcal{F}$ mentioned above.

It is known that the measure theory holds in a complete Banach space. Therefore, there is a normal measure μ in $\mathfrak{F}$, such that $\mu[\mathfrak{F}] = 1$ and $\mu[\mathcal{O}] > 0$ for any open set $\mathcal{O} \subset \mathfrak{F}$.

The following theorem asserts the scarcity of Peano phenomena among the planar differential equations (see [102]).

Theorem 1.9 *There is a normal measure ν in $\mathfrak{F}$, such that $\nu(\mathfrak{F}_P) = 0$.*

Proof. For brevity, we will give a sketch of the proof. Let
$$\mathfrak{F}_0 = \{g \in \mathfrak{F} : \ g(0,0) = 0\}.$$
Then $\mathfrak{F}_0$ is a linear subspace of $\mathfrak{F}$. It follows that, for given $f \in \mathfrak{F}$, we have
$$f(t,x) = f_0(t,x) + \lambda,$$
where $f_0 \in \mathfrak{F}_0$ and $\lambda = f(0,0) \in \mathbb{R}^1$. Hence,
$$\mathfrak{F} = \mathfrak{F}_0 \times \mathbb{R}^1. \tag{1.34}$$
Since $\mathfrak{F}_0$ is a closed subset in $\mathfrak{F}$, it is measurable.

On one hand, for any $g \in \mathfrak{F}_0$, we have
$$\{g\} \times \mathbb{R} = (\{g\} \times \Lambda_U(g)) \cup (\{g\} \times \Lambda_P(g)) \tag{1.35}$$
where the sets $\Lambda_U(g)$ and $\Lambda_P(g)$ are defined as follows:
$$\Lambda_U(g) = \{\lambda \in \mathbb{R} : \ g(t,x) + \lambda \in \mathfrak{F}_U\};$$
$$\Lambda_P(g) = \{\lambda \in \mathbb{R} : \ g(t,x) + \lambda \in \mathfrak{F}_P\}.$$

On other hand, for any positive integer n, let
$$\mathfrak{F}_P^{(n)} = \left\{ f \in \mathfrak{F}_P : \ \max_{t \in I} |\Phi(t) - \Psi(t)| \geq \frac{1}{n} \right\}.$$
It is clear that $\mathfrak{F}_P^{(n)}$ is a closed set and is thus measurable. Since
$$\mathfrak{F}_P = \bigcup_{n \geq 1} \mathfrak{F}_P^{(n)},$$
$\mathfrak{F}_P$ is also a measurable set.

Let μ_0 be the restricted measure of μ in the subspace $\mathfrak{F}_0$, and let σ be the Lebesgue measure in $\mathbb{R}^1$. Then, $\nu(\cdot) = \mu_0(\cdot) \times \sigma(\cdot)$ is a well-defined measure in the product space $\mathfrak{F}_0 \times \mathbb{R}^1 (= \mathfrak{F})$.

Besides, the initial-value problem (E_0) can be put in the form:
$$(E)_\lambda : \quad \frac{dx}{dt} = f^*(t,x) + \lambda, \quad x(t_0) = x_0,$$
where $f^* \in \mathfrak{F}_0$.

Lemma 1.4 *If $\lambda_1 \neq \lambda_2$, the Peano broom $\Omega((E)_{\lambda_1})$ does not intersect the Peano broom $\Omega((E)_{\lambda_2})$ except at the initial point (t_0, x_0).*

Proof. We claim that the solution of $(E)_{\lambda_1}$ does not intersect the solution of $(E)_{\lambda_2}$ except at the initial point (t_0, x_0).

For definiteness, let

$$\lambda_1 < \lambda_2, \tag{1.36}$$

and let $x = x_1(t)$ and $x = x_2(t)$ be the solutions of $(E)_{\lambda_1}$ and $(E)_{\lambda_2}$, respectively. It is noticed that

$$x_1(t_0) = x_2(t_0) = x_0.$$

It follows that there is a constant $\alpha > 0$, such that

$$x_2(t) > x_1(t), \qquad t_0 < t \leq t_0 + \alpha.$$

It suffices to prove that $t_0 + \alpha \geq 1$.

Otherwise, we have $t_1 = t_0 + \alpha < 1$ ($t_1 > t_0 \geq -1$), such that

$$x_2(t) > x_1(t) \quad (t_0 < t < t_1) \quad \text{and} \quad x_2(t_1) = x_1(t_1),$$

which implies $x_1'(t_1) \geq x_2'(t_1)$. That is,

$$f^*(t_1, x_1(t_1)) + \lambda_1 \geq f^*(t_1, x_2(t_1)) + \lambda_2.$$

It follows from $f^*(t_1, x_1(t_1)) = f^*(t_1, x_2(t_1))$ that

$$\lambda_1 \geq \lambda_2,$$

which contradicts the assumption (1.36).

Therefore, we have

$$x_2(t) > x_1(t) \qquad (t_0 < t \leq 1).$$

Similarly, we can prove

$$x_2(t) < x_1(t) \qquad (-1 \leq t < t_0).$$

The proof of Lemma 1.4 is thus completed. $\square$

Now, for given $f \in \mathfrak{F}_P$, we have $f = g + \lambda$ with $\lambda \in \Lambda_P(g)$. Using Corollary 1.2, we choose a fixed rational point p_λ^* in the interior of $\Omega((E)_\lambda)$. It follows from Lemma 1.4 that the Peano brooms $\{\Omega((E)_\lambda)\}$ corresponds, one by one, to the rational points $\{p_\lambda^*\}$. Hence, $\Lambda_P(g)$ is a countable set. It follows that $\sigma(\Lambda_P(g)) = 0$.

Finally, using Fubini theorem with (1.34) and (1.35) in mind, we get

$$\nu(\mathfrak{F}_P) = \int_{\mathfrak{F}_0} \int_{\Lambda_P(g)} d\lambda d\mu_0 = \int_{\mathfrak{F}_0} \sigma(\Lambda_P(g)) \, d\mu_0 = \int_{\mathfrak{F}_0} 0 \, d\mu_0 = 0,$$

which proves Theorem 1.9. □

Theorem 1.9 states that Peano phenomena are scarce among planar differential equations (i.e., differential equations of first order). Indeed, in this sense, it estimates the "upper bound" of the probability of planar Peano phenomena happening to the Cauchy problem (E_2) for $f \in \mathfrak{F}$.

However, we do not know whether or not the Peano phenomena are scarce among differential equations of higher-order too. Anyway, it can be seen from the proof of Theorem 1.9 and the Remark 1.1 that it is not trivial to prove the scarcity of Peano phenomena in higher dimensional space.

1.5 Convergence Theorem on Difference Methods

1.5.1 *Classical Difference Methods*

Consider the Cauchy problem

$$(E): \qquad \frac{dy}{dx} = f(x,y), \quad y(x_0) = y_0,$$

where $f(x, y)$ is a continuous n-vector valued function of (x, y) in a closed rectangular region $Q \subset \mathbb{R}^1 \times \mathbb{R}^n$. On the one hand, it is well-known in the theory that the Cauchy problem (E) has at least a solution $y = \varphi(x)$ on the interval $I = [x_0 - h, x_0 + h]$. On the other hand, it is a problem in the application to compute the solution. In literature, there is a large number of approximation methods in numerically solutions of differential equations (see [70] for example). The difference methods are widely used in numerical analysis. Since our treatment is admittedly elementary, we are only concerned with the convergence problem of difference methods.

Suppose that the integral curve

$$\Gamma: \qquad y = \varphi(x), \qquad x_0 - h \leq x \leq x_0 + h,$$

is interpolated by a finite number of points

$$S_m: \qquad (x_i, y_i), \qquad i = 0, \pm 1, \cdots, \pm m,$$

such that $y_i = y(x_i)$, where
$$x_0 - h = x_{-m} < \cdots < x_{-1} < x_0 < x_1 < \cdots < x_m = x_0 + h.$$
Let
$$\delta_m = \max_{0 \leq i \leq m-1} (x_{i+1} - x_i).$$
Assume S_m converges [6] to Γ_0 as $\delta_m \to 0$.

The main idea of difference methods is to find out such a series S_m through an approximation
$$\frac{\Delta y_i}{\Delta x_i} = F(x_i, y_i, y_{i+1}, \delta), \qquad 0 \leq i \leq m-1, \tag{1.37}$$
of the Cauchy problem (E), where
$$\frac{\Delta y_i}{\Delta x_i} = \frac{y_{i+1} - y_i}{x_{i+1} - x_i} \approx \frac{dy}{dx}$$
and
$$F(x_i, y_i, \Delta x_i) \approx f(x, y). \tag{1.38}$$
The *difference scheme* is given by the approximation (1.38), which is usually determined by the first few terms of the Taylor expansion
$$F(x_i, y_i, \Delta x_i) = f(x_i + \Delta x_i, y_i + \Delta y_i)$$
$$= f(x_i, y_i) + [f'_x(x_i, y_i)\Delta x_i + f'_y(x_i, y_i)\Delta y_i]$$
$$+ \frac{1}{2!}[f''_{x_i x_i}(x_i, y_i)\Delta^2 x_i + 2f''_{x_i y_i}\Delta x_i \Delta y_i + f''_{y_i y_i}(x_i, y_i)\Delta^2 y_i] + \cdots.$$

Based on the knowledge of the convergent series from calculus, we know that the accuracy of our approximation improves with the number of terms retained. On the other hand, the description of the higher-order terms gets more and more complicated and the associated calculations more profuse. The basic idea of the Runge-Kutta methods is to preserve the order of a Taylor approximation (in the sense of the error involved) while eliminating the necessity of calculating the various partial derivatives of $f(x, y)$ that are involved. The alternative proposed by these methods involves evaluating

---
[6] In fact, the point sequence S_m is understood here as a polygonal approximation with vertices (x_i, y_i) $(i = 0, \pm 1, \cdots, \pm m)$.

the function $f(x, y)$ at certain judicious points rather than evaluating the specific partial derivatives.

Example 1.4 The Euler polygonal approximation is in fact a simple difference method given by

$$y_{i+1} = y_i + F(x_i, y_i, \Delta x_i)\Delta x_i, \qquad (1.39)$$

where

$$F(x_i, y_i, \Delta x_i) = f(x_i, y_i).$$

Example 1.5 The Runge-Kutta approximation of order 3 is given by the following collection of formulas:

$$y_{i+1} = y_i + F(x_i, y_i, \Delta x_i)\Delta x_i, \qquad (1.40)$$

where

$$F(x_i, y_i, \Delta x_i) = \frac{u_i + 4v_i + w_i}{6},$$

and

$$\begin{cases} u_i = f(x_i, y_i), \\ v_i = f(x_i + \tfrac{1}{2}\Delta x_i, y_i + \tfrac{1}{2}u_i), \\ w_i = f(x_i + \Delta x_i, y_i + 2v_i - u_i). \end{cases}$$

Example 1.6 The Runge-Kutta approximation of order 4 is given by the following collection of formulas:

$$y_{i+1} = y_i + F(x_i, y_i, \Delta x_i)\Delta x_i, \qquad (1.41)$$

where

$$F(x_i, y_i, \Delta x_i) = \frac{a_i + 2b_i + 2c_i + d_i}{6},$$

and

$$\begin{cases} a_i = f(x_i, y_i), \\ b_i = f(x_i + \tfrac{1}{2}\Delta x_i, y_i + \tfrac{1}{2}a_i), \\ c_i = f(x_i + \tfrac{1}{2}\Delta x_i, y_i + \tfrac{1}{2}b_i), \\ d_i = f(x_i + \Delta x_i, y_i + c_i). \end{cases}$$

1.5.2 Generalized Difference Schemes

In 1971, W. Gear generalized the classical difference methods into a unified form as follows.

Let us approximate the solution of (E) by the difference equation:

$$\begin{cases} y_{i+1} - y_i = (x_{i+1} - x_i)F(x_i, y_i, y_{i+1}, \Delta x_i) & (\Delta x_i \neq 0), \\ x_{i+1} = x_i + \Delta x_i, \quad i = 0, \pm 1, \pm 2, \cdots, \pm(m-1), \\ \text{with } x_{\pm m} = x_0 \pm h, \end{cases} \quad (1.42)$$

where the function $F(x, y, z, \lambda)$ is continuous on the region

$$D^* : \quad |x - x_0| \leq a, \quad |y - y_0| \leq b, \quad |z - z_0| \leq b, \quad |\lambda| \leq c$$

and satisfies the consistency condition:

$$F(x, y, y, 0) = f(x, y). \quad (1.43)$$

The scheme (1.42) satisfying the consistency condition 1.43 is called the *Gear's difference equation* of (E), which unifies the classical difference schemes.[7]

In what follows, the consistency condition (1.43) will be tacitly assumed when we consider the scheme (1.42).

If the finite sequence

$$S_m : \begin{cases} (x_i, y_i) \in Q, \quad i = 0, \pm 1, \cdots, \pm(m-1), \\ \text{with } \Delta x_i = x_{i+1} - x_i \neq 0, \quad x_{\pm m} = x_0 \pm h, \end{cases}$$

satisfies (1.42), then S_m is called a solution of the difference equation (1.42).

The following result (due to Gear) is a general convergence theorem of difference methods.

Theorem 1.10 *If $F(x, y, z, \lambda)$ is a continuous function in the region D^* and satisfies the Lipschitz condition (with respect to y and z):*

$$|F(x, y_1, z_1, \lambda) - F(x, y_2, z_2, \lambda)| \leq L(|y_1 - y_2| + |z_1 - z_2|), \quad (1.44)$$

---
[7]Compared with the Runge-Kutta methods, the Gear's scheme is concerned with a more accurate approximation of $f(x, y)$ (for example, the difference computations of $f'_y(x, y)$).

then the solution of (1.42) converges to the solution of the initial-value problem (E) as the maximal steps

$$\delta_m = \max_{0 \le |i| \le |m-1|} |\Delta x_i| \to 0 \quad (\text{if } m \to \infty).$$

This theorem proves the convergence of the approximate solutions of Gear's difference equations. It can be seen that if $F(x, y, z, \lambda)$ satisfies the Lipschitz condition (1.44), then $f(x, y)$ satisfies the Lipschitz condition with respect to y. But, the converse is uncertain. Therefore, a natural question arises:

Does the solution of the Gear's differential equation (1.42) converge to the solution of (E) only if $f(x,y)$ is lipschitzian with respect to y?

The application of difference methods is really dependent on the answer of this question. The following theorem gives an affirmative answer to the question (see [28]).

Theorem 1.11 *If the solution $y = \Phi_0(x)$ of Cauchy problem (E) is unique, then the approximate solution of the difference equation (1.42) is convergent to $y = \Phi_0(x)$ as $\delta_m \to 0$.*

We will give the proof in the following several steps.

1.5.3 Preparatory Works

Take a constant

$$M \ge \max_{(x, y, z, \lambda) \in D^*} |F(x, y, z, \lambda)|.$$

Let

$$\alpha = \min\left(a, \frac{b}{M+1}\right), \quad \delta = \frac{\alpha}{m},$$

where m is a positive integer, satisfying $m > \alpha/c$.

Without destroying the generalization, we only consider the right-hand solution $[y_0, y_1, \cdots, y_m]$ for the points $x_{i+1} = x_i + \delta$, $i = 0, 1, \cdots, m-1$. The general case can be discussed in a similar manner.

Lemma 1.5 *The difference equation (1.42) has a right-hand solution $[y_0, y_1, \cdots, y_m]$, satisfying*

$$(x_k, y_k) \in Q \quad (k = 0, 1, \cdots, m). \tag{1.45}$$

Proof. It is clear that $(x_0, y_0) \in Q$. Now, assume $(x_j, y_j) \in Q$ for $j = 0, 1, \cdots, k$, $(k < m)$, and

$$|y_j - y_0| \leq \frac{jb}{m} \qquad (j = 0, 1, \cdots, k). \tag{1.46}$$

Then, we want to find the point y_{k+1} of (1.42) when $i = k$. It needs to prove the existence of the implicit solution $z = y_{k+1}$ of the equation

$$z = y_k + F(x_k, y_k, z, \delta)\delta.$$

For this aim, let

$$G(z) = y_k + F(x_k, y_k, z, \delta)\delta.$$

It is clear that $y = G(z)$ is continuous on the closed ball $B_b(y_0)$ (i.e., $|z - y_0| \leq b$), and satisfies the condition

$$\begin{aligned}|G(z) - y_0| &\leq |y_k - y_0| + \delta|F(x_k, y_k, z, \delta)| \\ &\leq \frac{kb}{m} + \frac{\alpha}{m}M \leq b.\end{aligned} \tag{1.47}$$

Therefore, $y = G(z)$ is a continuous mapping from the closed ball $B_b(y_0)$ into itself. Using the Brouwer fixed-point theorem (see the subsequent chapter of fixed point theorems), we assert that the mapping $y = G(z)$ has at least a fixed-point, denoted by y_{k+1}, in $B_b(y_0)$. It follows that

$$y_{k+1} = y_k + F(x_k, y_k, y_{k+1}, \delta)\delta, \quad |y_{k+1} - y_0| \leq b.$$

Then, it follows from the method of induction that (1.42) has a right-hand solution $[y_0, y_1, \cdots, y_m]$, satisfying (1.45).

In a similar way, we can prove that (1.42) has a left-hand solution. Lemma 1.5 is thus proved. □

Now, for a right-hand solution $[y_0, y_1, \cdots, y_k, y_{k+1}, \cdots, y_m]$ of (1.42), construct an interpolation as follows. Let

$$\Delta y_k = y_k - y_{k-1}, \qquad w_k = \frac{1}{\delta}\Delta y_k,$$

then

$$w_k = F(x_{k-1}, y_{k-1}, y_k, \delta), \qquad (k = 1, 2, \cdots, m);$$

in particular, let

$$w_0 = F(x_0, y_0, y_0, \delta).$$

Assume
$$\Delta w_{k+1} = w_{k+1} - w_k \quad (k = 0, 1, \cdots, m-1).$$
Then, on the interval
$$J_k = [x_0 + k\delta, \ x_0 + (k+1)\delta],$$
construct the following function
$$g_k(x) = y_k + (x - x_k)[w_{k+1} - (x - x_{k+1})^2 \delta^{-2} \Delta w_{k+1}]$$
$(k = 0, 1, \cdots, m-1)$.

Lemma 1.6   For each k $((0 \leq k \leq m-1))$, the function $g_k(x)$ has the following properties:

1. $g_k(x_k) = y_k, \quad g_k(x_{k+1}) = y_{k+1}$;
2. $g'_k(x_k) = w_k, \quad g'_k(x_{k+1}) = w_{k+1}$;
3. $|g'_k(x) - w_{k+1}| \leq |\Delta w_{k+1}|$.

Proof.   Property 1 is obvious. It follows from the definition of $g_k(x)$ that
$$g'_k(x) = w_{k+1} - (x - x_{k+1})(3x - x_{k+1} - 2x_k)\delta^{-2}\Delta w_{k+1},$$
which implies
$$g'_k(x_k) = w_{k+1} - \Delta w_{k+1} = w_k, \qquad g'_k(x_{k+1}) = w_{k+1}.$$
Property 2 is thus proved.

Finally, denoting by $(y)_i$ the i-th component of the vector y, we have
$$(g_k(x))_i = (y_k)_i + (x - x_k)[(w_{k+1})_i - (x - x_{k+1})^2 \delta^{-2}(\Delta w_{k+1})_i].$$
Because the function
$$(g_k(x))''_i = -(6x - 4x_{k+1} - 2x_k)\delta^{-2}(\Delta w_{k+1})_i,$$
changes its sign once only in the interval J_k, the parabola
$$u = (g_k(x))'_i = (w_{k+1})_i - (\Delta w_{k+1})_i(x - x_{k+1})(3x - x_{k+1} - 2x_k)\delta^{-2}$$
assumes its maximal and minimal value only possibly at the end-points of the interval J_k or at the point of vertex
$$x = \hat{x} = \frac{2x_{k+1} + x_k}{3},$$
corresponding respectively to the values:

(1) $(g'_k(x_k))_i = (w_k)_i = (w_{k+1})_i - (\Delta w_{k+1})_i,$
(2) $(g'_k(x_{k+1}))_i = (w_{k+1})_i,$
(3) $(g'_k(\hat{x}))_i = (w_{k+1})_i + \frac{1}{3}(\Delta w_{k+1})_i.$

It follows that $|(g'_k(x))_i - (w_{k+1})_i|$ has the maximal $|(\Delta w_{k+1})_i|$. We have thus

$$|g'_k(x) - w_{k+1}| \leq |\Delta w_{k+1}|,$$

which proves Property 3.

The proof of Lemma 1.6 is thus completed. □

Then, on the interval $J = [x_0, x_0 + \alpha]$, define the Spline function $Z_m(x)$, such that $Z_m(x) = g_k(x)$ for $x \in J_k$, $(0 \leq k \leq m-1)$.

Lemma 1.7 *The Spline function $Z_m(x)$ has the following properties:*

(1) $Z_m(x) \in C^1(J)$;
(2) $Z_m(x_k) = y_k$, $Z'_m(x_k) = w_k$, and

$$|Z'_m(x) - w_{k+1}| \leq |\Delta w_{k+1}| \qquad (x \in J_k);$$

(3) $(x, Z_m(x)) \in Q$, if $|\Delta w_{k+1}| \leq 1$.

Proof. Properties 1 and 2 are immediately derived from the definition of $Z_m(x)$ and Lemma 1.6.

Then, it follows from

$$|Z'_m(x) - w_{k+1}| \leq |\Delta w_{k+1}| \leq 1$$

that

$$|Z'_m(x)| \leq |w_{k+1}| + 1 \leq M + 1.$$

Using the Lagrange mean-value inequality for vector-function, we get

$$|Z_m(x) - y_0| = |Z_m(x) - Z_m(x_0)|$$

$$\leq \sup_{\xi \in J} |Z'_m(\xi)| \cdot |x - x_0| \leq (M+1)\alpha \leq b,$$

which implies the Property 3.

The proof of Lemma 1.7 is thus completed. □

Now, consider the remainder
$$R_m(x) := Z'_m(x) - F(x, Z_m(x), Z_m(x+\delta), \delta),$$
for $x \in [x_0, x_0 + (m-1)\delta]$; and
$$R_m(x) := R_m(x_0 + (m-1)\delta),$$
for $x \in (x_0 + (m-1)\delta, x_0 + \alpha]$.

Lemma 1.8 *The remainder $R_m(x)$ is continuous on the interval J, and converges uniformly to 0 as $m \to \infty$.*

Proof. Let $\varepsilon > 0$. Since the function F is continuous on a closed bounded region D, there is a constant $\sigma > 0$, such that if
$$|x - x^*| < \sigma, \quad |y - y^*| < \sigma, \quad |z - z^*| < \sigma,$$
the inequality
$$|F(x, y, z, \delta) - F(x^*, y^*, z^*, \delta)| < \frac{\varepsilon}{2}$$
is valid on D.

On the other hand, using (1.42), we conclude that if
$$m > N = \min\left\{n \in \mathbb{Z} : \quad n > \frac{(M+1)\alpha}{\sigma}\right\},$$
then
$$|\Delta y_{k+1}| = |\delta F(x_k, y_k, y_{k+1}, \delta)| \leq \frac{\alpha}{m}(M+1) < \sigma,$$
for $k = 1, \cdots, m-1$, and thus
$$\delta = \frac{\alpha}{m} \leq \frac{\alpha}{m}(M+1) < \sigma.$$
It follows that if $m > N$, we have
$$|\Delta w_1| = |F(x_0, y_0, y_1, \delta) - F(x_0, y_0, y_0, \delta)| < \frac{\varepsilon}{2},$$
and
$$|\Delta w_{k+1}| = |F(x_k, y_k, y_{k+1}, \delta) - F(x_{k-1}, y_{k-1}, y_k, \delta)| < \frac{\varepsilon}{2},$$
where $k = 1, \cdots, m-1$.

Now, for any given $x \in J$, choose the maximal x_k, such that $x_k \le x$. We get $|x - x_k| < \delta < \sigma$, and then

$$|Z_m(x) - y_k| = |Z_m(x) - Z_m(x_k)| \le (M+1)\delta < \sigma$$

and

$$|Z_m(x+\delta) - y_{k+1}| = |Z_m(x+\delta) - Z_m(x_k+\delta)| \le (M+1)\delta < \sigma.$$

It follows that if $m > N$ and $x_0 \le x \le x_0 + (m-1)\delta$, we have

$$|R_m(x)| = |Z'_m(x) - F(x, Z_n(x), Z_m(x+\delta), \delta)|$$

$$\le |Z'_m(x) - F(x_k, y_k, y_{k+1}, \delta)|$$

$$+ |F(x_k, y_k, y_{k+1}, \delta) - F(x, Z_m(x), Z_m(x+\delta), \delta)|$$

$$< |Z'_n(x) - w_{k+1}| + \frac{\varepsilon}{2} \le |\Delta w_{k+1}| + \frac{\varepsilon}{2} < \varepsilon.$$

On the other hand, when $x \in (x_0 + (m-1)\delta, x_0 + \alpha]$, we have

$$|R_m(x)| = |R_m(x_0 + (m-1)\delta)| < \varepsilon.$$

It follows that when $m > N$, the inequality $|R_m(x)| < \varepsilon$ holds on the interval J.

Lemma 1.8 is thus proved. □

It is noticed that although $\{R_m(x)\}$ converges uniformly to 0, we can not conclude that $\{Z_m(x)\}$ converges uniformly.

1.5.4 *Proof of Convergence*

Now, we need to prove the uniform convergence of the sequence $\{Z_m(x)\}$. From Lemma 1.7, we conclude that

$$|Z_m(x)| \le |y_0| + b, \quad x \in J.$$

It means that $\{Z_m(x)\}$ is uniformly bounded. Moreover, it follows from

$$|Z'_m(x)| \le M+1$$

that $Z_m(x)$ is equicontinuous. It follows from the Ascoli Lemma that there is a uniformly convergent subsequence $\{Z_{m_j}(x)\}$. Letting

$$\zeta(x) = \lim_{j \to \infty} Z_{m_j}(x),$$

we have

$$F(x, \zeta(x), \zeta(x), 0) = \lim_{j \to \infty} F\left(x, Z_{m_j}(x), Z_{m_j}(x + \frac{\alpha}{m_j}), \frac{\alpha}{m_j}\right).$$

It follows from the consistency condition that

$$F(x, \zeta(x), \zeta(x), 0) = f(x, \zeta(x)).$$

Then, using the definition of the remainder $R_m(x)$, we get

$$Z_{m_j}(x) = y_0 + \int_{x_0}^{x} \left[F\left(x, Z_{m_j}(x), Z_{m_j}(x + \frac{\alpha}{m_j}), \frac{\alpha}{m_j}\right) + R_{m_j}(x)\right] dx,$$

and letting $m_j \to \infty$ yields

$$\zeta(x) = y_0 + \int_{x_0}^{x} f(x, \zeta(x)) \, dx.$$

Therefore, $y = \zeta(x)$ is a solution of (E).

It is noticed that the Peano Existence Theorem (i.e., Theorem 1.1) can be proved by all the difference methods.

Now we are in a position to complete the proof of Theorem 1.11.

Because the solution of (E) is unique, we conclude that $y = \zeta(x)$ is equal to the unique solution $y = \Phi_0(x)$ of (E).

Claim: $y = Z_m(x)$ uniformly converges to $y = \Phi_0(x)$ of (E).

Assume the contrary. Then $\{Z_m(x)\}$ does not uniformly converge to $\Phi_0(x)$. It follows that there is a constant $\varepsilon_0 > 0$, such that for any positive integer s, there exists $x_s \in J$ satisfying

$$|Z_s(x_s) - \Phi_0(x_s)| \geq \varepsilon_0 > 0. \tag{1.48}$$

Without destroying the truth, assume $\{x_s\}$ is a convergent sequence. Let $x_s \to \hat{x}$. Then, similar to the above discussion, we can choose a uniformly convergent subsequence $\{Z_{s_j}(x)\}$ of the sequence $\{Z_s(x)\}$, which uniformly converges to a solution of (E). Since (E) has a unique solution $y = \Phi_0(x)$, we get

$$\lim_{j \to \infty} Z_{s_j}(x) = \Phi_0(x),$$

which is in conflict with (1.48).

We have thus proved Theorem 1.11. □

1.5.5 Necessary Condition of Convergence

Theorem 1.11 shows that the uniqueness of solutions of (E) is a sufficient condition for the convergence of Gear's difference approximation.

Now, we claim: *The uniqueness of solutions of (E) is also a necessary condition for the convergence of Gear's difference approximation.*

Proof. In fact, assume the contrary. Let $y = \Phi(x)$ and $y = \psi(x)$ be two different solutions of (E). Then, we claim that there are Gear's difference equations admitting of divergent solutions.

Without loss of generality, we assume
$$\Phi(x) \neq \psi(x), \quad x \in J = [x_0, x_0 + \alpha].$$

Now, construct a polygon $y = \Phi_m(x)$ passing through vertex-points $(x_i, \Phi(x_i))$ $(i = 0, 1, \cdots, m)$. Similarly, construct a polygon $y = \psi_m(x)$ passing the vertex-points $(x_i, \psi(x_i))$ $(i = 0, 1, \cdots, m)$.

It is clear that the polygonal approximations $y = \phi_m(x)$ and $y = \psi_m(x)$ tend to the solutions $y = \phi(x)$ and $y = \psi(x)$ of (E), respectively. Consider the positive constants
$$\sigma_m = \max_{x \in J} |\phi(x) - \phi_m(x)| + \frac{1}{m}$$
and
$$\tau_m = \max_{x \in J} |\psi(x) - \psi_m(x)| + \frac{1}{m}.$$

It follows that $\sigma_m \to 0$ and $\tau_m \to 0$ as $m \to \infty$.

Consider the tube-regions
$$T_{\sigma_m}(\phi): \quad |y - \phi(x)| \leq \sigma_m, \quad x \in J$$
and
$$T_{\tau_m}(\psi): \quad |y - \psi(x)| \leq \tau_m, \quad x \in J,$$
centered at the integral curves $y = \phi(x)$ and $y = \psi(x)$ $(x \in J)$, respectively. It can be seen that
$$\phi_m(x) \in T_{\sigma_m}(\phi) \quad \text{and} \quad \psi_m(x) \in T_{\tau_m}(\psi), \qquad (x \in J).$$

For the positive integer m, construct a continuous vector-function $F_m(x,y)$ on Q, such that:
1) When m is odd, let
$$F_m(x,y) = \begin{cases} f(x,y), & \text{as } (x,y) \in Q \setminus T_{\sigma_m}(\phi); \\ \dfrac{\phi_m(x_{i+1}) - \phi_m(x_i)}{x_{i+1} - x_i}, & \text{as } (x,y) = (x_i, \phi(x_i)), \end{cases}$$
for $i = 0, 1, \cdots, m-1$;
2) When m is even, let
$$F_m(x,y) = \begin{cases} f(x,y), & \text{as } (x,y) \in Q \setminus T_{\tau_m}(\psi); \\ \dfrac{\psi_m(x_{i+1}) - \psi_m(x_i)}{x_{i+1} - x_i}, & \text{as } (x,y) = (x_i, \psi(x_i)), \end{cases}$$
for $i = 0, 1, \cdots, m-1$.

Now, construct a continuous vector-function $F(x,y,\lambda)$ on Q with a parameter λ ($|\lambda| \leq \alpha$) as follows:
For $\lambda \leq 0$, let $F(x,y,\lambda) = f(x,y)$. In particular, we have
$$F(x,y,0) = f(x,y).$$
For $\lambda > 0$, there is a unique positive integer m, such that
$$\frac{\alpha}{m+1} < \lambda \leq \frac{\alpha}{m},$$
then define
$$F(x,y,\lambda) = \frac{m(m+1)}{\alpha}\left[\left(\frac{\alpha}{m} - \lambda\right) F_{m+1}(x,y) + \left(\lambda - \frac{\alpha}{m+1}\right) F_m(x,y)\right].$$
It is easy to verify that $F(x,y,\lambda)$ is continuous and satisfies
$$F\left(x, y, \frac{\alpha}{m}\right) = F_m(x,y).$$
Then, the initial-value problem (E) admits of a difference equation
$$y_{i+1} - y_i = \frac{\alpha}{m} F\left(x_i, y_i, \frac{\alpha}{m}\right), \quad (i = 0, 1, \cdots, m-1), \tag{1.49}$$
which belongs to the type (1.42), and has a solution $[y_0, y_1, \cdots, y_m]$ satisfying:
$$y_k = \begin{cases} \phi_m(x_k), & \text{as } m \text{ is odd}; \\ \psi_m(x_k), & \text{as } m \text{ is even}, \end{cases}$$

1.5. Convergence Theorem on Difference Methods

where $k = 0, 1, \cdots, m-1$. It can be seen that

$$(x_k, \phi(x_k)) \in T_{\sigma_m}(\phi), \quad (x_k, \psi(x_k)) \in T_{\tau_m}(\psi)$$

($k = 0, 1, \cdots, m-1$). Then, we have

1) *If m is odd and tends to $+\infty$, then $(x_m, y_m) \to (x, \phi(x))$;*
2) *If m is even and tends to $+\infty$, then $(x_m, y_m) \to (x, \psi(x))$.*

The condition $\phi(x) \neq \psi(x)$ implies that the approximate solutions $[y_0, y_1, \cdots, y_m]$ do not converge.

We have thus proved that when the solution of (E) is not unique, there exists a Gear's difference equation, such that the approximate solutions do not converge.

In this sense, the uniqueness of solutions of (E) is also a necessary condition for the convergence of the Gear's approximate solutions. □

Chapter 2

Global Behavior of Solution

2.1 Global Existence of Solution

2.1.1 Continuation of Solution

Consider the system of differential equations

$$\frac{dx}{dt} = f(t, x), \qquad (t, x) \in D, \tag{2.1}$$

where D is a connected region in the time-phase space $\mathbb{R} \times \mathbb{R}^n$, and the function $f(t, x)$ is n-vector valued and continuous in $(t, x) \in D$. Let (τ, ξ) be a given point in D. The Peano theorem implies that there is at least a solution $x = \varphi(t)$ of (2.1) on a local interval, satisfying the initial condition

$$x(\tau) = \xi. \tag{2.2}$$

For simplicity, assume the solution $x = \varphi(t, \tau, \xi)$ of the Cauchy problem (2.1)+(2.2) is unique for any given initial point $(\tau, \xi) \in D$.

Now, let $x = \Phi(t)$ be a solution of (2.1) on an interval I.

If there exists a solution $x = \widetilde{\Phi}(t)$ of (2.1) on an interval $\widetilde{I}$, where $\widetilde{I}$ contains I as a proper sub-interval, such that

$$\widetilde{\Phi}(t) = \Phi(t), \qquad t \in I,$$

then $x = \widetilde{\Phi}(t)$ is called a *continued solution* of $x = \Phi(t)$ (from I to $\widetilde{I}$).

If a solution $x = \Phi(t)$ has a continued solution, then it is said to be *continuable*; otherwise, it is called *non-continuable*.

Lemma 2.1 *If $x = \Phi(t)$ is a non-continuable solution of (2.1) on the interval I, then I is an open interval.*

Proof. Let $\tau_0 \in I$. Assume I is not an open interval. Then I contains at least an end-point, say the right end-point β. It follows that $[\tau_0, \beta] \subset I$. Letting $\eta = \Phi(\beta)$, we have a point $(\beta, \eta) \in D$.

It follows from the Peano theorem and the uniqueness assumption that there is a unique solution of $x = \phi(t)$ of (2.1) on a local interval $[\beta-h, \beta+h]$ for some constant $h > 0$, satisfying the initial condition $x(\beta) = \eta$. Now, define

$$\widetilde{\Phi}(t) = \begin{cases} \Phi(t), & t \in I; \\ \phi(t), & t \in [\beta, \beta+h], \end{cases}$$

which is a continuous function on the interval $\widetilde{I} = I \cup [\beta, \beta+h]$. It follows from the uniqueness assumption that $x = \widetilde{\Phi}(t)$ is a solution of (2.1) on the interval $\widetilde{I}$. Hence, it is a continued solution of $x = \Phi(t)$ (from I to $\widetilde{I}$). This is a contradiction to the assumption of the lemma. Hence, I is an open interval.

The proof of Lemma 2.1 is thus completed. $\square$

Let

$$\Gamma: \quad x = \Phi(t) \quad (t \in I)$$

be an integral curve of (2.1) in the region D.

Theorem 2.1 *If $x = \Phi(t)$ is a non-continuable solution on I, then Γ approaches to the boundary of D in the sense that each sequence $p_k = (t_k, x_k) \in \Gamma$ has a cluster point at the boundary of D whenever the endpoint of I is a cluster point of $\{t_k\}$.*

Proof. Assume the contrary. It follows that Γ lies in a compact region $K \subset D$. Hence, the continuous function $f(t, x)$ is bounded on Γ. In other words, there is a constant M, such that

$$|f(t, \Phi(t))| < M, \quad t \in I. \tag{2.3}$$

Using Lemma 2.1 yields that the interval I is open. Furthermore, since Γ lies in a compact set K, the interval I is finite. Therefore, I is an open finite interval, say $I = (\alpha, \beta)$, where α and β are constants.

Now, let $t_n \in I$ be a sequence, such that

$$\lim_{n \to \infty} t_n = \beta.$$

By
$$\Phi(t) = \xi_0 + \int_{\tau_0}^{t} f(t, \Phi(t))\, dt, \qquad t \in I,$$

together with (2.3), we get

$$|\Phi(t_l) - \Phi(t_k)| \le M|t_l - t_k|.$$

It follows that $\Phi(t_n)$ has a limit as $t_n \to \beta$, say

$$\eta = \lim_{t_n \to \beta} \Phi(t_n).$$

Since $(t_n, x_n) \in \Gamma \subset K$, we have $(\beta, \eta) \in \overline{\Gamma} \subset K \subset D$.

Using Peano theorem yields that there is a unique solution $x = \psi(t)$ of (2.1) on an interval $[\beta - h^*, \beta + h^*]$ for some constant $h^* > 0$, satisfying the initial condition $x(\beta) = \eta$. Then, the solution $x = \Phi(t)$ on I is prolonged to the solution

$$x = \widetilde{\Phi}(t) = \begin{cases} \Phi(t), & t \in I; \\ \psi(t), & t \in [\beta, \beta + h^*], \end{cases}$$

on the interval $\widetilde{I} = I \cup [\beta, \beta + h^*]$. However, it is in conflict with the assumption that $x = \Phi(t)$ is non-continuable on I.

Therefore, Theorem 11.8 is proved by the contradiction. □

Theorem 2.2 *Each solution $x = \Phi(t)$ of (2.1) on an interval I can be prolonged to a non-continuable solution $x = \widetilde{\Phi}(t)$.*

Proof. If the solution $x = \Phi(t)$ on an interval I is non-continuable, the conclusion of theorem holds naturally true.

If the solution $x = \Phi(t)$ on an interval I is continuable, then I contains at least a finite end-point, say the right-hand β. Let $\tau_0 \in I$. Then, applying a similar method used in the proof of Lemma 11.8, we can prove that the solution $x = \Phi(t)$ can be continuable to a solution $x = \Phi^*(t)$ on a prolonged interval $I^* \supset I$. Let $\widetilde{I}$ be the union of all such prolonged intervals $\{I^*\}$. Then $x = \widetilde{\Phi}(t)$ is a continued solution of $x = \Phi(t)$ (from the interval I to the interval $\widetilde{I}$), defined by

$$\widetilde{\Phi}(t) = \Phi^*(t), \qquad \text{as } t \in I^*.$$

It can be seen from the definition of $\widetilde{I}$ that the solution $x = \widetilde{\Phi}(t)$ is non-continuable on the interval $\widetilde{I}$.

Theorem 2.2 is thus proved. □

2.1.2 Non-Global Existence

Assume $D = I \times \mathbb{R}$. As shown above, each solution $x = \phi(t)$ of (2.1) is continuable to the boundary of the region D. In this case, the boundary is

$$\partial D = (\partial I \times \mathbb{R}^n) \cup (I \times \partial \mathbb{R}^n).$$

Let J be the maximal existence-interval of $x = \phi(t)$. It follows that $J \subset I$.
If $J = I$, then $x = \phi(t)$ is said to be *globally existent* on I.
If $J \neq I$, then $x = \phi(t)$ is said to be *non-globally existent* on I.

Example 2.1 Consider the differential equation of first order

$$\frac{dx}{dt} = 1 + x^2, \qquad (t, x) \in \mathbb{R} \times \mathbb{R}, \tag{2.4}$$

with $I = \mathbb{R}$.

On the one hand, using Theorems 11.8 and 2.2, we assert that every solution of (2.4) is continuable to the infinity (i.e., the boundary) of $\mathbb{R}^2$.

On the other hand, the elementary method of integration gives the general solutions of (2.4) as follows

$$x = \tan(t - c) \qquad \text{(where } c \text{ is an arbitrary constant)},$$

which has the maximal existence-interval $J = (c - \frac{\pi}{2}, c + \frac{\pi}{2})$, depending upon the solution itself. Since $J \neq \mathbb{R}$, each solution of (2.4) is non-globally existent (on $\mathbb{R}$).

Example 2.2 Consider a differential equation of second order

$$\frac{d^2 x}{dt^2} + q(t) x^m = 0, \qquad (t, x) \in I \times \mathbb{R}^1, \tag{2.5}$$

where m is a positive integer, and $q(t)$ is a continuous function on a given interval $I \subset \mathbb{R}$.

When $m = 1$, (2.5) is a linear differential equation. Therefore, its general solution exists globally on I.

When $m > 1$, (2.5) is a nonlinear differential equation, which has the following properties.

(P_1): If there is a positive constant κ, such that

$$q(t) \leq -\kappa^2 < 0 \qquad \text{for } t \in I,$$

then the differential equation (2.5) has non-globally existent solutions on I.

In fact, multiplying $\dfrac{dx}{dt}$ to (2.5), we get

$$\frac{dx}{dt}\frac{d^2x}{dt^2} + q(t)x^m \frac{dx}{dt} = 0.$$

Choose an initial point $(t_0, x(t_0), x'(t_0))$, such that

$$x(t_0) > 0 \quad \text{and} \quad x'(t_0) > 0.$$

Then there is an interval $J = [t_0, t_1] \subset I$, such that

$$x(t) > 0 \quad \text{and} \quad x'(t) \geq 0, \qquad t_0 \leq t \leq t_1. \tag{2.6}$$

By Integration, we obtain

$$\frac{1}{2}\left(\frac{dx(t)}{dt}\right)^2 = \frac{1}{2}(x'(t_0))^2 + \int_{t_0}^{t} -q(t)x^m(t)\frac{dx(t)}{dt}\,dt$$

$$\geq \frac{1}{2}(x'(t_0))^2 + \kappa^2 \int_{t_0}^{t} x^m(t)\frac{dx(t)}{dt}\,dt$$

$$= \frac{c}{2} + \frac{1}{2}\frac{2\kappa^2}{m+1}x^{m+1} \qquad (t \in J),$$

where the constant

$$c = \left[(x'(t_0))^2 - \frac{2\kappa^2}{m+1}x^{m+1}(t_0)\right] \geq 0$$

provided

$$x'(t_0) \geq \sqrt{\frac{2\kappa^2}{m+1}x^{m+1}(t_0)}. \tag{2.7}$$

Since the initial point $(t_0, x(t_0), x'(t_0))$ can be arbitrarily chosen for the differential equation of second order, the condition (2.7) is reasonable. It follows that

$$\frac{1}{2}\left(\frac{dx}{dt}\right)^2 \geq \frac{1}{2}\frac{2\kappa^2}{m+1}x^{m+1},$$

which implies

$$\frac{dx}{dt} \geq \sqrt{\frac{2}{m+1}}\,\kappa x^{\frac{m+1}{2}}, \qquad t \in J,$$

or

$$0 < x^{-\frac{m+1}{2}}\, dx \geq \frac{\sqrt{2}}{\sqrt{m+1}} \kappa\, dt, \qquad t \in I,$$

if $x > 0$. Integrating the inequality from t_0 to $t(> t_0)$ implies

$$\int_{x(t_0)}^{x(t)} x^{-\frac{m+1}{2}}\, dx \geq \frac{\sqrt{2}}{\sqrt{m+1}} \kappa\, (t - t_0), \qquad t > t_0.$$

Hence,

$$\frac{-2}{m-1}\left[x^{-\frac{m-1}{2}}(t) - x^{-\frac{m-1}{2}}(t_0) \right] \geq \frac{\sqrt{2}}{\sqrt{m+1}} \kappa\, (t - t_0), \qquad t > t_0.$$

It follows that

$$0 < x^{-\frac{m-1}{2}}(t) \leq \frac{1-m}{2} \frac{\sqrt{2}}{\sqrt{m+1}} \kappa\, (t - t_0) + x^{-\frac{m-1}{2}}(t_0). \tag{2.8}$$

Since $x(t_0) > 0$ and $1 - m < 0$, it follows that there is $t^* > t_0$, such that

$$\begin{cases} \dfrac{1-m}{2} \dfrac{\sqrt{2}}{\sqrt{m+1}} \kappa\, (t - t_0) + x^{-\frac{m-1}{2}}(t_0) > 0, & t_0 < t < t^*; \\[2mm] \dfrac{1-m}{2} \dfrac{\sqrt{2}}{\sqrt{m+1}} \kappa\, (t - t_0) + x^{-\frac{m-1}{2}}(t_0) = 0, & t = t^*, \end{cases}$$

which together with (2.8) yields

$$\lim_{t \to t_1^*} x(t) = +\infty,$$

where t_1^* is some point satisfying $t_0 < t_1^* \leq t^*$. It can be seen that if the initial value $x(t_0)$ is taken sufficiently large, then the constant t^* is sufficiently near t_0, and $t_0 \leq t < t_1^*$ is the (right-hand) maximal existence interval of the considered solution $x = x(t)$.

We have thus proved the property (P_1). □

The following property holds indeed as a consequence of (P_1).

(P_2): *If $q(t)$ changes sign on the interval I, then there are solutions of (2.5) which do not exist globally on on I.*

Finally, we suggest the reader to prove the following property.

(P_3): *If $q(t)$ is positive for $t \in I$, then the solution of (2.5) exists globally on I.*

Chapter 2. Global Behavior of Solution

2.1.3 Global Existence Theorem

Theorem 2.3 *If $f(t,x)$ is a continuous function, satisfying the condition*

$$|f(t,x)| \leq K_1|x| + K_0, \quad \text{for } (t,x) \in I \times \mathbb{R}^n, \tag{2.9}$$

where I is an open interval in $\mathbb{R}^1$, $K_1 > 0$ and $K_0 > 0$ are constants, then each solution $x = \phi(t)$ of (2.1) exists globally on I.

Proof. Let (t_0, x_0) be any given initial point in $I \times \mathbb{R}^n$, and consider the initial value problem

$$(E): \quad \frac{dx}{dt} = f(t,x), \quad x(t_0) = x_0.$$

Then Peano theorem implies that (E) has a solution $x = \phi(t)$ on a local interval. Assume J is the maximal existence-interval of $x = \phi(t)$. It follows from Lemma 2.1 that J is an open interval with $J \subset I$.

Let β be the right end-point of J.

We want to prove: *β is also the right end-point of I.*

Assume the contrary. Then, it follows from $J \subset I$ that β is an interior point of I.

Then $[t_0, \beta)$ is a proper sub-interval of J, and we have

$$\phi(t) = x_0 + \int_{t_0}^{t} f(t, \phi(t))\, dt, \quad \forall\, t_0 \leq t < \beta,$$

which implies

1) $|\phi(t)| \leq M_0 := |x_0| + M(\beta - t_0), \quad t_0 \leq t < \beta;$
2) $|\phi(t) - \phi(s)| \leq (K_1 M_0 + K_0)|t - s|, \quad t_0 \leq t, s < \beta.$

Then, using the properties 1) and 2), we prove that when $t \to \beta$, the limit of $\phi(t)$ exists, say

$$\eta = \lim_{t \to \beta} \phi(t).$$

Since β is an interior point of I, (β, η) is an interior point of $I \times \mathbb{R}^n$. Hence, the Peano theorem asserts that the initial value problem

$$\frac{dx}{dt} = f(t,x), \quad x(\beta) = \eta,$$

has at least a solution

$$x = \psi(t), \quad \beta \leq t \leq \beta + \sigma,$$

for some small constant $\sigma > 0$. It is clear that

$$x = \varphi^*(t) = \begin{cases} \phi(t), & t \in J; \\ \psi(t), & \beta \le t \le \beta + \sigma, \end{cases}$$

is a solution on the interval the interval $J^* = J \cup [\beta, \sigma]$, and is thus a prolonged solution of $x = \phi(t)$ (from the interval J to J^*). This is in conflict with the fact that J is the maximal existence-interval of the solution $x = \phi(t)$.

Hence, the right end-point of J agrees with the right end-point of I.

In a similar manner, we can prove that the left end-point of J also agrees with the left end-point of I.

Since both the intervals I and J are open, we have $J = I$. Theorem 2.3 is thus proved. □

Corollary 2.1 *The solution of a linear system of differential equations*

$$\frac{dx}{dt} = A(t)x + b(t), \qquad t \in J,$$

exists globally on J if the coefficient $n \times n$-matrix $A(t)$ and the perturbation n-vector $b(t)$ are continuous in the interval J.

Examples 2.1 and 2.2 express that it is impossible in general to diminish the severe assumption of Theorem 2.3. Usually, the global existence of solution of differential equation is studied through special consideration of the equation.

2.1.4 Differential Equation of Second Order

The following differential equation of second order

$$\frac{d^2x}{dt^2} + f(x)\frac{dx}{dt} + g(x) = e(t), \qquad (t,x) \in \mathbb{R} \times \mathbb{R}, \tag{2.10}$$

plays an important role in the theory of non-linear oscillations. It is a mathematical model of, for example, the oscillation of mechanical spring (of unit-mass), with restoration force $g(x)$, resistance coefficient $f(x)$, and forced term $e(t)$.

Theorem 2.4 *Assume $g(x)$ and $f(x)$ are continuous functions, such that*

$$f(x) \ge -N, \qquad x \cdot g(x) > 0 \qquad (\text{for } |x| > d), \tag{2.11}$$

where $N > 0$ and $d \geq 0$ are constants, and assume $e(t)$ is a continuous function in $\mathbb{R}^1$. Then each solution $x = x(t)$ of (2.10) exists globally for $t \geq t_0$ if the initial condition of $x = x(t)$ is given at $t = t_0$.

Proof. We first put (2.10) into its equivalent system

$$\frac{dx}{dt} = y, \quad \frac{dy}{dt} = -f(x)y - g(x) + e(t). \tag{2.12}$$

It follows from the Peano theorem that the solution $x = x(t)$, $y = y(t)$ of (2.12), satisfying the initial condition

$$x(t_0) = x_0, \quad y(t_0) = y_0,$$

exists on a local interval. Let J be its maximum interval of existence.

It suffices to prove that $[t_0, +\infty) \subset J$.

In fact, assume the contrary. Then, we have

$$J \cap [t_0, +\infty) = [t_0, \beta),$$

with some constant $\beta > 0$. If there is a sequence $t_j \to \beta$, such that

$$\lim_{j \to \beta} (x(t_j), y(t_j)) \to (\xi, \eta) \in \mathbb{R}^2,$$

then using Theorem 1.1 yields that the existence interval of the solution $x = x(t)$, $y = y(t)$ can be prolonged to $[t_0, \beta]$. This is in conflict with the assumption that $[t_0, \beta)$ is the maximum interval of existence. It follows that

$$\lim_{t \to \beta} [|x(t)| + |y(t)|] = \infty. \tag{2.13}$$

On the other hand, it follows from the condition (2.11) that the function $g(x)$ satisfies $x \cdot g(x) > 0$ when $|x| \geq d$. Hence, its primitive has a lower bound $-K_1$ in $\mathbb{R}$; that is,

$$G(x) = \int_0^x g(x)\,dx \geq -K_1, \quad x \in \mathbb{R},$$

with some constant $K_1 > 0$. Again, using (2.11) yields:

$$f(x) \geq -K_2, \quad x \in \mathbb{R},$$

with some constant $K_2 \geq N > 0$.

Multiplying y to the equation (2.10), we get

$$y\frac{dy}{dt} + f(x)y^2 + g(x)\frac{dx}{dt} = e(t)y,$$

and then integrating it on the interval $(\sigma \leq t < \beta)$, we have

$$\frac{1}{2}y^2 + \int_\sigma^t f(x)y^2\, dt + G(x) = \int_\sigma^t e(t)y\, dt + C,$$

with the solution $x = x(t)$, $y = y(t)$, and the integral constant

$$C = \frac{1}{2}y^2(\sigma) + G(x(\sigma)).$$

It follows that

$$\frac{1}{2}y^2 - K_2 \int_\sigma^t y^2\, dt - K_1 \leq \int_\sigma^t |e(t)||y|\, dt + |C|,$$

that is,

$$\frac{1}{2}y^2 \leq K_2 \int_\sigma^t y^2\, dt + \int_\sigma^t |e(t)||y|\, dt + |C| + K_1.$$

Let

$$v(t) = \sup_{\sigma \leq s \leq t} |y(s)|.$$

Then

$$\frac{1}{2}y^2(t) \leq K_2(\beta - \sigma)v^2(t) + v(t) \int_\sigma^t |e(t)|\, dt + |C| + K_1.$$

Let the number $\sigma \in (t_0, \beta)$ be near β enough, so that

$$\beta - \sigma < \frac{1}{4K_2}.$$

We have thus

$$\frac{1}{2}y^2(t) \leq \frac{1}{4}v^2(t) + C_1 v(t) + C_0 \qquad (\sigma \leq t < \beta), \tag{2.14}$$

with constants

$$C_0 = |C| + K_1, \quad \text{and} \quad C_1 = \int_\sigma^\beta |e(t)|\, dt.$$

If $y(t)$ is bounded on the finite interval $[\sigma, \beta)$, then its primitive $x(t)$ is bounded too. However, this is in conflict with (2.13). Hence, $y(t)$ must be unbounded. It follows that there is a sequence

$$\sigma < t_1 < t_2 < \cdots < t_k < \cdots (\to \beta)$$

such that $v(t_k) = |y(t_k)| \to +\infty$. Then, using (2.14), we get

$$\frac{1}{4}v^2(t_k) \leq C_1 v(t_k) + C_0,$$

or

$$\frac{1}{4} \leq \frac{C_1}{v(t_k)} + \frac{C_0}{v^2(t_k)},$$

which implies $\frac{1}{4} \leq 0$. This contradiction proves the desired conclusion: $[t_0, \infty) \subset J$.

The proof of Theorem 2.4 is thus completed. □

If the coefficient function $f(x)$ satisfies the condition

$$f(x) > 0, \quad \text{for } |x| > d > 0,$$

and there is a point $x_0 \in (-d, d)$ satisfying $f(x_0) < 0$, then (2.10) is called a *Liénard equation*. For example, the van der Pol equation

$$\frac{d^2 x}{dt^2} + \mu(x^2 - 1)\frac{dx}{dt} + x = e(t), \quad \text{(with parameter } \mu > 0) \quad (2.15)$$

is a special Liénard equation.

If the coefficient function $f(x)$ is a constant, then (2.10) is called a *Duffing equation*. Therefore, the Duffing equation reads as

$$\frac{d^2 x}{dt^2} + c\frac{dx}{dt} + g(x) = e(t), \quad \text{(where } c \text{ is a constant}). \quad (2.16)$$

It follows from Theorem 1.5 that each solution of Liénard equation exists globally in the positive direction. In particular, each solution of the van der Pol equation and the Duffing equation as well exists globally in the positive direction. Moreover, using Theorem 1.5 together with a substitution $-t$ to t in the equation (2.16), we conclude that the solutions of the Duffing equation exist globally in both positive and negative directions.

However, the solution of the van der Pol equation

$$\frac{d^2 x}{dt^2} + \mu(x^2 - 1)\frac{dx}{dt} + x = 0$$

does not exist globally in the negative direction if its initial amplitude $|(x(t_0), x'(t_0))|$ is sufficiently large.

Indeed, it can be seen from [86] that the van der Pol equation has some solution satisfying the condition:

$$x(t) > x(t_0) > K > 1, \quad \delta > y(t) > y(t_0) > 0 \quad (t < t_0),$$

where $y(t) = x'(t)$, t_0 is an initial instant, and $K > 1$ and $\delta > 0$ are constants. Then, we have

$$\frac{dy}{dt} = -x - \mu(x^2 - 1)y < -\mu(x^2 - 1)y < -\mu(K^2 - 1)y \quad (t < t_0),$$

which together with $y(t) > 0$ implies

$$\int_t^{t_0} \frac{dy(t)}{y(t)} < -\mu(K^2 - 1)(t_0 - t) < \mu(K^2 - 1)t.$$

It follows that

$$t > \frac{\log y(t_0) - \log y(t)}{\mu(K^2 - 1)} > \frac{\log y(t_0) - \log \delta}{\mu(K^2 - 1)} = \text{constant}.$$

Therefore, the above-mentioned solution of the van der Pol equation does not exist globally in the negative direction.

2.2 Predictability of Solution

In history, people (for example, Laplace) considered that the process defined by differential equations is *deterministic* or *predictable* in the sense that both the future $(t > t_0)$ and past states $(t < t_0)$ are uniquely determined by its initial conditions at $(t = t_0)$. This view point comes naturally from the existence and uniqueness theorem of solution for the Cauchy problem.

2.2.1 The View Point of Poincaré

However, H. Poincaré pointed a different point of view in his essay *Science and Methods* (1882), which can be sketched as follows.

"The difficulty lay not in the rules of differential equations, but rather in specifying the initial conditions of solutions.
For example, let $x = \Phi(t)$ be a solution of the system

$$\frac{dx}{dt} = f(t, x) \tag{2.17}$$

on the infinite interval $[t_0, \infty)$. Assume that, for any given constant $\varepsilon > 0$, there is a constant $\delta = \delta(t_0, \varepsilon) > 0$, such that the inequality

$$|\varphi(t) - \Phi(t)| < \varepsilon \quad (t_0 \leq t < \infty)$$

holds for any solution $x = \varphi(t)$ of (2.17) whenever the initial condition $|\varphi(t_0) - \Phi(t_0)| < \delta$ is satisfied. Then the solution $x = \Phi(t)$ is said to be *regular*. Otherwise, it is said to be *irregular*.

Therefore, it may happen that a small error in the initial condition of an irregular motion [1] will produce very great ones in the final phenomena. In this case, prediction becomes impossible, and we have the *fortuitous* or *unpredictable* phenomenon."

Now, let $x = \Phi(t)$ be a motion (i.e., solution) of the system of (2.17). A substitution $x = y + \Phi(t)$ in (2.17) yields

$$\frac{dy}{dt} = g(t, y), \tag{2.18}$$

where

$$g(t, y) = f(t, y + \Phi(t)) - f(t, \Phi(t)).$$

It follows that $x = \Phi(t)$ is a solution of (2.17) if and only if $y = 0$ is a trivial solution of (2.18).

Therefore, the study of the solution $x = \Phi(t)$ of (2.17) is equivalent to the study of a trivial solution (or, an equilibrium point) $y = 0$ of (2.18).

2.2.2 Lagrange Principle

Lagrange was the first person in history to study the stability of equilibrium points. A system of differential equations is called a *Lagrange system* if it is given by a potential function $G(q)$ ($q \in \mathbb{R}^n$) as follows:

$$\frac{d^2 q_i}{dt^2} = -\frac{\partial G}{\partial q_i} \quad (i = 1, \cdots, n). \tag{2.19}$$

Then the system (2.19) of second order differential equations is equivalent to the following system of first order differential equations

$$\frac{dq_i}{dt} = p_i, \quad \frac{dp_i}{dt} = -\frac{\partial G}{\partial q_i} \quad (i = 1, \cdots, n). \tag{2.20}$$

If the potential function $G(q)$ has an isolated minimum point $q = q^*$, then $(q, p) = (q^*, 0)$ is an equilibrium point of the system (2.20). It follows that

---
[1] In this book, the term 'motion' will be sometimes used instead of 'solution'.

the kinematic energy

$$v(q,p) = \frac{1}{2}\sum_{i=1}^{n} p_i^2 + [G(q) - G(q^*)]$$

is positively definite in a small neighborhood of the point $(q^*, 0)$. It is noticed that $v(q^*, 0) = 0$, and the directional derivative of the function $v(q,p)$ with respect to the solution of (2.20) is equal to

$$\left.\frac{dv}{dt}\right|_{(2.20)} = \sum_{i=1}^{n}\left(\frac{\partial v}{\partial q_i}\frac{dq_i}{dt} + \frac{\partial v}{\partial p_i}\frac{dp_i}{dt}\right) = \sum_{i=1}^{n}\left(p_i\frac{\partial G}{\partial q_i} - \frac{\partial G}{\partial q_i}p_i\right) = 0.$$

Hence, any solution $q = q(t)$, $p = p(t)$ of (2.20) satisfies

$$v(q(t), p(t)) \equiv c,$$

where $c = v(q(0), p(0))$ is a constant. It can be seen that $c > 0$ is sufficiently small whenever $(q(0), p(0))$ is sufficiently near $(q^*, 0)$. Using the positive definiteness of $v(q, p)$ near $(q^*, 0)$, we conclude that $|p(t)|^2 + |q(t) - q^*|^2$ is sufficiently small if c (or $|p(0)|^2 + |q(0) - q^*|^2$) is small enough. This proves the first part of **Lagrange Principle**:

The equilibrium point $(q^*, 0)$ of (2.20) is stable, if q^* is an isolated minimum point of the potential function $G(q)$.

Similarly, we can study the case when $q = q^*$ is an isolated maximum point of the potential $G(q)$.

In fact, if $q = q^*$ is an isolated maximum point of $G(q)$, we have

$$\sum_{i=1}^{n} -(q_i - q_i^*)\frac{\partial G}{\partial q_i}(q) > 0 \qquad (0 < |q - q^*| \ll 1). \tag{2.21}$$

Consider the function

$$w(q, p) = \sum_{i=1}^{n} (q_i - q_i^*)p_i,$$

which changes signs in a small neighborhood of the point $(q^*, 0)$. On the other hand, it follows from (2.21) that the directional derivative of $w(q,p)$ with respect to the solution of (2.20), i.e.,

$$\left.\frac{dw}{dt}\right|_{(2.20)} = \sum_{i=1}^{n}\left(-(q_i - q_i^*)\frac{\partial G}{\partial q_i} + p_i^2\right),$$

is positively definite near the point $(q^*, 0)$. Therefore, if the initial point (q_0, p_0) of a solution $(q(t), p(t))$ is sufficiently near the equilibrium point

$(q^*, 0)$, satisfying $w(q_0, p_0) > 0$, $w(q(t), p(t))$ is monotonously increasing for $t > 0$. It follows that the solution $(q(t), p(t))$ will leave from the equilibrium point $(q^*, 0)$. This proves the second part of the Lagrange principle:

The equilibrium point $(q^*, 0)$ of (2.20) is unstable if q^* is an isolated minimum point of the potential function $G(q)$.

Example 2.3 The equation of simple pendulum

$$\frac{d^2\theta}{dt^2} + a^2 \sin\theta = 0$$

is equivalent to the following Lagrange system

$$\frac{d\theta}{dt} = \varphi, \qquad \frac{d\varphi}{dt} = -\frac{\partial G}{\partial \theta},$$

where the potential function $G(\theta) = -a^2 \cos\theta$. It is noted that $G(\theta)$ has an isolated minimum point $\theta = 0$ and an isolated maximum point $\theta = \pm\pi$. It follows from the Lagrange principle that the simple pendulum has a stable equilibrium point $(\theta, \varphi) = (0, 0)$ and an unstable equilibrium point $(\theta, \varphi) = (\pi, 0) = (-\pi, 0)$.

It is clear that the Lagrange principle is available to Lagrange systems only. Liapunov generalized later the Lagrange principle to the stability theory of motion for general differential equations; that is, the so-called second method of Liapunov (see, for examples, [21], [86] and [100], etc.).

2.3 Liapunov Stability

2.3.1 *General Definitions*

It can be seen in literature that A. Liapunov (1980's) developed the stability theory of motion from the concepts of the Poincaré's regular solution and the Lagrange's principle.

Consider the system of differential equations

$$\frac{dx}{dt} = f(t, x), \qquad (t, x) \in \mathbb{R} \times \mathbb{R}^n, \tag{2.22}$$

where $f(t, x)$ is continuous in $(t, x) \in \mathbb{R} \times \mathbb{R}^n$.

Denote by $x = \phi(t, p)$ the unique solution of (2.22) satisfying the initial condition $p = \phi(\tau, p)$ for $p \in \mathbb{R} \times \mathbb{R}^n$.

The theory of Liapunov stability is concerned with the equi-continuous dependence of the solution $x = \phi(t, p)$ on initial condition p as t varies in the infinite interval $[t_0, \infty)$.

Definition 2.1 Assume $x = \Phi(t)$ is a motion (i.e., a solution) of (2.22) existing on the infinite interval $I_0 = [t_0, \infty)$. Given $\varepsilon > 0$, if there exists a $\delta = \delta(t_0, \varepsilon) > 0$, such that the motion $x = \varphi(t)$ of (2.22) satisfies

$$|\varphi(t) - \Phi(t)| < \varepsilon, \qquad \text{for } t \in I_0, \tag{2.23}$$

whenever $|\varphi(t_0) - \Phi(t_0)| < \delta$, then the motion $x = \Phi(t)$ is said to be *Liapunov stable* (in positive direction). Otherwise, it is called *Liapunov unstable*.

The Liapunov stable (or unstable) motion in negative direction can be defined in a similar manner.

It is evident that a Liapunov stable motion is just a Poincaré regular solution.

The definition of a Liapunov stable motion $x = \Phi(t)$ depends on the initial time t_0. However, we have the following result:

Assume the motion of (2.22) passing through any point $p \in \mathbb{R} \times \mathbb{R}^n$ is unique. If the motion $x = \Phi(t)$ is Liapunov stable on the interval $I_0 = [t_0, \infty)$, then it is also Liapunov stable on the interval $I_\sigma = [t_0 + \sigma, \infty)$, where σ is any constant.

In fact, the result is a direct consequence of the continuous dependence of solution on initial condition in a finite interval $[t_0, t_0 + \sigma]$ if $\sigma > 0$ (or $[t_0 + \sigma, t_0]$ if $\sigma < 0$).

Now, let α be a fixed constant, and let $t_0 \geq \alpha$.

Definition 2.2 If the constant $\delta = \delta(t_0, \varepsilon)$ in Definition 2.1 does not depend on t_0 (i.e., $\delta = \delta(\varepsilon)$ depends only on ε), then the motion $x = \Phi(t)$ is called (Liapunov) *uniformly stable*. Otherwise, it is *not uniformly stable*.

Definition 2.3 Let $x = \Phi(t)$ be a motion of (2.22). If there is a constant $\eta = \eta(t_0) > 0$, such that the motion $x = \varphi(t)$ of (2.22) satisfies

$$\lim_{t \to +\infty} |\varphi(t) - \Phi(t)| = 0,$$

provided $|\varphi(t_0) - \Phi(t_0)| < \eta$, then $x = \Phi(t)$ is called an *attractive motion*. Otherwise, it is called *non-attractive*.

Assume the motion $x = \Phi(t)$ is attractive, and $x = \varphi(t)$ is a motion satisfying Definition 2.3. Let $\xi = \varphi(t_0)$. Then, for given $\sigma > 0$, there is a constant $T = T(t_0, \sigma, \xi) > 0$, such that

$$|\varphi(t) - \Phi(t)| < \sigma, \qquad \text{for } t > t_0 + T.$$

Definition 2.4 If the above constant T is independent on t_0 and ξ, then $x = \Phi(t)$ is called a *uniformly attractive* motion. Otherwise, it is called *non-uniformly attractive*.

Definition 2.5 If $x = \Phi(t)$ is Liapunov stable and attractive, then it is called an *asymptotically stable* motion. Otherwise, it is called *non-asymptotically stable*.

Definition 2.6 If $x = \Phi(t)$ is Liapunov uniformly stable and uniformly attractive, then it is called a *uniformly asymptotically stable* motion.

The following diagram expresses a relationship of various stabilities in the sense of Liapunov.

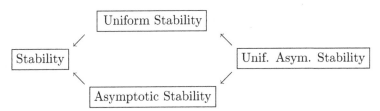

where Unif.=Uniform and Asym.=Asymptotic.

On the other hand, it is not hard to construct counter examples to show that all the implications in the diagram are not invertible (see [100]).

Let $x = \Phi(t)$ be an attractive motion. Then, given t_0, there is an *attractive domain* of the motion $\Phi(t)$; that is,

$$\mathcal{A}(t_0) = \left\{ \xi \in \mathbb{R}^n : \quad \xi = \varphi(t_0) \quad \text{such that} \quad \lim_{t \to +\infty} |\varphi(t) - \Phi(t)| = 0 \right\}.$$

In general, the attractive domain $\mathcal{A}(t_0)$ depends on the initial time t_0.

If $\mathcal{A}(t_0) = \mathbb{R}^n$ for all $t_0 \geq \alpha$, then $x = \Phi(t)$ is called a *globally attractive* motion. Moreover, it is said to be *globally asymptotically stable* if it is a globally attractive Liapunov stable motion.

It is obvious that for linear differential equations a globally attractive motion is globally attractive Liapunov stable. However, in the case of nonlinear differential equations, there are examples of globally attractive Liapunov unstable motion. Since those examples in literature are rather artificial (see [108]), we are interested in the following simple polynomial system (see [69]):

$$\frac{dx}{dt} = y, \quad \frac{dy}{dt} = -x(x-1)y - \frac{x^3}{8}, \quad (x,y) \in \mathbb{R}^2, \qquad (2.24)$$

which has a Liapunov unstable but globally attractive equilibrium point $\mathfrak{o} = (0,0)$ as shown in the Fig. 2.

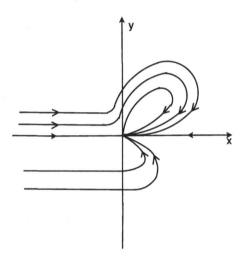

Fig. 2

2.3.2 Particular Examples

It is usually suggestive to consider particular examples in the analysis of Liapunov stability theory.

Example 2.4 Consider the differential equation

$$\frac{d^2x}{dt^2} + \mu \frac{dx}{dt} + x - x^3 = 0,$$

which is equivalent to the system

$$\frac{dx}{dt} = y, \quad \frac{dy}{dt} = -x - \mu y + x^3, \qquad (2.25)$$

where $\nu > 0$ is a constant. It is not a Lagrangian system since no function $G(x)$ satisfies the condition $\partial G/\partial x = x + \mu y - x^3$.

System (2.25) has three equilibrium points: $(0,0)$, $(1,0)$ and $(-1,0)$. In the neighborhood of the equilibrium point $(0,0)$,

$$V(x,y) = \frac{y^2}{2} + \frac{x^2}{2} - \frac{x^4}{4}$$

is a positive definite function (i.e., $V(x,y) > 0$ if the point $(x,y) \neq (0,0)$ is sufficiently near to $(0,0)$ and $V(x,y) = 0$ if the point $(x,y) = (0,0)$). Compute the directional derivative of $V(x,y)$ with respect to the system (2.20); that is,

$$\left.\frac{dV}{dt}\right|_{(2.25)} = v'_x \frac{dx}{dt} + v'_y \frac{dy}{dt} \qquad (2.26)$$

$$= (x - x^3)y + y(-x - \mu y + x^3) = -\mu y^2 \leq 0,$$

which implies

$$V(x(t), y(t)) \leq V(x_0, y_0), \qquad (t > t_0) \qquad (2.27)$$

where $x = x(t)$, $y = y(t)$ is the solution of equation (2.25) satisfying the initial condition $x(t_0) = x_0$, $y(t_0) = y_0$. Since $(0,0)$ is an isolated minimum point of the function $V = V(x,y)$, the inequality (2.27) implies that the motion $(x(t), y(t))$ will stay in a neighborhood of the point $(0,0)$ for $t > t_0$ provided the initial point (x_0, y_0) is sufficiently near the point $(0,0)$. This proves that the equilibrium point $(0,0)$ is Liapunov stable.

In a neighborhood of the equilibrium point $(1,0)$, consider the function

$$W(x,y) = \frac{y^2}{2} - \frac{(x^2 - 1)^2}{4},$$

which has the property that $W(1,0) = 0$ and $W(x,y)$ changes signs in the neighborhood of $(1,0)$. The directional derivative along the motion $x = x(t)$, $y = y(t)$ of the system (2.25) is

$$\left.\frac{dW}{dt}\right|_{(2.25)} \leq -\mu y^2(t) \leq 0, \qquad (t > t_0), \qquad (2.28)$$

which implies that $W(x(t), y(t))$ is a monotone decreasing function for $t > t_0$. It follows that

$$W(x(t), y(t)) \leq W(x_0, y_0) \qquad (t > t_0),$$

where $(x_0, y_0) = (x(t_0), y(t_0))$. On the other hand, in the neighborhood of $(1, 0)$ there exist initial points

$$(x_0^*, y_0^*) = (1 + \xi, 0) \qquad (0 < |\xi| \ll 1),$$

such that $W(x_0^*, y_0^*) < 0$. Let $(x^*(t), y^*(t))$ be the motion passing through the initial point (x_0^*, y_0^*). Then $W(x^*(t), y^*(t))$ is monotone decreasing for $t > t_0$ and satisfies

$$W(x^*(t), y^*(t)) < W(x_0^*, y_0^*) < W(1, 0) = 0, \qquad t > t_0.$$

Hence, the motion $(x^*(t), y^*(t))$ will leave from the equilibrium point $(1, 0)$. We have thus proved that the equilibrium point $(1, 0)$ is unstable.

In a similar manner, it can be proved that the equilibrium point $(-1, 0)$ is also unstable.

In the theory of stability, the above functions $V(x, y)$ and $W(x, y)$ are called *Liapunov functions*, which play the similar roles as the kinematic energy $v(p, q)$ and the function $w(p, q)$ in the above Lagrange system (2.20), respectively.

2.3.3 General Linearized System

Assume $x = 0$ is an equilibrium point of the system (2.22), where $f(t, x)$ is continuously differentiable with respect to x. Then we can put the system into the following form

$$\frac{dx}{dt} = A(t)x + h(t, x), \qquad (2.29)$$

where $A(t)$ is a continuous $n \times n$-matrix and $h(t, x)$ is continuous function in (t, x), satisfying $h(t, x) = o(|x|)$ for $t > 0$. Indeed, we can consider (2.29) as a perturbed system of the linear system

$$\frac{dx}{dt} = A(t)x \qquad (x \in \mathbb{R}^n). \qquad (2.30)$$

It is natural to think that the stability property of the equilibrium point $x = 0$ of (2.29) will be the same as that of (2.30) if the perturbation $h(t, x)$ is sufficiently small (for example, $h(t, x) = o(|x|)$). However, the following examples express the complexity of the problem.

Example 2.5 Consider a perturbed differential equation

$$\frac{du}{dt} = u \sin t + u^3 \sin t, \qquad (t, u) \in \mathbb{R}^1 \times \mathbb{R}^1, \qquad (2.31)$$

of the linear differential equation

$$\frac{du}{dt} = u \sin t. \tag{2.32}$$

When $|u| \ll 1$, the perturbation

$$h(t, u) = u^3 \sin t = o(|u|^2) \quad \text{(is uniformly for } t \in \mathbb{R}^1\text{).}$$

The general solution of (2.32) is

$$u = ce^{1-\cos t}, \quad \text{where } c = u(0) \text{ is an arbitrary constant.}$$

Hence, the equilibrium point $u = 0$ of (2.32) is Liapunov stable.

On the other hand, the Bernoulli equation (2.31) has a general solution

$$u = \frac{ce^{-\cos t}}{\sqrt{1 - c^2 e^{-2\cos t}}}$$

near the equilibrium point $u = 0$ with $0 \leq c < e^{-1}$. It implies that the equilibrium point $u = 0$ of (2.31) is also Liapunov stable.

Example 2.6 Consider the perturbed differential equation

$$\frac{du}{dt} = u \sin t + u^3 e^{2\cos t}, \quad (t, u) \in \mathbb{R}^1 \times \mathbb{R}^1, \tag{2.33}$$

of the linear differentil equation (2.32). When $|u| \ll 1$, the perturbation

$$h(t, x) = u^3 e^{2\cos t} = o(u^2) \quad \text{(is uniformly small for } t \in \mathbb{R}^1\text{).}$$

The general solution of the Bernoulli equation (2.33) is

$$u = \frac{u(0)e^{1+\cos t}}{\sqrt{1 - 2te^2 u^2(0)}}, \quad 0 \leq t < \tau = \frac{1}{2e^2 u^2(0)},$$

which yields

$$\lim_{t \to \tau} u(t) = \epsilon \cdot \infty, \quad \epsilon = \mathrm{sgn}(u(0)).$$

Hence, the equilibrium point $u = 0$ of (2.33) is unstable for $t > 0$, although the equilibrium point $u = 0$ of (2.32) is stable for $t > 0$.

Example 2.7 Consider the linear differential equation

$$\frac{du}{dt} = \frac{u}{1+t} \quad (t \geq 0, \ u \in \mathbb{R}^1), \tag{2.34}$$

which has the general solution

$$u = \frac{u(0)}{1+t}, \quad t \geq 0.$$

Hence, the equilibrium point $u = 0$ is Liapunov uniformly asymptotically stable.

On the other hand, consider the perturbed differential equation

$$\frac{du}{dt} = \frac{u}{1+t} + \frac{u^3}{(1+t)^2} \quad (t \geq 0, \ u \in \mathbb{R}^1), \tag{2.35}$$

of the linear equation (2.34). When $|u|$ is small, the perturbation

$$h(t, u) = \frac{u^3}{(1+t)^2} = o(u^2)$$

is uniformly small with respect to $t > 0$.

The general solution of the Bernoulli differential equation (2.35) can be found as follows

$$u = \frac{(1+t)u(0)}{\sqrt{1 - 2u^2(0)t}}, \quad 0 \leq t < b = \frac{1}{2u^2(0)},$$

which yields

$$\lim_{t \to b} \frac{u(t)}{u(0)} = +\infty \quad (u(0) \neq 0).$$

Hence, the equilibrium point $u = 0$ of (2.35) is Liapunov unstable for $t > 0$ although the equilibrium point $u = 0$ of its linearized system (2.34) is Liapunov uniformly asymptotically stable for $t > 0$.

Hence, in general, the stability (or instability) of equilibrium of the linearized system does not suffice to determine the stability (or instability) of equilibrium of the non-linear perturbed system.

2.3.4 *Linearized System with Constant Coefficients*

Now, consider the perturbed system

$$\frac{dx}{dt} = Ax + h(t, x), \quad (t, x) \in \mathbb{R}^1 \times \mathbb{R}^n \tag{2.36}$$

of a linear system

$$\frac{dx}{dt} = Ax, \qquad (2.37)$$

where A is a coefficient matrix A and $h(t,x)$ is a continuous perturbation. In this case, we have the following linearization theorems of stability [21].

Theorem 2.5 *The equilibrium point of (2.36) is asymptotically stable if the equilibrium point of (2.37) is asymptotically stable and the perturbation*

$$h(t,x) = o(|x|) \qquad (|x| \to 0) \qquad (2.38)$$

holds uniformly with respect to $t \geq 0$.

Proof. It follows from Peano theorem that the solution $x = \varphi(t)$ of (2.36) exists in a local interval $|t| \leq \alpha$ when $|\varphi(0)|$ is small. Hence, we have

$$\varphi(t) = e^{At}\varphi(0) + \int_0^t e^{(t-s)A} h(s, \varphi(s))\, ds \qquad (2.39)$$

so long as $\varphi(t)$ exists. Because the real parts of the characteristic roots of A are negative, there are positive constants K and σ, such that

$$|e^{At}| \leq K e^{-\sigma t}, \qquad t \geq 0. \qquad (2.40)$$

It follows from (2.39) and (2.40) that

$$|\varphi(t)| \leq K|\varphi(0)|e^{-\sigma t} + K \int_0^t e^{-\sigma(t-s)} |h(s, \varphi(s))|\, ds.$$

On the other hand, the condition (2.38) implies that given $\varepsilon > 0$, there exists a δ, such that $|h(t,x)| \leq \varepsilon |x|/K$ for $|x| \leq \delta$. Hence, we have

$$e^{\sigma t}|\varphi(t)| \leq K|\varphi(0)| + \varepsilon \int_0^t e^{\sigma s}|\varphi(s)|\, ds,$$

which together with the Gronwall lemma implies

$$e^{\sigma t}|\varphi(t)| \leq K|\varphi(0)|e^{\varepsilon t}$$

or

$$|\varphi(t)| \leq K|\varphi(0)|e^{-(\sigma-\varepsilon)t}. \qquad (2.41)$$

If ε is so chosen that $\varepsilon < \sigma$, then (2.41) implies that $|\varphi(t)| \leq K|\varphi(0)|$ so long as $|\varphi(t)| \leq \delta$ for $t > 0$. If $|\varphi(0)| < \varepsilon/K$, it follows from (2.41) that

$$|\varphi(t)| < \varepsilon e^{-(\sigma-\varepsilon)t}, \qquad \text{for } t > 0,$$

which completes the proof of Theorem 2.5 because of $(\sigma - \varepsilon) > 0$. □

Theorem 2.6 *Assume the coefficient matrix A in (2.36) has at least a characteristic value having positive real part and let $h(t,x)$ satisfy (2.38). Then the equilibrium point $x = 0$ of (2.36) is Liapunov unstable.*

Proof. To prove the theorem, we use a substitution $x = Py$, where P is a constant matrix, such that the system (2.36) is transformed into the following form

$$\frac{dy}{dt} = By + g(t, y), \qquad (2.42)$$

with

$$B = P^{-1}BP = \begin{pmatrix} B_1 & 0 \\ 0 & B_2 \end{pmatrix} \quad \text{and} \quad g(t,y) = P^{-1}h(t, Py),$$

where B_1 is a diagonal $k \times k$-matrix ($1 \leq k \leq n$) having characteristic roots with positive real parts, while B_2 is a diagonal $l \times l$-matrix ($l = n - k$) having characteristic roots with non-positive real parts. The characteristic values are in the main diagonal of B. Those elements off the main diagonal can be made smaller than any assigned positive quantity through a proper choice of P.

Let the components of y be y_i and let

$$u = \sqrt{\sum_{i=1}^{k} |y_i|^2} \quad \text{and} \quad v = \sqrt{\sum_{i=k+1}^{n} |y_i|^2}.$$

Let the real parts of the characteristic roots of B_1 be larger than some constant $\sigma > 0$. Then choose a positive constant $\varepsilon < \sigma/10$ and choose $\eta > 0$ and $T > 0$ so that

$$|g(t,y)| \leq \varepsilon |y| \qquad (t \geq T) \qquad (2.43)$$

for $|y| \leq \eta$.

Suppose the equilibrium point of (2.42) is stable. Therefore, for η and T chosen above, there exists a $\delta > 0$ such that, if $y = \varphi(t)$ is a solution of (2.42) satisfying $u(T) + v(T) < \delta$, then $u(t) + v(t) < \eta$ for $t \geq T$. Moreover, let $v(T) < u(T)/2$.

It follows from (2.42) and (2.43) that

$$\sum_{i=1}^{k} (\varphi_i' \varphi_i + \varphi_i \varphi_i') = 2uu' \geq 2\sigma u^2 - 2\gamma u^2 - 2\varepsilon u(u+v) \qquad (t \geq T),$$

Chapter 2. Global Behavior of Solution

where $\gamma > 0$ is a sufficiently small constant, such that
$$\frac{du}{dt} \geq \frac{1}{2}\sigma u - \varepsilon v.$$

In a similar way, we have
$$\frac{dv}{dt} \leq \varepsilon(u+v) + \frac{1}{2}\sigma v.$$

Hence, when $\varepsilon > 0$ is small enough, we have
$$\frac{d(u-v)}{dt} \geq \frac{1}{4}\sigma(u-v),$$

which yields
$$u(t) - v(t) \geq (u(T) - v(T))e^{\frac{1}{4}\sigma(t-T)} \geq \frac{1}{2}u(T)e^{\frac{1}{4}\sigma(t-T)},$$

thus
$$u(t) \geq \frac{1}{2}u(T)e^{\frac{1}{4}\sigma(t-T)} \quad \to \quad \infty, \quad \text{as} \quad t \to \infty.$$

However, this is in conflict with the stability assumption:
$$u(t) + v(t) < \eta, \quad \text{for} \quad t \geq T.$$

The proof of Theorem 2.6 is thus completed. $\square$

For a concise survey of Liapunov stability, the reader is referred to the book [114], for example.

2.4 Liapunov Unstability

We begin with the following examples.

Example 2.8 Consider the differential equation
$$\frac{du}{dt} = cu, \quad u \in \mathbb{R}, \tag{2.44}$$

where c is a constant. It is clear that $u = 0$ is an equilibrium point of the equation.

If $c \leq 0$, the equilibrium point $u = 0$ is Liapunov stable, and its adjacent motions are all Liapunov stable and bounded in $\mathbb{R}$; if $c > 0$, the equilibrium point $u = 0$ is Liapunov unstable, and its adjacent motions are all Liapunov unstable and unbounded in $\mathbb{R}$.

Example 2.9 Consider the differential equation

$$\frac{du}{dt} = f(u), \quad u \in \mathbb{R}, \tag{2.45}$$

where

$$f(u) = \begin{cases} u^3 \sin \dfrac{\pi}{u}, & u \neq 0; \\ 0, & u = 0, \end{cases}$$

which is C^1-differentiable.

It can be seen that for an integer $k \neq 0$, $u = \dfrac{1}{k}$ is an equilibrium point of the equation. It follows that the equilibrium point $u = 0$ is Liapunov stable.

On the other hand, the trivial solution $u = \dfrac{1}{k}$ is a Liapunov stable (or unstable) equilibrium point whenever k is an odd (or even) integer. Besides, a nontrivial solution approaches to some equilibrium point. It follows that the nontrivial solutions of the equation are Liapunov stable and bounded.

In summary, $u = 0$ is a Liapunov stable equilibrium point of (2.45), with the property that there are infinite many Liapunov stable motions as well as infinite many Liapunov unstable motions in the small neighborhood of $u = 0$.

The Liapunov unstable motion is sensitive to the initial condition since a small disturbance on the initial condition may cause certain 'large' deviation of the motion. It usually causes the complexity of motion. But, in the theory of Liapunov stability, the unstable motion is considered in the time-phase space, which has no compactness to focus the behavior of motion. To explore the complexity of Liapunov unstability, we will consider the problem later in a compact invariant set in the phase space (that is, the theory of chaotic motions).

Chapter 3

Autonomous Systems

The system of differential equations

$$\frac{dx}{dt} = f(t,x), \qquad (t,x) \in \mathfrak{D} \subset \mathbb{R}^1 \times \mathbb{R}^n,$$

is called *autonomous* if $f(t,x)$ is independent of the variable t. Otherwise, it is called *non-autonomous*.

Therefore, the autonomous system can be in general written in the following form

$$\frac{dx}{dt} = V(x), \qquad (3.1)$$

where $V(\cdot)$ is a vector field defined in the region $G \subset \mathbb{R}^n$.

Suppose $x = \phi(t)$ is a solution of differential equation, existing globally on the real axis $\mathbb{R}^1$. It is obvious that $x = x(t)$ is not bounded in a compact set of the time-phase space $\mathbb{R}^1 \times \mathbb{R}^n$. The bound of the solution $x = x(t)$ is indeed defined by the bound of the projection of solution in the phase-space $\mathbb{R}^n$. The problem is that the projections in the phase-space are usually very complicated.

However, we are benefited from the simple property of autonomous system, such that the orbit passing through an initial point in the phase-space is unique.

3.1 Phase Portrait

Let us begin with an elementary analysis of autonomous systems, which provides indeed the foundation of dynamical systems.

In what follows, assume $V(x)$ satisfies the local Lipschitz condition with respect to x in the region $G \subset \mathbb{R}^n$.

Lemma 3.1 *If $x = \varphi(t)$ is a solution of (3.1) on the interval (α, β), then $x = \varphi(t - c)$ is also a solution of (3.1) on the interval $(\alpha + c, \beta + c)$, where c is any given (real) constant.*

Proof. Since $x = \varphi(t)$ is a solution of (3.1) on (α, β), we have

$$\varphi'(t) = V(\varphi(t)), \qquad t \in (\alpha, \beta).$$

Then, replacing t by $t - c$ for a constant c, we get

$$\varphi'(t - c) = V(\varphi(t - c)), \qquad t \in (\alpha + c, \beta + c),$$

which is equivalent to

$$u'(t) = V(u(t)), \qquad t \in (\alpha + c, \beta + c),$$

where $u(t) = \varphi(t - c)$. It means that $x = u(t) = \varphi(t - c)$ is a solution of (3.1) on the interval $(\alpha + c, \beta + c)$.

The proof of Lemma 3.1 is thus completed. □

It follows from Lemma 3.1 that an integral curve of the autonomous system remains to be an integral curve of the system under the translation along the t-axis (with a finite displacement). It is known that this property is a key-point in the qualitative theory of ordinary differential equations.

Given a point $p \in G$, let $x = \varphi(t, p)$ be the solution of (3.1) satisfying the initial condition

$$x(0) = \varphi(0, p) = p. \tag{3.2}$$

Assume J is the maximal existence interval of $x = \varphi(t, p)$. Then

$$\Gamma_p := \{x \in \mathbb{R}^n : \quad x = \varphi(t, p), \quad \forall\, t \in J\}$$

is a differentiable curve in $\mathbb{R}^n$, which is called the *orbit* of the solution $x = \varphi(t, p)$. Following a tradition in mathematical physics, we will optionally call a *motion* instead of the solution of differential equation.

It is clear that the orbit Γ_p of motion $x = \varphi(t, p)$ is different from the integral curve

$$C_p = \{(t, x) \in \mathbb{R}^1 \times \mathbb{R}^n : \quad x = \varphi(t, p), \quad \forall\, t \in J\,\}.$$

In fact, the orbit Γ_p is a projection of the integral curve C_p from the time-phase space $\mathbb{R}^1 \times \mathbb{R}^n$ to the phase space $\mathbb{R}^n$.

It follows from Lemma 3.1 that, given a constant c, $x = \varphi(t - c, p)$ is also a solution of (3.1), such that it satisfies the initial condition $x(0) = q$ with

$q = \varphi(-c, p)$. Since the integral curve C_q is a translation of the integral curve C_p along the t-axis with a displacement c, the integral curves C_p and C_q have the same projection to the phase space $\mathbb{R}^n$. Hence, the orbit Γ_q of the motion $x = \varphi(t-c, p)$ is identical to the orbit Γ_p of the motion $x = \varphi(t, p)$ (i.e., $\Gamma_q = \Gamma_p$).

Lemma 3.2 *If the orbit Γ_u intersects the orbit Γ_v (i.e., $\Gamma_u \cap \Gamma_v \neq \emptyset$), then $\Gamma_u = \Gamma_v$.*

Proof. Let Γ_u and Γ_v be the orbits of the motions $x = \varphi(t, u)$ and $x = \varphi(t, v)$, respectively. If they intersect at a point p_0, then there are t_1 and t_2, satisfying

$$\varphi(t_1, u) = p_0 = \varphi(t_2, v).$$

Hence, the solution $x = \varphi(t, u)$ satisfies the condition

$$x(t_1) = \varphi(t_1, u) = p_0. \tag{3.3}$$

Now, using Lemma 3.1, we conclude that $x = \varphi(t+(t_2-t_1), v)$ is a solution of (3.1) satisfying the condition

$$x(t_1) = \varphi(t_2, v) = p_0. \tag{3.4}$$

It follows from (3.3) and (3.4) that the solutions $x = \varphi(t, u)$ and $x = \varphi(t+(t_2-t_1), v)$ satisfy the same initial condition: $x(t_1) = p_0$. Hence, the uniqueness of solution implies

$$\varphi(t, u) \equiv \varphi(t+(t_2-t_1), v).$$

On the other hand, the orbit of $x = \varphi(t+(t_2-t_1), v)$ is equal to the orbit Γ_v. Therefore, we conclude that $\Gamma_u = \Gamma_v$.

Lemma 3.2 is thus proved. □

Assume $x = \varphi(t, q)$ is a constant solution of (3.1); that is,

$$\varphi(t, q) = q, \qquad \forall\, t \in \mathbb{R}. \tag{3.5}$$

It follows that the orbit of a constant solution $x = \varphi(t, q)$ is degenerate since the considered orbit is a single point q. A constant solution of (3.1) is usually called an *equilibrium state* of (3.1).

It follows from (3.5) that

$$\frac{d\varphi(t, q)}{dt} = 0, \qquad \forall\, t \in \mathbb{R}.$$

which together with (3.1) implies

$$0 = \frac{d\varphi(t,q)}{dt} = V(\varphi(t,q)) = V(q),$$

that is, $V(q) = 0$. Hence, an equilibrium state of (3.1) corresponding to a singular point of the vector field $V(\cdot)$ and vice versa. For simplicity, the singular point of $V(\cdot)$ is also called the *singular point* of (3.1). It is noted that, for the autonomous system of differential equation, a singular point is the degenerate orbit of an equilibrium motion. Hence, a singular point is also called an equilibrium point.

On the other hand, a point p in G is called an *ordinary point* of (3.1) if it is not a singular point. Hence, p is an ordinary point if and only if $V(p) \neq 0$.

It can be seen from Lemma 3.2 that if p is an ordinary point of (3.1), then the orbit Γ_p contains no singular point of (3.1), and Γ_p is thus said to be *non-degenerate*. It follows that if Γ_p is a non-degenerate orbit, $V(z)$ is a non-singular tangent vector of Γ_p at $z \in \Gamma_p$. Therefore, the vector field $V(z)$ defines a natural direction of Γ_p, which is called the *positive direction* of the orbit.

For any given point $p \in G$, there is a unique orbit Γ_p passing through p. When p is not a singular point, Γ_p is a non-degenerate orbit with a natural positive direction.

The qualitative theory of differential equations is mainly concerned with the *phase-portrait*, which describes the key picture of orbits.

3.2 Orbital Box

Let $p \in G$ be an ordinary point of the system (3.1). Then there is a neighborhood $\mathcal{N}(p)$ of p, such that $V(z) \neq 0$ for $z \in \mathcal{N}(p)$. It follows that there exists an $(n-1)$-dimensional small (closed) disk $\bar{D}$ in $\mathcal{N}(p)$, centered at p, such that the vector field $V(\cdot)$ is *transversal* to the disk $\bar{D}$ (i.e., given $z \in \bar{D}$, the vector $V(z)$ intersects $\bar{D}$ non-tangentially at the point z). It follows that the orbits meeting the disk $\bar{D}$ will pass across $\bar{D}$ from one side to the other side (for example, from the outside to the inside).

In this case, there is a small constant $h > 0$, such that the solution $x = \varphi(t, \xi)$ starting from $\xi \in \bar{D}$ has an orbital arc

$$\Gamma_\xi[-h, h] := \{\, x = \varphi(t, \xi), \quad \forall\, t \in [-h, h] \,\} \subset \mathcal{N}(p).$$

Now, set

$$T(p) = \{\Gamma_\xi[-h, h] : \quad \forall\, \xi \in \bar{D}\},$$

which is called an *orbital box* of the system (3.1) across the point p. It is remarked that the orbital box is different from the integral box in definition. The orbital box $T(p)$ consists of the orbital arcs $\Gamma_\xi[-h, h]$ for all $\xi \in \bar{D}$. Sometimes, an orbital tube $T(p)$ is also considered as a point-set; that is,

$$T(p) = \{\, x \in \mathcal{N}(p) : \quad x = \varphi(t, \xi),\ -h \le t \le h,\ \ \forall\, \xi \in \bar{D}\,\}.$$

To analyze $T(p)$, consider the following solid box

$$\mathcal{H}(p) = \{\, (t, \xi) \in \mathbb{R} \times \bar{D} \subset \mathbb{R}^n : \quad |t| \le h,\ \xi \in \bar{D}\,\}$$

in the phase-space $\mathbb{R}^n$. Notice that $\mathcal{H}(p)$ consists of the family of parallel axial line-segments

$$L_\eta := \{\, (t, \xi) \in \mathbb{R}^n : \quad -h \le t \le h,\ \xi = \eta\,\}, \qquad \forall\, \eta \in \bar{D}.$$

Consider the mapping

$$\Phi : \quad \mathcal{H}(p) \;\to\; T(p); \qquad (t, \xi) \;\mapsto\; x = \varphi(t, \xi).$$

Since the system (3.1) satisfies the local Lipschitz condition, the solution $x = \varphi(t, \xi)$ is uniquely determined by the initial condition $x(0) = \xi$, and $x = \varphi(t, \xi)$ is continuous in (t, ξ). It follows that Φ is a topological mapping from $\mathcal{H}(p)$ onto $T(p)$, such that

$$\mathcal{H}(p) = \Phi^{-1}(T(p)) \qquad \text{and} \qquad L_\eta = \Phi^{-1}(\Gamma_\eta[-h, h]).$$

The above argument leads to the following result.

Proposition 3.1 *The orbits in the orbital box $T(p)$ are topologically equivalent to the parallel (axial) line-segments in the box $\mathcal{H}(p)$.*

Therefore, the family of local orbits around an ordinary point p of (3.1) is as simple as a family of local parallel line-segments. However, the above method of orbital box does not hold in the neighborhood of a singular point.

3.3 Types of Orbits

We have shown that a singular point q of the system (3.1) has a degenerate orbit Γ_q (i.e., $\Gamma_q = q$) and for a non-singular (ordinary) point p, there is a unique non-degenerate orbit Γ_p passing through p. What is the geometry of a non-degenerate orbit?

Assume p is an ordinary point of (3.1). Then there is a unique solution $x = \varphi(t, p)$ satisfying the initial condition $x(0) = p$. Let J be the maximal existence-interval of this solution. Without loss of generality, let $J = \mathbb{R}$.

Then we have the following alternatives:

(A_1) There are constants t_2 and t_1 ($t_2 > t_1$), such that
$$\begin{cases} \varphi(t_2, p) = \varphi(t_1, p); \\ \varphi(t, p) \neq \varphi(t_1, p), \quad (t_1 < t < t_2). \end{cases}$$

(A_2) For any constants t_2 and t_1 ($t_2 \neq t_1$), we have
$$\varphi(t_2, p) \neq \varphi(t_1, p).$$

In the first case (A_1), we have a positive constant $T = t_2 - t_1$. It follows from Lemma 3.1 that $x = v(t) = \varphi(t + T, p)$ is a solution of (3.1) satisfying the initial condition
$$v(t_1) = \varphi(t_2, p).$$

On the other hand, $x = u(t) = \varphi(t, p)$ is a solution of (3.1), satisfing the initial condition
$$u(t_1) = \varphi(t_1, p).$$

It is clear that (A_1) implies $u(t_1) = v(t_1)$. It follows from the uniqueness of solution that $v(t) \equiv u(t)$. We have thus proved
$$\varphi(t + T, p) = \varphi(t, p), \quad \forall\, t \in \mathbb{R},$$

which means that $x = \varphi(t, p)$ is a *periodic* solution of period T. In this case, we have a topological map
$$\Phi: \quad \mathbb{S}^1 \to \Gamma_p; \quad s \mapsto x = \varphi(s, p),$$

where $\mathbb{S}^1 = \mathbb{R}^1/\mathrm{mod}(T)$ is a circle. It concludes that the orbit Γ_p of a periodic solution is topologically equivalent to a circle $\mathbb{S}^1$. In other words, Γ_p is a simple closed curve in the phase-space $\mathbb{R}^n$, which is called a *closed orbit* of (3.1).

It follows from the periodicity that
$$\varphi(t + kT, p) = \varphi(t, p), \quad \forall\, t \in \mathbb{R},$$

for any integer k. Hence, given an integer $m \geq 1$, mT is also a period of $x = \varphi(t,p)$, and $T = (t_2 - t_1) > 0$ is the least period determined by (A_1).

In the final the case (A_2), $x = \varphi(t,p)$ is a non-periodic solution. It can be seen that

$$\Psi : \quad t \quad \mapsto \quad x = \varphi(t,p)$$

is a topological transformation from $\mathbb{R}^1$ onto the orbit Γ_p. It follows that Γ_p is topologically equivalent to the open interval $\mathbb{R}^1$. In this case Γ_p is called an *open orbit*.

In summary, an orbit of autonomous system belongs to one type of the following three categories:

(1) Singular Points;   (*singular case*)
(2) Closed Orbits;   (3) Open Orbits;   (*ordinary case*)

A singular point is a degenerate orbit of a constant solution, and a (non-degenerate) closed orbit represents the orbit of some periodic solution. The first two categories of orbits are simple in geometry since the orbits themselves are closed in the sense that the closure of the singular point and the closed orbit as well is the orbit itself.

The most significant feature of the open orbit is that the closure of the orbit is different from the orbit itself. Usually, it is not trivial to determine the closure $\widehat{\Gamma}_p$ of the open orbit Γ_p. For an open orbit, the orbit Γ_p is a proper subset of its closure $\widehat{\Gamma}_p$, which describes the asymptotical behavior of Γ_p (or the corresponding motion $x = \varphi(t,p)$).

3.4 Singular Points

Assume z_0 is a singular point of (3.1). Hence, $V(z_0) = 0$. Furthermore, assume the singular point z_0 is isolated. For example, the singular point z_0 is isolated if $\det [V'(z_0)] \neq 0$.

The singular point z_0 is said to be *simple*, if each eigenvalue of the matrix $V'(z_0)$ does not vanish.

The singular point z_0 is said to be *critical*, if there is at least an eigenvalue of $V'(z_0)$ with vanishing real part. Therefore, the singular point z_0 is *non-critical*, if the real part of each eigenvalue of $V'(z_0)$ does not vanish.

It is noticed that a non-critical singular point is simple, but the converse is in general not true.

For brevity, assume the singular point z_0 is the origin $\mathbf{o}$ of the space $\mathbb{R}^n$.

Then the autonomous system (3.1) can be put in the form

$$\dot{z} = Az + g(z) \qquad (|z| \ll 1), \tag{3.6}$$

where Az is the linear part of $V(z)$ with coefficient matrix $A = V'(z_0)$, and $g(z)$ is the non-linear part of $V(z)$ (i.e., $g(z) = V(z) - Az$). The linear part

$$\dot{z} = Az \qquad (z \in \mathbb{R}^n) \tag{3.7}$$

is called the *linearized system* of (3.6).

Without loss of generality, let the matrix A be taken in the Jordan form.

3.4.1 Noncritical Singular Point

Let the singular point $\mathfrak{o}$ of (3.6) be noncritical. Then the real part of each eigenvalue of A is non-vanishing.

Denote by $\mathfrak{F}_n$ the phase portrait of (3.7) in the neighborhood of $\mathfrak{o} \in \mathbb{R}^n$.

Example 3.1 When $n = 1$, the linear system (3.7) becomes

$$\dot{x} = \lambda x \qquad (x \in \mathbb{R}),$$

where $\lambda \neq 0$. Then we have the general solution

$$x = x_0 e^{\lambda t} \qquad (x_0 \text{ is an arbitrary constant}),$$

which gives the phase portrait $\mathfrak{F}_1$ as shown in Figures 3 and 4.

$(\lambda < 0)$ \qquad\qquad $(\lambda > 0)$

Fig. 3 A Sink \qquad\qquad Fig. 4 A Source

Example 3.2 When $n = 2$, we have

$$z = \begin{pmatrix} x \\ y \end{pmatrix}, \qquad A = \begin{pmatrix} a & b \\ c & d \end{pmatrix}.$$

Let λ_1 and λ_2 be the eigenvalues of A. Since $\mathfrak{o}$ is noncritical, we have $\lambda_1 \lambda_2 \neq 0$, with the following possible cases.

Chapter 3. Autonomous Systems   81

Case 1 : $A = \begin{pmatrix} \lambda_1 & 0 \\ 0 & \lambda_2 \end{pmatrix}$ (with real eigenvalues $\lambda_1 \neq \lambda_2$).

Then (3.7) has the general solution
$$x = x_0 e^{\lambda_1 t}, \quad y = y_0 e^{\lambda_2 t}, \tag{3.8}$$
where $z_0 = (x_0, y_0)$ is an arbitrary initial point, and the phase portraits are shown in Figures 5-10:

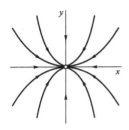

$(\lambda_1 < \lambda_2 < 0)$

Fig. 5   Improper node

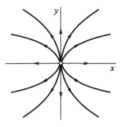

$(0 < \lambda_1 < \lambda_2)$

Fig. 6   Improper node

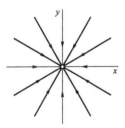

$(\lambda_1 = \lambda_2 < 0)$

Fig. 7   Proper node

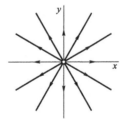

$(\lambda_1 = \lambda_2 > 0)$

Fig. 8   Proper node

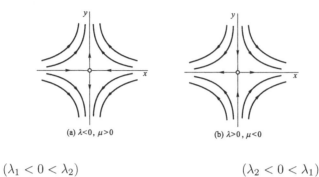

(a) $\lambda<0, \mu>0$ (b) $\lambda>0, \mu<0$

$(\lambda_1 < 0 < \lambda_2)$ $(\lambda_2 < 0 < \lambda_1)$

Fig. 9 Saddle point   Fig. 10 Saddle point

Case 2: $\quad J_2 = \begin{pmatrix} \lambda & 0 \\ 1 & \lambda \end{pmatrix} \quad (\lambda \neq 0).$

Then (3.7) has the general solution

$$x = x_0 e^{\lambda t}, \quad y = y_0 e^{\lambda t} + x_0 t e^{\lambda t},$$

which yields the phase portraits as shown in Figures 12-13.

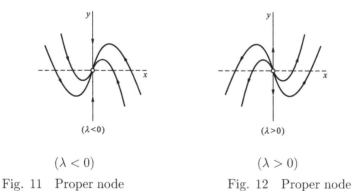

($\lambda<0$)   ($\lambda>0$)

$(\lambda < 0)$ $(\lambda > 0)$

Fig. 11 Proper node   Fig. 12 Proper node

Case 3: $\quad A = \begin{pmatrix} \alpha & -\beta \\ \beta & \alpha \end{pmatrix} \quad (\alpha \neq 0, \ \beta \neq 0);$

(in this case, $\lambda_1 = \alpha + i\beta$ and $\lambda_2 = \alpha - i\beta$ are complex eigenvalues).
It follows from (3.7) that

$$\dot{x} = \alpha x - \beta y, \qquad \dot{y} = \beta x + \alpha y,$$

which implies

$$\dot{r} = \alpha r, \qquad \dot{\theta} = \beta,$$

where r and θ are the polar coordinates of (x, y). Then the general solution

$$r = r_0 e^{\alpha t}, \qquad \theta = \beta t + \theta_0,$$

yields the phase portraits as shown in the Figures 14-15. Correspondingly, the singular point is called a *spiral point* (or *focus*).

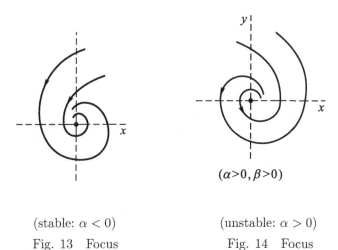

($\alpha > 0, \beta > 0$)

(stable: $\alpha < 0$) (unstable: $\alpha > 0$)

Fig. 13 Focus   Fig. 14 Focus

In summary, the phase portraits $\mathfrak{F}_2$ are shown in Fig.6 - Fig.14.

In fact, the phase-portrait $\mathfrak{F}_n$ can be analyzed in a similar manner in principle. Moreover, it is not hard to prove the following result.

Theorem 3.1 *In the noncritical case, the phase-portrait of the non-linear system (3.6) nearby the singular point* o *is topologically equivalent to a corresponding phase-portrait listed in* $\mathfrak{F}_n$ *under small perturbation* $g(z)$ *of order* $o(|z|)$.

3.4.2 Critical Singular Point

Compared with the non-critical singular points (see Theorem 3.1), it is rather difficult to determine the phase-portraits of critical singular points. The following is a short discussion about the simple critical singular points in the plane.

Example 3.3 When

$$A = \begin{pmatrix} 0 & -\beta \\ \beta & 0 \end{pmatrix} \qquad (\beta \neq 0),$$

the singular point o of the non-linear system (3.6) is critical and simple. The corresponding linearized system (3.7) is

$$\dot{x} = -\beta y, \qquad \dot{y} = \beta x,$$

which is equivalent to

$$\dot{r} = 0, \qquad \dot{\theta} = \beta.$$

Then we have the general solution

$$r = r_0, \qquad \theta = \beta t + \theta_0,$$

with arbitrary constants $r_0 \geq 0$ and θ_0. It gives the phase portraits of the singular point o as shown in Fig. 16 and Fig. 17, where the singular point o is called a *center*. Note that the closed orbits have a clockwise direction when $\beta < 0$ and a counter clockwise direction when $\beta > 0$.

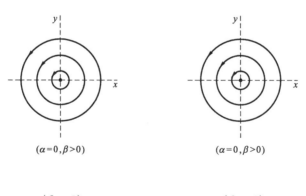

($\alpha=0, \beta>0$) ($\alpha=0, \beta>0$)

($\beta > 0$)　　　　　　　　($\beta < 0$)

Fig. 15. Center　　　Fig. 16. Center

Then consider the following perturbed system.

Example 3.4 Assume the nonlinear system

$$\begin{cases} \dot{x} = -y + \sigma(x^2 + y^2)^m - y(x^2 + y^2)^m, \\ \dot{y} = x + \sigma(x^2 + y^2)^m + x(x^2 + y^2)^m, \end{cases} \quad (3.9)$$

where σ is a constant, and $m \geq 1$ is an integer. Note that the perturbation (i.e., the part of nonlinear terms) is a small term of type $o[(x^2 + y^2)^m]$ near the center o. The order of smallness (i.e., $2m$) can be assumed to be higher as we like.

Using polar coordinates, (3.9) is written in the form

$$\dot{r} = \sigma r^{2m-1}, \qquad \dot{\theta} = -1 - r^{2m},$$

which implies:

(1) the singular point o is a *center* if $\sigma = 0$;

(2) the singular point o is a *stable focus* [1] if $\sigma < 0$;

(3) the singular point o is an *unstable focus* [2] if $\sigma > 0$.

Hence, the phase-portrait of the perturbed system for a center cannot be determined by the order of perturbation. This is indeed the central difficulty of the *Poincaré center problem* concerning about the phase portrait of a center under small perturbations.

The following problem is called the *Liapunov stability problem in second critical case*, which involves essentially the same difficulty of the Poincaré center problem:

How to determine the stability of the equilibrium state o of (3.6) when $\lambda_1 = i\beta$, $\lambda_2 = -i\beta$ ($\beta \neq 0$), and $Re(\lambda_j) < 0$ ($3 \leq j \leq n$)?

Meanwhile, the *Liapunov stability problem in the first critical case* states as follows:

How to determine the stability of the equilibrium state o of (3.6) when $\lambda_1 = 0$ and $Re(\lambda_j) < 0$ ($2 \leq j \leq n$)?

For the planar system (i.e., $n = 2$), the present author gave an answer to this problem through the analysis of the phase-portrait of the perturbed singular point o in the first critical case (see [104] and [123]).

---

[1] That is, $r \to 0$ and $\theta \to -\infty$ as $t \to +\infty$.
[2] That is, $r \to 0$ and $\theta \to +\infty$ as $t \to -\infty$.

Generally speaking, the analysis of critical point is not trivial. For example, the singular point o of the planar system (2.24) is critical. The book [123] contains a study of critical points in more details.

3.5 General Property of Singular Points

The above examples provides an abundance of singular points. However, they possess the common feature as stated in the following.

Theorem 3.2 *Assume o is an isolated singular point of (3.6). Then, given a neighborhood $\mathcal{N}$ of o, there exists at least a non-degenerate semi-orbit of of (3.6) contained in $\mathcal{N}$.*

Proof. Let $\widehat{B}_\sigma$ be the closed ball of radius $\sigma > 0$ centered at the singular point o. Since the singular point o is isolated, we can take $\sigma > 0$, such that o is a unique singular point in $\widehat{B}_\sigma$.

We want to prove that for given δ $(0 < \delta \leq \sigma)$ there exists at least a non-degenerate (positive or negative) semi-orbit, which stays everlasting in the ball $\widehat{B}_\delta$.

Assume the contrary. Then, given a point $p \in (\widehat{B}_\delta \setminus \{o\})$, there are constants $\alpha \leq 0$ and $\beta \geq 0$, such that the solution $\varphi(t,p)$ of (3.6) has a closed orbit-arc

$$\Gamma_p[\alpha, \beta] = \{ x = \varphi(t,p) : \quad \alpha \leq t \leq \beta \}$$

staying in $\widehat{B}_\delta$, but for given small constant $\varepsilon > 0$, both the open orbit-arcs

$$\Gamma_p(\alpha - \varepsilon, \alpha) = \{ x = \varphi(t,p) : \quad \alpha - \varepsilon \leq t \leq \alpha \}$$

and

$$\Gamma_p(\beta, \beta + \varepsilon) = \{ x = \varphi(t,p) : \quad \beta \leq t \leq \beta + \varepsilon \}$$

are located at the outside of $\widehat{B}_\delta$.

Denote by $\mu(p)$ the distance from the singular point o to the orbit-arc $\Gamma_p[\alpha, \beta]$. It can be seen that $\mu(p)$ is a continuous function for $p \in (\widehat{B}_\delta \setminus \{o\})$, satisfying

$$\mu(p) > 0, \quad \text{for } p \in S_\delta,$$

where S_δ is the boundary of $\widehat{B}_\delta$ (i.e., $S_\delta = \partial B_\delta$). It follows that the restriction of $\mu(p)$ to the sphere S_δ is continuous and positive. Hence, we have a constant

$$\sigma = \min_{p \in S_\delta} \mu(p) > 0.$$

It is clear that the non-degenerate orbit Γ_z ($0 < |z| < \sigma$) cannot reach the sphere S_σ. Hence, we have $\Gamma_z \subset B_\sigma$. This is a contradiction to the assumption that there is no non-degenerate semi-orbit staying in B_δ ($0 < \delta \leq \sigma$) forever.

Theorem 3.2 is thus proved by contradiction. □

According to Theorem 3.2, there arises a natural problem as follows:

Assume o is an isolated singular point in $\mathbb{R}^n$. Does there exist at least a non-degenerate orbit tending positively or negatively to o?

When o is a singular point in the pase space $\mathbb{R}^1$, it is trivial to prove that the answer is 'yes'.

When o is a singular point in the pase space $\mathbb{R}^2$, it can be seen from the Fig. 16-17 that the answer is 'no'.

When o is a singular point in the pase space $\mathbb{R}^3$, the answer is not trivial. It leads to the following Reeb's problem:

Is there an isolated singular point o in the phase space $\mathbb{R}^3$, such that there is no (non-degenerate) orbit tending positively or negatively to o?

T. Ding gave an affirmative answer to this problem by constructing an example (see [25]), which will be discussed in details later.

With such an example in mind, it will be easy, by using the method of topological product, to construct the example of isolated singular point o in the phase space $\mathbb{R}^n$ (for $n \geq 4$), such that each non-degenerate orbit is kept a positive distance from the singular point o.

3.6 Closed Orbit

In this section, we are concerned with the orbits of non-trivial periodic motions.[3]

Example 3.5 The small oscillation of a spring-oscillator is approximated by the solution of a linear system

$$\dot{x} = \sqrt{\kappa}y, \qquad \dot{y} = -\sqrt{\kappa}x, \qquad (3.10)$$

---
[3] A non-trial periodic motion means that the motion is periodic but not equilibrium.

where $\kappa > 0$ is some constant. It is known that
$$x = x_0 \cos \sqrt{\kappa} t + y_0 \sin \sqrt{\kappa} t, \qquad y = y_0 \cos \sqrt{\kappa} t - x_0 \sin \sqrt{\kappa} t$$
is a periodic motion of (3.10), satisfying the initial condition
$$x(0) = x_0, \qquad y(0) = y_0.$$
Moreover, the orbit of the periodic motion is given by the circle
$$\Gamma_c: \quad x^2 + y^2 = c^2, \qquad \text{where } c = \sqrt{x_0^2 + y_0^2}.$$
The orbit Γ_c is degenerated to the singular point o when $c = 0$. It can be seen that the closed orbits $\{\Gamma_c\}$ are not isolated.

In general, when $x = \Phi(t)$ is a periodic solution of a linear system of differential equations, $x = c\Phi(t)$ is also a periodic solution for any constant c. Therefore, the (non-degenerate) closed orbits of a linear system are not isolated.

In history, Poincaré first found an isolated orbit in a polynomial system of differential equations. This is an interesting non-linear phenomenon.

Example 3.6 Consider the following polynomial system
$$\begin{cases} \dot{x} = -\sigma y - x(x^2 + y^2 - 1), \\ \dot{y} = +\sigma x - y(x^2 + y^2 - 1), \end{cases} \quad (x, y) \in \mathbb{R}^2, \qquad (3.11)$$
where σ is a positive constant.

Using the polar coordinates, we have
$$r\dot{r} = x\dot{x} + y\dot{y}, \qquad r^2 \dot{\theta} = x\dot{y} - y\dot{x},$$
which together with (3.11) implies
$$\dot{r} = -r(r^2 - 1), \qquad \dot{\theta} = \sigma.$$
It follows from $\dot{\theta} = \sigma > 0$ that all non-degenerate orbits move in counter clockwise direction. More exactly, it can be seen that the circle $r = 1$ is an isolated closed orbit Γ and the origin $r = 0$ is an equilibrium state o. The non-degenerate orbits starting from the inside of Γ will tend positively to Γ and negatively to o, while the orbits starting from the outside of Γ will tend positively to Γ and negatively to the infinity as shown in the following Fig.17.

Roughly speaking, a closed orbit Γ is called *orbital stable* (in positive or negative direction) if the nearby orbits stay nearby Γ forever (in positive or negative direction, respectively). On the other hand, Γ is called *asymptotically orbital stable* (in positive or negative direction) if the nearby orbits approach to Γ (in positive or negative direction, respectively). It is obvious that an asymptotically orbital stable closed orbit is orbital stable.

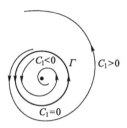

Fig. 17. Poincaré limit cycle

For example, the closed orbit Γ in the above example is asymptotically orbital stable in positive direction since $r \to 1$ as $t \to \infty$.

A planar closed orbit is called a *limit-cycle* if it is the limit-set of the nearby orbits as $t \to \pm\infty$. Therefore, the closed orbit Γ in Fig. 17 is a limit-cycle of (3.11).

However, a motion on an orbitally stable orbit may be not Liapunov stable as shown in the following.

Example 3.7 The planar system

$$\begin{cases} \dot{x} = -y(x^2 + y^2) - x(x^2 + y^2 - 1)^3, \\ \dot{y} = +x(x^2 + y^2) - y(x^2 + y^2 - 1)^3, \end{cases} \quad (3.12)$$

is equivalent to the system

$$\dot{r} = -r(r^2 - 1)^3, \qquad \dot{\theta} = r^2, \quad (3.13)$$

in polar coordinates. We can thus assert that $r = 1$ is an isolated closed orbit, and the nearby orbits approach monotonically to it (i.e., $r \to 1$ as $t \to \infty$). Hence, $r = 1$ is a limit-cycle, which is asymptotically orbitally stable as $t \to \infty$.

Note that

$$r = R_1(t) = 1, \qquad \theta = \Theta_1(t) = t + \theta_0 \quad (3.14)$$

is a solution of (3.13) on the closed orbit $r = 1$ satisfying the condition
$$R_1(0) = 1, \qquad \Theta_1(0) = \theta_0.$$
Let
$$r = r(t), \qquad \theta = \theta(t)$$
be a solution of (3.13) satisfying the initial condition
$$r(0) = r_0 > 0, \qquad \theta(0) = \alpha \qquad (r_0 \neq 1).$$
Then, using (3.13), we get the argument deviation
$$\theta(t) - \Theta_1(t) = (\alpha - \theta_0) + \int_0^t [r^2(t) - 1] \, dt$$
$$= (\alpha - \theta_0) - \int_{r_0}^r \frac{dr}{r(r^2 - 1)^2}.$$

It follows from
$$\int_{r_0}^r \frac{dr}{r(r^2 - 1)^2} \to \pm\infty, \qquad \text{as } r \to 1$$
that
$$\lim_{t \to \infty} [\theta(t) - \Theta_1(t)] = \pm\infty.$$
It follows that the motion (3.14) on the closed orbit $r = 1$ is not Liapunov stable nevertheless it is orbitally stable.

Finally, we would like to give a few words to the following famous Hilbert's 16-th problem, which is concerned about the number of limit-cycles for the planar system of differential equations
$$\dot{x} = P_n(x, y), \qquad \dot{y} = Q_n(x, y), \qquad (3.15)$$
where $P_n(x, y)$ and $Q_n(x, y)$ are polynomials in (x, y), such that
$$n = \max\{\deg[P_n(x, y)], \deg[Q_n(x, y)]\}.$$

What is the upper bound $N(n)$ of the number of limit-cycles for all of the systems (3.15) ?

No doubt, this is one of the most difficult problems in the theory of differential equations. Although there is a great amount of papers published for this problem, people are still far from the answer. For $n = 2$, it is proved that $N(2) \geq 4$. For a detailed survey, the interested readers are referred to the books (for examples, [119] and [123]).

In 1920's, van der Pol discovered an important fact that the self-excited oscillation of an electronic triode can be described by the limit cycle of the differential equation

$$\frac{d^2x}{dt^2} + \mu(x^2 - 1)\frac{dx}{dt} + x = 0, \tag{3.16}$$

in the phase plane (i.e., the (x,y)-plane with $y = \dot{x}$). Since then, the interest on limit cycles was greatly inspired not only by the theoretic Hilbert's 16-th problem but also by the applications in nonlinear oscillations. The study on limit-cycles for the Lienárd's type of differential equations is especially fruitful (see, for example, [123]).

On the other side, the study of closed orbit in the space is more challengeable in topology. The root lies in the fact that a spatial closed orbit cannot divide the space into two disjoint parts while a planar closed orbit separates the plane because of the Jordan's theorem.

3.7 Invariant Torus

Let us give a heuristic description at first. Roughly speaking, a tyre-like surface is called a *torus*. The important feature of torus is that the local structure is planar, but the global structure is non-planar. For example, the Jordan theorem does not hold on torus (i.e., in general, a simple closed curve on the torus does not separate the torus into disjoint parts). Therefore, the torus has a local planar structure meantime it admits certain space-like structure, which provides a simple method to exhibit spatial orbits.

However, for the calculus on torus, we have to describe the geometry of torus with more exactness. Let us define the torus by

$$\mathbb{T}^2 = \mathbb{R}^2/\mathrm{mod}\,(2\pi\mathbb{Z}^2),$$

which is an analogue of the definition of circle

$$\mathbb{S}^1 = \mathbb{R}^1/\mathrm{mod}\,(2\pi\mathbb{Z}^1).$$

It means that the torus $\mathbb{T}^2$ is topologically equivalent to the topological product of a pair of circles; that is, $\mathbb{T}^2 = \mathbb{S}^1 \times \mathbb{S}^1$.

Hence, a point (x^*, y^*) in $\mathbb{T}^2$ is in fact the projection of the point (x, y) in $\mathbb{R}^2$, such that

$$x^* = x, \qquad y^* = y \qquad (\text{mod } 2\pi).$$

Without loss of generalization, assume

$$0 \le x^* < 2\pi, \qquad 0 \le y^* < 2\pi.$$

In other words, a point (x^*, y^*) in $\mathbb{T}^2$ represents the (congruent) class of points $\{(x, y)\}$ in $\mathbb{R}^2$, where $(x, y) = (x^* + l, y^* + k)$ for all $(l, k) \in \mathbb{Z}^2$. On the other hand, for a given point $(x, y) \in \mathbb{R}^2$, there is $(l, k) \in \mathbb{Z}^2$, such that $x^* = x - l$, $y^* = y - k$ satisfy $0 \le x^* < 2\pi$, $0 \le y^* < 2\pi$, respectively. Then we have the map

$$P: \quad (x, y) \quad \mapsto \quad (x^*, y^*),$$

which is called the *projection* from $\mathbb{R}^2$ to $\mathbb{T}^2$.

The plane $\mathbb{R}^2 = P^{-1}(\mathbb{T}^2)$ is called the *covering space* of the torus $\mathbb{T}^2$. In general, if Ω is a set in $\mathbb{T}^2$, then the set $\tilde{\Omega} = P^{-1}(\Omega)$ in $\mathbb{R}^2$ is called the *lift* of Ω. The covering space $\mathbb{R}^2$ of $\mathbb{T}^2$ is the same as the lift of $\mathbb{T}^2$. Image that the torus $\mathbb{T}^2$ is horizontally seated. Note that, under the projection P, a vertical line in $\mathbb{R}^2$ is projected to a meridian circle on $\mathbb{T}^2$, and a horizontal line in $\mathbb{R}^2$ is projected to a latitude circle on $\mathbb{T}^2$, and so on.

Now, consider the system of differential equations

$$\frac{dx}{dt} = g(x, y), \qquad \frac{dy}{dt} = h(x, y), \tag{3.17}$$

in the plane $\mathbb{R}^2$, where the functions $f(x, y)$ and $g(x, y)$ are 2π-periodic and satisfy the Lipschitzian condition with respect to x and y, respectively. The vector field $V(x, y) = (f(x, y), g(x, y))$ is both 2π-periodic with respect to x and y.

It can be seen that a vector field V^* on $\mathbb{T}^2$ is induced by the planar vector field V through the tangential map P' of P. In fact, through the tangential map

$$P': \quad \mathbb{R}^2 \times \mathbb{R}^2 \quad \to \quad \mathbb{T}^2 \times \mathbb{R}^2$$

we obtain

$$P'[(x, y) \times V(x, y)] = (x^*, y^*) \times W(x^*, y^*),$$

where the vector field W on $\mathbb{T}^2$ is defined by $W(x^*, y^*) = V(x, y)$ (it will be simply denoted by $W = P'(V)$). Therefore, a periodic system (3.17) of differential equations on $\mathbb{R}^2$ is induced to the system

$$\frac{dx^*}{dt} = g(x^*, y^*), \qquad \frac{dy^*}{dt} = h(x^*, y^*), \tag{3.18}$$

on the torus $\mathbb{T}^2$. Naturally, we can analyze the system (3.18) on torus by means of the formula (3.17).

Example 3.8 The system of differential equations

$$\frac{dx}{dt} = \cos y, \qquad \frac{dy}{dt} = \sin x$$

on $\mathbb{R}^2$ induces a system on $\mathbb{T}^2$, which has four singular points

$$S_1 = \left(0, \frac{\pi}{2}\right), \quad S_2 = \left(\pi, \frac{\pi}{2}\right), \quad S_3 = \left(0, \frac{3\pi}{2}\right), \quad S_4 = \left(\pi, \frac{3\pi}{2}\right).$$

According to the Jacobian matrix of the system

$$\frac{\partial(\cos y, \sin x)}{\partial(x, y)} = \begin{pmatrix} 0 & -\sin y \\ \cos x & 0 \end{pmatrix},$$

it is easy to verify that S_1 and S_4 are centers and S_2 and S_3 are saddles.

Example 3.9 Consider the system

$$\frac{dx}{dt} = 1, \qquad \frac{dy}{dt} = \lambda, \tag{3.19}$$

on the torus $\mathbb{T}^2$, where λ is a constant. Note that there is no singular point of (3.19). When the the system (3.19) is considered as a planar system, the straight line

$$\Gamma : \qquad y = \lambda x + y_0$$

is an orbit in $\mathbb{R}^2$ passing through the initial point $(0, y_0)$. Therefore, the projection $\Gamma^* = P(\Gamma)$ is an orbit of (3.19) on the torus $\mathbb{T}^2$ passing through the point $(0, y_0^*)$.

Let $x_n = 2n\pi$ and $y_n = \lambda x_n + y_0$. It follows that

$$P(x_n, y_n) = (x_n^*, y_n^*) \in \Gamma^* \cap (\{0\} \times \mathbb{S}^1),$$

which means

$$x_n^* = 0, \qquad y_n^* \in \mathbb{S}^1, \qquad (n \in \mathbb{Z}).$$

Remembering the following basic knowledge in analysis:
  (a) $\{y_n^*\}_{n\in\mathbb{Z}}$ is a finite set in $\mathbb{S}^1$ if λ is a rational number;
  (b) $\{y_n^*\}_{n\in\mathbb{Z}}$ is a dense set in $\mathbb{S}^1$ if λ is an irrational number,
we conclude that
  (1) Γ^* is a closed orbit on the torus $\mathbb{T}^2$ if λ is rational;
  (2) Γ^* is a dense orbit on the torus $\mathbb{T}^2$ if λ is irrational.
Therefore, each orbit of (3.19) is closed on $\mathbb{T}^2$ when λ is a rational number, and is dense on $\mathbb{T}^2$ when λ is an irrational number.

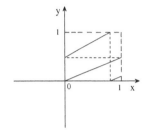
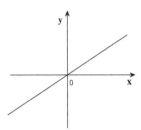

Fig.18   Fig.19

Example 3.10   Consider the system

$$\frac{dx}{dt} = F(x,y), \qquad \frac{dy}{dt} = \lambda F(x,y), \qquad (3.20)$$

on the torus $\mathbb{T}^2$, where λ is a constant and $F(x,y)$ is a real continuous function in $(x,y) \in \mathbb{T}^2$. Assume $F(x,y) \not\equiv 1$.

It can be seen that if $F(x,y) > 0$, the orbit Γ_p of (3.20) passing through the point p agrees with the orbit of (3.19) passing through p. But the motions are different since the corresponding velocities $(F(x,y), \lambda F(x,y))$ and $(1, \lambda)$ are distinct from each other.

Assume

$$\begin{cases} F(x,y) = 0, & (x,y) = \mathfrak{o}, \\ F(x,y) > 0, & (x,y) \neq \mathfrak{o}. \end{cases}$$

It can be seen that the system (3.20) has a singular point $\mathfrak{o}$. If λ is an irrational number, the dense orbit $\Gamma_\mathfrak{o}$ of (3.19) is divided into three parts:

$$\Gamma_\mathfrak{o} = \Gamma_\mathfrak{o}^- \cup \{\mathfrak{o}\} \cup \Gamma_\mathfrak{o}^+,$$

where Γ_o^- is the negative semi-orbit of Γ_o and Γ_o^+ is the positive semi-orbit of Γ_o, respectively. On the other hand, both Γ_o^- and Γ_o^+ stand, respectively, for complete orbits of (3.20) which do not intersect the singular point o. The orbit Γ_o^- of (3.20) approaches positively to o and is negatively dense in $\mathbb{T}^2$, while the orbit Γ_o^+ of (3.20) approaches negatively to o and is positively dense in $\mathbb{T}^2$.

3.8 Limit-Point Set

Let $x = \varphi(t,p)$ be the motion of the autonomous system (3.1) starting from the initial point $p \in \mathbb{R}^n$ at $t = 0$. Denote by Γ_p the orbit of the motion $x = \varphi(t,p)$ for $t \in (-\infty, \infty)$. Let Γ_p^+ (Γ_p^-) be the positive (negative) semi-orbit of the motion $x = \varphi(t,p)$ for $t > 0$ ($t < 0$).

Let q be a point in $\mathbb{R}^n$. If there is a sequence of $\{t_k\}$ tending to $+\infty$, such that the points $p_k = \varphi(t_k, p)$ approaches to the point q as $t_k \to \infty$; namely,

$$\lim_{t_k \to \infty} \text{dist}\,[p_k, q] = 0,$$

then q is called a *positive limit-point* of the motion $x = \varphi(t,p)$ (or the orbit Γ_p). A *negative limit-point* of the motion $x = \varphi(t,p)$ (or the orbit Γ_p) can be defined in a similar manner. In literature, the positive limit-point and the negative limit-point are occasionally called the ω-limit point and the α-limit point respectively. Let

$$\Omega_p = \{q \in \mathbb{R}^n : \quad q \text{ is a positive limit-point of } \Gamma_p\}$$

be the set of ω-limit points of Γ_p, and let

$$A_p = \{q \in \mathbb{R}^n : \quad q \text{ is a negative limit-point of } \Gamma_p\}$$

be the set of α-limit points of Γ_p.

3.8.1 *General Property of Limit-Point Set*

The following properties of limit-point sets are elementary in the theory of dynamical systems.

Theorem 3.3 *Both the sets Ω_p and A_p are closed (possibly empty).*

Proof. The proof is a direct consequence of the definitions of Ω_p and A_p together with the property that the solution of (3.1) is continuous with respect to the initial condition. □

Theorem 3.4 Ω_p (A_p) is a bounded and connected set if the orbit Γ_p is positively (negatively) bounded.

Proof. It follows from the positive (negative) bounded-ness of Γ_p that Ω_p (A_p) is bounded.

It remains to prove the connect-ness. Assume Ω_p is not connected. Then we have

$$\Omega_p = A \cup B, \tag{3.21}$$

where A and B are disjoint closed bounded sets in $\mathbb{R}^n$. Denote the distance between A and B by $d_0 > 0$. Since the points of A and B are ω-limit-points of Ω_p, there are sequences α_k and β_k ($\alpha_k < \beta_k$) tending to ∞, such that

$$\text{dist}\,[\varphi(\alpha_k, p), A] < \frac{d_0}{3}, \qquad \text{dist}\,[\varphi(\beta_k, p), B] < \frac{d_0}{3}.$$

It follows that there a sequence t_k ($\alpha_k < t_k < \beta_k$), such that

$$\text{dist}\,[\varphi(t_k, p), A] = \frac{d_0}{2},$$

which implies that $\varphi(t_k, p)$ is a bounded sequence. It follows that $\varphi(t_k, p)$ has a subsequence converging to a point w^* with $\text{dist}\,[w^*, A] = d_0/2$. Hence, we obtain a point $w^* \in \Omega_p$, such that $w^* \notin (A \cup B)$. However, it contradicts to (3.21).

Theorem 3.4 is thus proved. □

Examples:

(**1**) If $x = \varphi(t, q)$ is an equilibrium motion, then Ω_q and A_q are identical to the singular point q;

(**2**) If $x = \varphi(t, q)$ is a non-trivial periodic motion $x = \varphi(t, p)$, then Ω_q and A_q are identical to the closed orbit Γ_p;

(**3**) If Γ_p is a non-degenerate orbit of Example 3.6, then Ω_p is the limit-cycle C_1 ($r = 1$), and

$$A_p = \begin{cases} \mathfrak{o}, & \text{if dist}\,[p, \mathfrak{o}] < 1, \\ C_1, & \text{if dist}\,[p, \mathfrak{o}] = 1, \\ \infty, & \text{if dist}\,[p, \mathfrak{o}] > 1 \end{cases} \quad \text{(where } \infty \text{ is the infinite point of } \mathbb{R}^2\text{)};$$

(4) To find a disconnected limit-set, consider first the planar system

$$\dot{x} = y, \quad \dot{y} = -x - y, \tag{1}$$

having the phase-portrait $\mathfrak{F}_1$ with a unique unstable focus $\mathfrak{o}$ (i.e., all the non-degenerate orbits spirally departure from the focus $\mathfrak{o}$ and tend to the infinity of $\mathbb{R}^2$ in positive direction). Then, consider the mapping

$$\mathcal{T}: \quad u = x, \quad v = \arctan y,$$

which transforms $\mathfrak{F}_1$ into the phase-portrait $\mathfrak{F}_2$ of the system

$$\dot{u} = \tan v, \quad \dot{v} = -u\cos^2 v - \sin v \cos v, \tag{2}$$

in the strip

$$S: \quad -\infty < u < \infty, \quad -\frac{\pi}{2} < v < \frac{\pi}{2}.$$

It follows that an orbit Γ_z ($z \neq \mathfrak{o}$) of (2) in the strip S approaches spirally in positive direction to the horizontal lines H_1 ($v = \pi/2$) and H_2 ($v = -\pi/2$). Hence, we obtain a disconnected limit-set $\Omega_z = H_1 \cup H_2$.

(5) For the orbit Γ_p of Example 3.9, the limit-sets Ω_p and A_p are identical to the torus $\mathbb{T}^2$ if λ is irrational; and they are identical to the closed orbit Γ_p if λ is rational;

(6) When λ is irrational, Example 3.10 has a unique orbit $\Upsilon_1 = \Gamma_\mathfrak{o}^-$ positively tending to the singular point $\mathfrak{o}$ and negatively dense in $\mathbb{T}^2$, as well as, a unique orbit $\Upsilon_2 = \Gamma_\mathfrak{o}^+$ negatively tending to the singular point $\mathfrak{o}$ and positively dense in $\mathbb{T}^2$. For all the other orbits Γ_p, the limit-point sets Ω_p and A_p are identical to the torus $\mathbb{T}^2$. Finally, we leave the phase portrait of Example 3.10 to the reader when λ is rational.

3.9 Poincaré-Bendixson Theorem

Assume (3.1) is a planar system. Let Γ_p^+ be a positive semi-orbit of (3.1) contained in a compact set K. Then the positive limit set Ω_p is a non-empty closed set in K.

The following theorem is called the *Poincaré-Bendixson Theorem* (or, for brevity, the *P-B Theorem*), which is a basic tool in the planar qualitative theory of differential equations.

Theorem 3.5 *If Ω_p contains no singular point, then either*

(i) Γ_p *is a closed orbit; or*

(ii) Ω_p is a closed orbit.

Proof. There are various proofs in literature (see, for examples, [21] and [86]), where the essential point is the Jordan theorem.[4]

If Γ_p is a closed orbit, we have $\Gamma_p = \Omega_p$.

Now, assume Γ_p is not a closed orbit. Then, we want to prove that Ω_p is a closed orbit.

We can first prove with ease that Ω_p is compact. Moreover, it is invariant (i.e., $z \in \Omega_p$ implies $\Gamma_z \subset \Omega_p$).

Assume $q \in \Omega_p$. Then $\Gamma_q \subset \Omega_p$. It follows that Γ_q is a non-degenerate orbit, with the possibilities:

(1) Γ_q is a closed orbit;

(2) Γ_q is not a closed orbit.

Assume (2) holds.

Since $\Gamma_q \subset \Omega_p$, we have $\Omega_q \subset \Omega_p$. It follows that Ω_q contains no singular point. Hence, there is a non-singular point $\xi \in \Omega_q$. Let L_ξ be a line-segment centered at ξ, such that L_ξ perpendicular to the planar vector $V(\xi) \neq 0$. Since ξ is an ω-limit point of Γ_q, there are points $q_1 = \varphi(t_1, q) \in L_\xi$ and $q_2 = \varphi(t_2, q) \in L_\xi$ with $t_2 > t_1$, such that the orbital arc $\Gamma_q[q_1, q_2]$ of Γ_q from q_1 to q_2 and the interval $L_\xi[q_2, q_1]$ of L_ξ from q_2 to q_1 form a Jordan closed curve; namely,

$$J = \Gamma_q[q_1, q_2] \cup L_\xi[q_2, q_1]$$

is a Jordan closed curve. We can take the line-segment L_ξ small enough, such that the vector $V(x)$ on L_ξ directs from the outside to the inside of J (or the converse). In any case, Γ_q is separated by J. Since $\xi \in \Omega_p$, ξ lies in the side (in the outside). It follows that the orbit Γ_p will come into the inside (or the outside) of J forever. Therefore, we conclude that $\Gamma_q \subset \Omega_p$ will stay in the inside (or the outside) of J. This contradicts to the fact Γ_q is separated by J. Hence, the contradiction proves that the possibility (2) does not hold.

Therefore, the possibility (1) holds. We have thus conclude that Ω_p contains a closed orbit Γ_q. Since Γ_p is not a closed orbit but Γ_q is a closed orbit, we have $\Gamma_p \neq \Gamma_q$. It follows that (a) Γ_p lies in the inside of the closed orbit Γ_q and spirally approaches to Γ_q; (b) Γ_p lies in the outside of the closed orbit Γ_q and spirally approaches to Γ_q.

---

[4]That is, a simple closed curve in plane separates the plane into two disjoint parts called, respectively, the *interior part* and the *exterior part*.

In both the cases (a) and (b), it can be proved that $\Omega_p = \Gamma_q$ is a closed orbit.

The proof is thus complete. □

Finally, we remark that the following theorem (i.e., the *generalized Poincaré-Bendixson Theorem*) can be proved in a similar manner (see [86]).

Theorem 3.6 *Assume Γ_p^+ is a positive (non-degenerate) semi-orbit of a planar system in a bounded and closed set K, then Ω_p is either a singular point, a closed orbit, or a closed polygon Δ_p.*[5]

Example 3.11 Consider the system

$$\dot{x} = -x, \qquad \dot{y} = y, \qquad (x, y) \in \mathbb{R}^2.$$

It is obvious that $\mathbf{o} = (0, 0)$ is a singular point.

Let $p = (\xi, \eta) \neq \mathbf{o}$. It can be seen that the orbit Γ_p is given by

$$x = \xi e^{-t}, \qquad y = \eta e^t,$$

which yields the following conclusions:
(1) $\Omega_p = A_p = \emptyset$ whenever $\xi \neq 0$ and $\eta \neq 0$;
(2) $\Omega_p = \{\mathbf{o}\}$ and $A_p = \emptyset$ whenever $\xi \neq 0$ and $\eta = 0$;
(3) $\Omega_p = \emptyset$ and $A_p = \{\mathbf{o}\}$ whenever $\xi = 0$ and $\eta \neq 0$.

Example 3.12 Consider the system

$$\dot{x} = y, \qquad \dot{y} = -x, \qquad (x, y) \in \mathbb{R}^2,$$

which has a unique singular point $\mathbf{o} = (0, 0)$.

Let $p = (\xi, \eta) \neq \mathbf{o}$. It can be seen that Γ_p is a closed orbit given by the circle

$$x^2 + y^2 = c^2, \qquad \text{with the constant } c = \sqrt{\xi^2 + \eta^2} > 0.$$

It can be seen from the equation $\dot{x} = y$ that the direction of Γ_p is clockwise.
Moreover, we have $\Gamma_p = \Omega_p = A_p$ for all $p \in \mathbb{R}^2$.

Example 3.13 Consider the system

$$\dot{x} = y + x(x^2 + y^2 - 1), \qquad \dot{y} = -x + y(x^2 + y^2 - 1), \qquad (x, y) \in \mathbb{R}^2,$$

which has a unique singular point $\mathbf{o} = (0, 0)$.

---
[5]A closed polygon Δ_p is a contour, consisting of some non-degenerate orbits and singular points, such that there is an orbit Γ_p spirally approaching to Δ_p on one side.

Let $p = (\xi, \eta) \neq \mathfrak{o}$. Using the polar coordinates $(x, y) = \langle r, \theta \rangle$, we transform the original system into the following form
$$\dot{r} = r(r^2 - 1), \qquad \dot{\theta} = -1.$$
It follows that
(1) If $|p| = 1$, Γ_p is the closed orbit Δ ($r = 1$) with clockwise direction;
(2) If $0 < |p| < 1$, $\Omega_p = \mathfrak{o}$ and $A_p = \Delta$;
(3) If $1 < |p|$, $\Omega_p = \Delta$ and $A_p = \emptyset$.

Example 3.14 Consider the system
$$\begin{cases} \dot{x} = [(x-1)^2 + y^2][y + x(x^2 + y^2 - 1)], \\ \dot{y} = [(x-1)^2 + y^2][-x + y(x^2 + y^2 - 1)], \end{cases} (x, y) \in \mathbb{R}^2,$$
which has the singular points $S_1 = (0, 0)$ and $S_2 = (1, 0)$.

Let $p = (\xi, \eta) \neq \mathfrak{o}$. It is noted that the unit-circle Δ is a closed polygon since $S_2 \in \Delta$. It is clear that
(1) If $|p| = 1$ and $p \neq S_2$, $\Gamma_p = (\Delta \setminus S_2)$ is a bounded open orbit with clockwise direction, such that $\Omega_p = A_p = S_2$;
(2) If $0 < |p| < 1$, $\Omega_p = \mathfrak{o}$ and $A_p = \Delta$;
(3) If $1 < |p|$, $\Omega_p = \Delta$ and $A_p = \emptyset$.

Finally, we emphasize again the important role of Jordan theorem in the proof of Poincaré-Bendixson theorem. In fact, the Poincaré-Bendixson theorem does not hold in the space where the Jordan theorem is invalid. For example, let us consider the system
$$\dot{x} = 1, \qquad \dot{y} = \lambda \qquad \text{(where } \lambda \text{ is an irrational number)}$$
on the torus $\mathbb{T}^2$, where the Jordan theorem fails.

It is obvious that the above system has no singular point. For a given point $p \in \mathbb{T}^2$, the orbit Γ_p is dense in $\mathbb{T}^2$ and the limit-set Ω_p is thus identical to $\mathbb{T}^2$. Hence, neither Γ_p nor Ω_p is a closed orbit. It follows that the Poincaré-Bendixson theorem does not hold on the torus.

Chapter 4

Non-Autonomous Systems

4.1 General Systems

The qualitative theory of differential equations is concerned with not only the autonomous differential equations but also with the non-autonomous differential equations. Sometimes, it is termed as the theory of *nonlinear oscillations* since they have no significant difference.

In what follows, we consider the general system of differential equations

$$\frac{dx}{dt} = f(t, x), \tag{4.1}$$

where $f(t, x)$ is a continuous function in $(t, x) \in \mathbb{R}^1 \times \mathbb{R}^n$ with value in $\mathbb{R}^n$. If $f(t, x)$ is periodic in t with period $T > 0$; i.e.,

$$f(t + T, x) = f(t, x), \quad \forall\ (t, x) \in \mathbb{R}^1 \times \mathbb{R}^n,$$

then (4.1) is called a *periodic system* of period T or a T-*periodic system* (with respect to the variable t). It can be seen that if (4.1) is T-periodic, then it is kT-periodic too, where k any given positive integer.

An autonomous system can be also considered as a T-periodic system for any constant $T > 0$. Therefore, an autonomous system is a periodic system without minimal period. On the other hand, it is not hard to prove that a non-autonomous periodic system has the minimal period.

4.1.1 Linear System

As we know, the stability of the solution $x = \phi(t)$ of (4.1) depends heavily upon the stability of the trivial solution for the linear homogeneous system of differential equations

$$\frac{dx}{dt} = A(t)x, \qquad x \in \mathbb{R}^n, \qquad (4.2)$$

where

$$A(t) = f'_x(t, \phi(t)), \qquad t \in \mathbb{R}^1,$$

is an $n \times n$-matrix. Let $U(t)$ be the fundamental matrix-solution $U(t)$ of (4.2), satisfying the initial condition $U(0) = I$, where I is the identity.

The following results are known in literature:

(**1**) If $A(t)$ is a constant matrix B, the stability of (4.2) is determined by the fundamental matrix solution

$$U(t) = e^{Bt} \qquad (t \in \mathbb{R}^1) \qquad (4.3)$$

of (4.2). Hence, the stability of (4.2) is completely solved by the Jordan form of B (see, for example, [21] and [56]).

(**2**) If $A(t)$ is a T-periodic matrix, the fundamental matrix solution $U(t)$, generally speaking, can not be found by the elementary method via the coefficient matrix $A(t)$. However, there holds the *Floquet theorem* as follows.

Theorem 4.1 *If $A(t)$ is T-periodic, then we have*

$$U(t) = P(t)e^{Kt} \qquad (4.4)$$

where $P(t)$ is a T-periodic $n \times n$-matrix and K is a constant $n \times n$-matrix.

Although the Floquet theorem is more or less theoretic, it concludes that the stability of the linear T-periodic system (4.2) is solved by a constant matrix K. In this sense, the T-periodic system (4.2) is called *reducible*.

It is an open question about the reducibility of the almost periodic system (4.2). In other words, if $A(t)$ is an almost periodic matrix, we do not know if the fundamental matrix solution $U(t)$ of (4.2) can be put in the following form

$$U(t) = Q(t)e^{Mt},$$

where $Q(t)$ is an almost periodic matrix $Q(t)$ and M is a constant matrix.

Nothing can be said about the reducibility of (4.2) in other cases.

4.1.2 Poincaré Map

Assume (4.1) is T-periodic, and has a unique solution $x = \varphi(t, \xi)$ satisfying the initial condition

$$x(0) = \xi \qquad (\forall\, \xi \in \mathbb{R}^n).$$

Assume $x = \varphi(t, \xi)$ exists globally on the real axis $\mathbb{R}^1$ for all $\xi \in \mathbb{R}^n$, then

$$\Phi: \quad \xi \mapsto \varphi(T, \xi) \qquad (\forall\, \xi \in \mathbb{R}^n)$$

defines a continuous map from $\mathbb{R}^n$ to itself, which is called the *Poincaré map* of the periodic system (4.1). Moreover, we have the iteration map $\Phi^k(\xi) = \varphi(kT, \xi)$ for $k \in \mathbb{Z}_+$.

It can be seen that the map Φ is one-to-one, but it may be not 'onto' (i.e., it is possible that $\Phi(\mathbb{R}^n) \neq \mathbb{R}^n$). If $x = \varphi(t, \xi)$ exists globally on the real axis $(-\infty < t < \infty)$ for all $\xi \in \mathbb{R}^n$, then the map Φ is 'onto', and it is thus a homeomorphism of $\mathbb{R}^n$.

Obviously, the solution $x = \varphi(t, \xi)$ of (4.1) is positively (or negatively) bounded if and only if the orbit $\Delta_\xi = \{\Phi^k(\xi) : k \in \mathbb{Z}\}$ of Φ passing through the initial point ξ is positively (or negatively) bounded. Moreover, there is an important relationship between the periodic solutions of (4.1) and the fixed points of the Poincaré map Φ.

Theorem 4.2 $x = x(t)$ *is a T-periodic solution of (4.1) if and only if the initial point $\xi = x(0)$ is a fixed point of Φ.*

Proof. Let ξ be a fixed point of Φ (i.e., $\xi = \Phi(\xi) = \varphi(T, \xi)$). Since $x = x(t)$ satisfies the initial condition $x(0) = \xi$, the uniqueness theorem of solution implies that $x = x(t) = \varphi(t, \xi)$. Let $x = y(t) = \varphi(t + T, \xi)$. It can be seen that $x = y(t)$ is still a solution of (4.1) and satisfies the initial condition

$$y(0) = \varphi(T, \xi) = \Phi(\xi) = \xi.$$

Hence, $x(0) = y(0)$. It follows from the uniqueness theorem that $y(t) = x(t)$; namely, $x(t + T) = x(t)$ for $t \in \mathbb{R}^1$. In other words, $x = x(t)$ is a T-periodic solution of (4.1).

Conversely, assume $x = x(t)$ is a T-periodic solution of (4.1). Then we have $x(t+T) = x(t)$ for all $t \in \mathbb{R}^1$. Setting $\xi = x(0)$, we have $x(t) = \varphi(t, \xi)$. It follows that $\varphi(t + T, \xi) = \varphi(t, \xi)$. Then, letting $t = 0$, we have

$$\Phi(\xi) = \varphi(T, \xi) = \varphi(0, \xi) = \xi.$$

It proves that the initial point $\xi = x(0)$ of the T-periodic solution $x = x(t)$ is a fixed point of Φ.

The proof of Theorem 4.2 is thus completed. □

In a similar manner, we can prove the following result.

Corollary 4.1 *The solution $x = x(t)$ of (4.1) is mT-periodic if and only if its initial point $\xi = x(0)$ is a m-periodic point of Φ (i.e., $\xi = \Phi^m(\xi)$).*

4.2 Conservative Systems

Assume Φ is a continuous map from $\mathbb{R}^n$ to itself. If for any measurable set $E \subset \mathbb{R}^n$, the Lebesgue measure $\mu(E)$ is invariant under Φ (i.e., $\mu(\Phi(E)) = \mu(E)$), then Φ is called a *measure-preserving* map. It is remarked that a measure-preserving map is also called an area-preserving map when it is defined on some two-dimensional manifold.

Consider the system of differential equations

$$\frac{dx}{dt} = f(t, x), \tag{4.5}$$

where

$$x = \begin{pmatrix} x_1 \\ \vdots \\ x_n \end{pmatrix} \quad \text{and} \quad f(t, x) = \begin{pmatrix} f_1(t, x) \\ \vdots \\ f_n(t, x) \end{pmatrix}$$

are n-dimensional vectors. Assume $f(t, x)$ is continuous in $(t, x) \in \mathbb{R}^1 \times \mathbb{R}^n$ and differentiable with respect to x.

If the divergence $\operatorname{div}_x[f]$ of f is identically equal to zero; that is,

$$\operatorname{div}_x[f](t, x) := \sum_{j=1}^{n} \frac{\partial f_j}{\partial x_j}(t, x) = 0, \quad (t, x) \in \mathbb{R}^1 \times \mathbb{R}^n, \tag{4.6}$$

then the system (4.5) is said to be *conservative*.

Theorem 4.3 *If the system (4.5) is conservative and T-periodic, then its Poincaré map Φ is measure-preserving.*

Proof. Assume $x = \varphi(t, \xi)$ is the solution of (4.5) satisfying the initial condition $\varphi(0, \xi) = \xi$. Differentiating to the identity
$$\frac{d\varphi(t, \xi)}{dt} = f(t, \varphi(t, \xi))$$
with respect to ξ, we obtain a linear system
$$\frac{du}{dt} = f'_x(t, \varphi(t, \xi))u,$$
which has the fundamental matrix of solutions
$$u = \varphi'_\xi(t, \xi) = \begin{pmatrix} \frac{\partial \varphi_1}{\partial \xi_1}(t, \xi) & \cdots & \frac{\partial \varphi_1}{\partial \xi_n}(t, \xi) \\ \vdots & \vdots & \vdots \\ \frac{\partial \varphi_n}{\partial \xi_1}(t, \xi) & \cdots & \frac{\partial \varphi_n}{\partial \xi_n}(t, \xi) \end{pmatrix}.$$

It follows from (4.6) that the Wronskii determinant
$$\det [\varphi'_\xi(t, \xi)] = \exp\left(\int_0^t \mathrm{div}_x[f](t, \varphi(t, \xi))\, dt\right) = 1.$$
In particular, letting $t = T$, we have
$$\det [\varphi'_\xi(T, \xi)] = 1.$$
Now, let $x = \varphi(T, \xi)$ for $\xi \in E$, where $E \subset \mathbb{R}^n$ is a given measurable set. Then we have
$$\mu(\Phi(E)) = \int_{\Phi(E)} dx = \int_E \det[\varphi'_\xi(T, \xi)]\, d\xi = \int_E d\xi = \mu(E),$$
which implies that the Poincaré map Φ is measure-preserving. Theorem 4.3 is thus proved. □

Now, consider the system of differential equations
$$\begin{cases} \dfrac{du}{dt} = H'_v(t, v, u), \\ \dfrac{dv}{dt} = -H'_u(t, u, v), \end{cases} \quad (t, u, v) \in \mathbb{R}^1 \times \mathbb{R}^n \times \mathbb{R}^n, \quad (4.7)$$
where $H(t, u, v)$ is a continuous scalar function in (t, u, v) and is C^1-differentiable with respect to (u, v). Then (4.7) is called a *Hamiltonian system* having the *Hamiltonian function* $H(t, u, v)$. It is obvious that a Hamiltonian system is conservative. Hence, for a T-periodic Hamiltonian

system (4.7), the corresponding Poincaré map Φ is measure-preserving in the phase space $\mathbb{R}^n \times \mathbb{R}^n$.

4.3 Dissipative Systems

Assume $\mathfrak{D}$ is a compact (simply connected) domain in $\mathbb{R}^n$ (i.e., it is a topological n-dimensional ball). If for each solution $x = x(t)$ of the system

$$\frac{dx}{dt} = g(t,x), \qquad (t,x) \in \mathbb{R}^1 \times \mathbb{R}^n, \tag{4.8}$$

there is a constant $\tau \geq 0$, such that

$$x(t) \in \mathfrak{D}, \qquad \forall\, t \geq \tau,$$

then (4.8) is called a *dissipative system*.

Assume system (4.8) has a unique solution $x = \varphi(t, t_0, \xi)$ satisfying the initial condition $\varphi(t_0, t_0, \xi) = \xi$.

If (4.8) is dissipative, then, for any given initial point $\xi \in \mathbb{R}^n$, there is a time $s = s(t_0, \xi) > 0$, such that

$$\varphi(t, t_0, \xi) \in \mathfrak{D}, \qquad \forall\, t \geq s.$$

It means that the solution $x = \varphi(t, t_0, \xi)$ starting from $\xi \in \mathbb{R}^n$ at the time $t = t_0$ will comes into the domain $\mathfrak{D}$ at some time $t = s$ and stays there forever. The domain $\mathfrak{D}$ is called an *attractor* of the dissipative system (4.8).

Moreover, assume the dissipative system (4.8) is T-periodic, then, for any initial point $\xi \in \mathfrak{D}$, there is a time $\tau = \tau(\xi) > 0$, such that

$$\varphi(t, 0, \xi) \in \mathfrak{D}, \qquad \forall\, t \geq \tau.$$

Obviously, we have a constant

$$\tau_* = \inf_{\xi \in \mathfrak{D}} \tau(\xi) \geq 0,$$

such that

$$\varphi(t, 0, \xi) \in \mathfrak{D}, \qquad \forall\, t \geq \tau_*, \tag{4.9}$$

for any $\xi \in \mathfrak{D}$. Hence, there is a positive integer m, such that $mT \geq \tau_*$, and (4.9) implies that $\varphi(mT, 0, \xi) \in \mathfrak{D}$; that is,

$$\Phi^m(\xi) \in \mathfrak{D}, \qquad \forall\, \xi \in \mathfrak{D},$$

where $\Phi(\cdot) = \varphi(T, 0, \cdot)$ is the Poincaré map of (4.8). Then, the Brouwer fixed point theorem asserts that Φ^m has at least a fixed point $\xi_0 \in \mathfrak{D}$. It follows that the system (4.8) has a mT-periodic solutin $x = \varphi(t, 0, \xi_0)$. However, this solution may be kT-periodic, where k is some positive integer k less than mT. In this case, the integer m must be divisible by the integer k. As we know, it is very hard to verify the minimal period of a periodic solution in the theory of differential equations.

A dissipative system is apparently not conservative and conversely, a conservative system is not dissipative. The major subjects of nonlinear oscillations are concerned with the properties of these systems. We are usually encountered with the following problems:

(Q_1) *How to identify the dissipative system?*

(Q_2) *Is there a T-periodic solution for dissipative T-periodic system ?*

These problems are studied by Cartwright-Littlewood [18] and Levinson [88] in details for the nonlinear differential equations of second order

$$\ddot{x} + f(x, \dot{x})\dot{x} + g(x) = p(t) \qquad (4.10)$$

where $x \in \mathbb{R}^1$.

Theorem 4.4 *If the following conditions*

(A_1) *There exist constant $a > 0$, $m > 0$ and $M > 0$, such that*

$$f(x, y) \begin{cases} \geq M, & \text{if } |x| \geq a, \\ \geq -m, & \text{if } |x| \leq a; \end{cases}$$

(A_2) *There exists a constant $A > 0$, such that $xg(x) > 0$ if $|x| \geq A$, and $g(x)$ is monotonically increasing for $x \geq A$;*

(A_3) *$g(x) \to \infty$ as $|x| \to \infty$; moreover, $g(x)/G(x) = \mathcal{O}(1/|x|)$, where $G(x) = \int_0^x g(x)\, dx$;*

(A_4) *$p(t)$ is a bounded continuous function;*

are satisfied, then (4.10) is a dissipative differential equation.

Corollary 4.2 *The Duffing equation*

$$\ddot{x} + c\dot{x} + g(x) = p(t), \qquad \text{where } c \text{ is a positive constant}, \qquad (4.11)$$

is dissipative whenever $p(t)$ satisfies the condition (A_4) and $g(x)$ satisfies the conditions (A_2) and (A_3).

4.3.1 Duffing Equation in Higher Dimensional Space

The above Corollary 4.2 will be partially generalized to Duffing equation in higher dimensional space as follows.

$$\frac{d^2x}{dt^2} + C\frac{dx}{dt} + G'_x(x) = p(t), \qquad (t,x) \in \mathbb{R}^1 \times \mathbb{R}^n, \qquad (4.12)$$

where C is the coefficient $(n \times n)$-matrix of friction and $G'_x(x)$ denotes the restoring force of the equation with the potential (differentiable) function $G(x)$ from $\mathbb{R}^n$ to $\mathbb{R}^1$. Note that (4.12) is the *scalar Duffing equation* if $n = 1$.

When $C = 0$, we have the Duffing equation

$$\frac{d^2x}{dt^2} + G'_x(x) = p(t), \qquad (t,x) \in \mathbb{R}^1 \times \mathbb{R}^n, \qquad (4.13)$$

without the friction term. In this case, setting

$$E(t,x,y) = \frac{1}{2}\sum_{j=1}^{n} y_j^2 + G(x) - \sum_{j=1}^{n} x_j p_j(t),$$

we write (4.13) into the equivalent Hamiltonian system

$$\frac{dx}{dt} = E'_y(t,x,y), \qquad \frac{dy}{dt} = -E'_x(t,x,y), \qquad (4.14)$$

which is T-periodic if the perturbed term $p(t)$ of (4.13) is T-periodic. Let Φ be the Poincaré map of (4.14) (i.e., (4.13)). Then Φ is a measure-preserving map in $\mathbb{R}^n$. Generally speaking, it is not an easy task to study the existence of fixed-point of Φ (or the existence of periodic solution of (4.13)) in the higher dimensional space (see [4]) although it is fruitful to analyze the Duffing equation of second order in the phase-plane $\mathbb{R}^2$.

When $C \neq 0$, we assume the following conditions:

(H_1): $\langle Cy, y \rangle \geq a \langle y, y \rangle, \quad \forall\, y \in \mathbb{R}^n$;

(H_2): $\langle x, G'_x(x) \rangle \geq b \langle x, x \rangle, \quad \forall\, |x| > d$,

where a, b and d are some positive constants, and $\langle \cdot, \cdot \rangle$ denotes the inner product of n-vectors.

Condition (H_1) means that C is a positively definite matrix (i.e., the Duffing system (4.12) has a positively definite coefficient-matrix of friction), and Condition (H_2) means that the system has a sufficiently large restoring force $G'_x(x)$.

Chapter 4. Non-Autonomous Systems

It is known that the Duffing equation (4.12) is equivalent to the system

$$\frac{dx}{dt} = y, \quad \frac{dy}{dt} = -Cy - G'_x(x) + p(t). \tag{4.15}$$

Moreover, consider the auxiliary family of differential equations

$$\frac{dx}{dt} = y, \quad \frac{dy}{dt} = -Cy - G'_x(x) + \sigma p(t), \tag{4.16}$$

where σ is a parameter ($0 \leq \sigma \leq 1$). Note that the system (4.16) agrees with (4.15) when $\sigma = 1$. Let

$$x = x(t, \xi, \eta, \sigma), \quad y = y(t, \xi, \eta, \sigma) \tag{4.17}$$

be the solution of (4.16) satisfying the initial condition

$$x(0) = \xi, \quad y(0) = \eta, \quad (\xi, \eta) \in \mathbb{R}^n \times \mathbb{R}^n.$$

Using the condition (H_2), we have the global existence of the solution on the interval $-\infty < t < \infty$. Then the uniqueness theorem of solution implies that the solution (4.17) is continuous with respect to

$$(t, \xi, \eta, \sigma) \in \mathbb{R}^1 \times \mathbb{R}^n \times \mathbb{R}^n \times [0, 1].$$

Then, for given $\sigma \in [0, 1]$, we obtain the family of vector fields

$$F_\sigma(z) = \begin{pmatrix} x(T, u, v, \sigma) - u \\ y(T, u, v, \sigma) - v \end{pmatrix},$$

where $z = (u, v)$ and $F_\sigma(z)$ are vectors in $\mathbb{R}^n \times \mathbb{R}^n$.

Without loss of generality, assume $G(0) = 0$. Let

$$h(x, y) = \frac{1}{2} \langle y, y \rangle + G(x).$$

Lemma 4.1 *If the condition (H_2) is satisfied, then for sufficiently large constant $E > 0$, the equation*

$$h(x, y) = E, \quad (x, y) \in \mathbb{R}^n \times \mathbb{R}^n, \tag{4.18}$$

defines a manifold $\mathfrak{M}_E$ of dimension $(n-1)$, which is starlike with respect to the origin of $\mathbb{R}^n \times \mathbb{R}^n$. Moreover, $\mathfrak{M}_E$ encloses any given bounded set whenever $E > 0$ is large enough.

Proof. It can be seen that $w = h(x, y)$ is a hyper-surface of dimension $2n$ in the (x, y, w)-space $\mathbb{R}^{2n+1}$. Since $h(0, 0) = 0$ and

$$h(x, y) \to \infty, \qquad \text{as } \sqrt{|x|^2 + |y|^2} \to \infty,$$

the equation (4.18) determines a compact set $\mathfrak{M}_E$ in the hyperplane $w = 0$.

Let $(x, y) \in \mathfrak{M}_E$ and let $\mathfrak{r} = (x, y, 0)$ be the coordinates-vector of the point $(x, y, 0)$. Note that

$$\mathfrak{n} = (G'_x(x),\ y,\ -1)$$

is the normal vector of $w = h(x, y)$ at the point (x, y), which yields the inner product

$$\langle \mathfrak{n}, \mathfrak{r} \rangle = \langle x, G'_x(x) \rangle + \langle y, y \rangle \geq b\langle x, x \rangle + \langle y, y \rangle > 0, \qquad \forall\ (x, y) \in \mathfrak{M}_E.$$

It follows that any ray on the hyperplane $w = E$ issuing from the origin $\mathfrak{o}$ intersects the hyper-surface $w = h(x, y)$ at one and only one point; hence $\mathfrak{M}_E$ is starlike with respect the origin of $\mathbb{R}^n \times \mathbb{R}^n$.

Lemma 4.1 is thus proved. $\square$

Lemma 4.2 *For any point $(u, v) \in \mathfrak{M}_E$, the vectors $F_0(u, v)$ and (u, v) are different in direction when $E > 0$ is large enough.*

Proof. Differentiating the function $w = h(x, y)$ with respect to (4.15) with $\sigma = 0$, we have

$$\frac{dw}{dt} = \langle y, \frac{dy}{dt} \rangle + \langle G'_x(x), y \rangle$$

$$= \langle y, -Cy - G'_x(x) \rangle + \langle y, G'_x(x) \rangle = -\langle y, y \rangle \leq 0.$$

It follows that the function $w = h(x, y)$ is decreasing along the orbit of (4.15) with $\sigma = 0$. Let $\mathfrak{D}_E$ be the domain bounded by $\mathfrak{M}_E$. It follows that

$$(x(2\pi, \xi, \eta, 0), y(2\pi, \xi, \eta, 0)) \in \mathfrak{D}_E,$$

for $(\xi, \eta) \in \mathfrak{M}_E$. Since $\mathfrak{D}_E$ is starlike, the vector

$$F_0(\xi, \eta) = (x(2\pi, \xi, \eta, 0) - \xi,\ y(2\pi, \xi, \eta, 0) - \eta)$$

does not agree the direction of the coordinate vector (ξ, η) whenever $F_0(\xi, \eta) \neq 0$. In the next Lemma, which is independent on this Lemma, we prove that $F_0(\xi, \eta) \neq 0$ for $(\xi, \eta) \in \mathfrak{M}_E$ if E is large enough.

Hence, Lemma 4.2 is proved. $\square$

As a consequence, the index of the vector field F_0 on $\mathfrak{M}_E$

$$\text{Ind}\,[F_0, \mathfrak{M}_E] = 1, \quad \text{if } E > 0 \text{ is large enough.} \tag{4.19}$$

Before proving the main theorem of this section, we need to prove the following result.

Lemma 4.3 *Assume $(x(t), y(t))$ is any T-periodic solution of the system (4.15) with $\sigma \in [0, 1]$. Then*

$$(x(t), y(t)) \in \mathfrak{D}_E, \quad t \in \mathbb{R}^1,$$

if $E > 0$ is large enough.

Proof. We will proceed the proof in the following steps:

Step 1. Let $(x(t), y(t))$ be any T-periodic solution of the system (4.15) with $\sigma \in [0, 1]$. Then $h = h(x(t), y(t))$ is a T-periodic function, satisfying

$$\frac{dh}{dt} = \langle y(t), \sigma p(t) \rangle - \langle Cy(t), y(t) \rangle$$

which implies

$$h(x(t), y(t)) = h_0 + \int_0^t \langle y(t), \sigma p(t) \rangle \, dt - \int_0^t \langle Cy(t), y(t) \rangle \, dt,$$

where $h_0 = h(x(0), y(0))$ is a constant. The property of periodic function yields

$$\int_0^t \langle y(t), \sigma p(t) \rangle \, dt = m_1 t + P(t),$$

and

$$\int_0^t \langle Cy(t), y(t) \rangle \, dt = m_2 t + Q(t),$$

where m_1 and m_2 are the corresponding mean values of integration, and $P(t)$ and $Q(t)$ are T-periodic functions. Hence, we have

$$h(x(t), y(t)) = h_0 + (m_1 - m_2)t + (P(t) - Q(t)).$$

Since $h(x(t), y(t))$ is T-periodic, we get

$$m_1 = m_2. \tag{4.20}$$

It is clear that
$$m_1 = \frac{\sigma}{T} \int_0^T \langle y(t), p(t) \rangle \, dt \leq \|y\| \cdot \|p\|$$

and
$$m_2 = \frac{1}{T} \int_0^T \langle Cy(t), y(t) \rangle \, dt \geq \frac{a}{T} \int_0^T \langle y(t), y(t) \rangle \, dt = a\|y\|^2,$$

where
$$\|y\| = \sqrt{\left\{ \frac{1}{T} \int_0^T \langle y(t), y(t) \rangle \, dt \right\}}$$

is the norm of $y(t)$ in the Hilbert space $L^2[0,T]$. It follows from (4.20) that

$$\|y\| \leq c_1, \qquad \text{where the constant } c_1 = \frac{\|p\|}{a} > 0. \qquad (4.21)$$

Step 2. Now, we claim that *there is a constant $c_2 > 0$, such that any T-periodic solution $(x(t), y(t))$ of (4.15) for $\sigma \in [0,1]$ satisfies*

$$|x(t^*)| < c_2 \qquad (4.22)$$

at some time $t^* \in [0,T]$.

In fact, assume the contrary. Then we have a sequence $\{d_j\}$ tending to $+\infty$ and a sequence $\{\sigma_j\}$ in $[0,1]$, such that there is a T-periodic solution $(x_j(t), y_j(t))$ of (4.15) for $\sigma = \sigma_j$, satisfying

$$|x_j(t)| > d_j, \qquad \forall \, t \in [0,T] \qquad (j = 1, 2, \cdots). \qquad (4.23)$$

Without destroying the generality, assume $d_j > d$, where the constant d is given in (H_2). For simplicity, let (x, y) denote the above mentioned T-periodic solutions. Then we have

$$\langle x, \frac{dy}{dt} \rangle = \langle x, \sigma_j p(t) \rangle - \langle x, Cy \rangle - \langle x, G'_x(x) \rangle. \qquad (4.24)$$

Note that
$$\int_0^T \langle x, \frac{dy}{dt} \rangle \, dt = -\int_0^T \langle \frac{dx}{dt}, y \rangle \, dt = -\int_0^T \langle y, y \rangle \, dt = -T\|y\|^2,$$

and
$$\int_0^T \langle C\frac{dx}{dt}, x \rangle \, dt = -\int_0^T \langle x, C\frac{dx}{dt} \rangle \, dt,$$

which implies
$$\int_0^T \langle Cy, x\rangle\, dt = \int_0^T \langle x, C\frac{dx}{dt}\rangle\, dt = 0.$$

It follows from (4.21) and (4.24) that
$$\int_0^T \langle x, G'_x(x)\rangle\, dt = T\|y\|^2 + \int_0^T \langle \sigma_j p(t), x\rangle\, dt \leq Tc_1^2 + T\|p\| \cdot \|x\|.$$

It follows from (H_2) that
$$Tb\|x\|^2 \leq Tc_1^2 + T\|p\| \cdot \|x\|,$$

which yields
$$\|x\| = \|x_j\| \leq c_3 := \frac{\|p\| + \sqrt{\|p\|^2 + 4bc_1^2}}{2b}.$$

This inequality is obviously in conflict with (4.23). The inequality (4.22) is thus proved.

Step 3. Using
$$x(t) = x(t^*) + \int_{t^*}^t y(t)\, dt$$

together with (4.22) and (4.21) implies that
$$|x(t)| \leq |x(t^*)| + \int_0^T |y(t)|\, dt \leq c_2 + \frac{1}{2}\int_0^T (1 + |y(t)|^2)\, dt \leq c_4,$$

for $t \in [0, T]$, where c_4 is equal to the constant $(c_2 + T/2) + Tc_1^2/2$.

On the other hand, it can be easily seen from (4.21) that there is some time $t_0 \in [0, T]$, such that
$$|y(t_0)| \leq c_1. \tag{4.25}$$

Letting
$$|G'_x(x)| < c_5, \qquad \text{for } |x| \leq c_4$$

and
$$|p(t)| < c_6, \qquad \forall\, t \in \mathbb{R}^1,$$

where c_5 and c_6 are some constants. Then, using

$$y(t) = y(t_0) - C[x(t) - x(t_0)] - \int_{t_0}^{t} G'_x(x(t))\, dt + \sigma \int_{t_0}^{t} p(t)\, dt,$$

we obtain

$$|y(t)| \leq c_7 := c_1 + 2|C|c_4 + T(c_5 + c_6), \qquad 0 \leq t \leq T.$$

It follows that

$$|x(t)| + |y(t)| \leq c_8 := c_4 + c_7, \qquad \forall\, t \in \mathbb{R}^1,$$

where c_8 is a positive constant. Therefore, we assert that

$$(x(t), y(t)) \in \mathfrak{D}_E, \qquad \forall\, t \in \mathbb{R}^1,$$

whenever E is large enough.

The proof of Lemma 4.3 is thus complete. $\square$

Now, we are in position to prove the main result.

Theorem 4.5 *If the conditions* (H_1) *and* (H_2) *are valid, then the Duffing equation (4.12) has at least a T-periodic solution.*

Proof. Using Lemma 4.3, we have

$$F_\sigma(x, y) \neq 0, \quad (x, y) \in \mathfrak{M}_E, \quad \text{for all } \sigma \in [0, 1],$$

which implies that the vector fields F_1 and F_0 are homotopic on $\mathfrak{M}_E$. Hence, the corresponding Poincaré indices are equal; that is,

$$\mathrm{Ind}[F_1, \mathfrak{M}_E] = \mathrm{Ind}[F_0, \mathfrak{M}_E],$$

which together with (4.19) implies

$$\mathrm{Ind}\,[F_1, \mathfrak{M}_E] = 1.$$

It follows that the Poincaré map Φ of (4.12) has at least a fixed point $(\xi, \eta) \in \mathfrak{D}_E$, and the corresponding solution $x = x(t, \xi, \eta)$, $y = y(t, \xi, \eta)$ of (4.16) is T-periodic. Therefore, $x = x(t, \xi, \eta)$ is a T-periodic solution of (4.12).

We have thus proved Theorem 4.5. $\square$

4.4 Planar Periodic Systems

4.4.1 Various Statements of Poincaré-Bendixson's Theorem

In the qualitative theory of differential equations, the Poincaré-Bendixson theorem is a powerful tool for proving the existence of limit cycles. However, it is a pity that this theorem is merely valid for the autonomous differential equations in the plane (or sphere). Massera first generalized the Poincaré-Bendixson theorem to the planar periodic systems. We will explain the matter with details.

Let us take a careful review of planar autonomous system

$$\frac{dz}{dt} = V(z), \qquad z = (x,y) \in \mathbb{R}^2, \qquad (4.26)$$

where the vector field $V(z)$ satisfies a local Lipschitz condition.

On the one hand, it is clear that the Poincaré-Bendixson's theorem (i.e., Theorem 3.5) can be restated as follows.

Theorem 4.6 *Let γ be a positively bounded orbit of (4.26), and denote by $\Omega(\gamma)$ the positive limit-set of γ. Then we have the following result:*

1) The system (4.26) has at least a singular point in $\Omega(\gamma)$; otherwise

2) $\Omega(\gamma)$ is a closed orbit Δ enclosing at least a singular point.

On the other hand, let $x = \varphi(t,\xi)$ be the solution of (4.26) satisfying the initial condition $x(0) = \xi$. Assume the Poincaré map of (4.26)

$$\Phi: \quad \xi \quad \mapsto \quad \varphi(\sigma,\xi) \qquad (\xi \in \mathbb{R}^2),$$

is defined for some fixed constant $\sigma > 0$. It is noted that the map Φ is direction-preserving.

For a given point $p \in \mathbb{R}^2$, then

$$\Sigma_p = \{\, z \in \mathbb{R}^2 : \quad z = \Phi^k(p) \qquad (k \in \mathbb{Z}^1)\,\}$$

is the (discrete) orbit of Φ passing through the initial point p. It can be seen that if p is a point of the closed orbit Δ of (4.26), then Σ_p surrounds the fixed point q of Φ in the sense that q is surrounded by $\in \Delta$. Moreover, we have

(1) the orbit Σ_p is periodic when T/σ is rational;

(2) the orbit Σ_p is (nontrivial) recurrent when T/σ is irrational.

It follows that if q is a fixed point of Φ, then q is either a singular point or a recurrent point on some closed orbit that is just Γ_q.

Finally, we restate Theorem 4.6 in the term of Poincaré map Φ.

Theorem 4.7 *If there is a positively bounded orbit Σ of Φ, then*

1) *The Poincaré map Φ has at least a fixed point in $\Omega(\Sigma)$; or*
2) *$\Omega(\Sigma)$ is a recurrent set [1] of Φ containing no fixed point, and Φ has at least a fixed $\xi_0 \notin \Omega(\Sigma)$,*

where $\Omega(\Sigma)$ is the positive limit-set of the orbit Σ.

4.4.2 Massra's Theorem

Instead of the autonomous (planar) system (4.26), assume the periodic system

$$\frac{dz}{dt} = V(t, z), \qquad (t, z) \in \mathbb{R}^1 \times \mathbb{R}^2, \tag{4.27}$$

where the vector field $V(t, z)$ is continuous in (t, x) and periodic in t with period $T > 0$, and satisfies a local Lipschitz condition with respect to z.

Let $z = \psi(t, z_0)$ be the solution of (4.27) satisfying the initial condition $z(0) = z_0 \in \mathbb{R}^2$. Then the Poincaré map

$$\Psi : \quad z_0 \mapsto \psi(T, z_0),$$

is defined on some subset G of $\mathbb{R}^2$ (possibly, $G = \emptyset$).

In 1950, Massera proved the following interesting theorem [93].

Theorem 4.8 *Assume f is a continuous direction-preserving map from $\mathbb{R}^2$ to itself. If f has a positively bounded orbit, then it has at least a fixed point.*

As a direct consequence of Massera Theorem (i.e., Theorem 4.8), we obtain the following result.

Corollary 4.3 *If the periodic system (4.27) admits a positively bounded solution and its Poincaré map Ψ is defined on $\mathbb{R}^2$, then it has at least a periodic solution.*

It is clear that the value of Massera Theorem (i.e., Theorem 4.8) is the generalization of the Poincaré-Bendixson Theorem from the autonomous planar systems to the periodic planar systems. In the review [MR: 2000d: 34088], R. Ortega gives a helpful remark to the proof of Massera theorem.

---
[1] That is, each point of the set is a recurrent point of Φ.

4.4.3 Further Generalization

In 1989, the author proved the following theorem (see [41]), such that the Massera theorem is furthermore generalized.

Theorem 4.9 *If the Poincaré map Ψ of (4.27) is defined on the plane $\mathbb{R}^2$ and has a positively bounded orbit Σ, then*

(1) The positive limit-set Ω of Σ contains at least a fixed point of Ψ; or

(2) If Ω contains no fixed point of Ψ, there is at least a fixed point $\xi_0 \notin \Omega_\Sigma$.

Proof. Since Σ is a positively bounded orbit of Ψ, the positive limit-set $\Omega(\Sigma)$ is a nonempty compact invariant set. It follows that Ψ has at least a minimal set $M \subset \Omega(\Sigma)$, and each point of M is recurrent. Therefore, $\Omega(\Sigma)$ contains at least a recurrent point of Ψ.

If Ψ has a fixed point in $\Omega(\Sigma)$, then the first part of the theorem is valid.

If $\Omega(\Sigma)$ contains no fixed point, then there is at least a recurrent (non-fixed) point q, such that any sufficiently small neighborhood $N(q)$ satisfies $\Psi(N(q)) \cap N(q) = \emptyset$ and $\Psi^s(q) \in N(q)$ for some integer $s \geq 2$.

Therefore, we have a point $q \in \Omega(\Sigma)$ and a neighborhood $U(q)$ of q with some integer $s \geq 2$, such that

$$\Psi(U(q)) \cap U(q) = \emptyset, \qquad \Psi^s(U(q)) \cap U(q) \neq \emptyset. \qquad (4.28)$$

Then, given a point $\xi_0 \in \Psi^s(U(q)) \cap U(q)$, we obtain a point $\eta_0 \in U(q)$ satisfying $\xi_0 = \Psi^s(\eta_0)$. Let

$$\Upsilon_0 = \{ \Psi^j(\eta_0) : \quad j = 1, \cdots, s-1 \}.$$

Since Υ_0 contains only finitely many points, we have a continuous curve C connecting the points ξ_0 and η_0, such that $C \cap \Upsilon_0 = \emptyset$ and $C \subset U(q)$. Then there is a neighborhood W of C, such that $W \subset U(q)$ and $W \cap \Upsilon_0 = \emptyset$.

Now, construct the homotopy

$$h_\lambda: \quad \mathbb{R}^2 \;\to\; \mathbb{R}^2 \quad (0 \leq \lambda \leq 1),$$

such that

1) $h_\lambda(W) \subset W$, $\quad 0 \le \lambda \le 1$;

2) $h_\lambda(z) = z$, $\forall\, z \in (\mathbb{R}^2 \setminus W)$, $\quad 0 \le \lambda \le 1$;

3) $h_1(\xi_0) = \eta_0$;

4) h_0 is the identity map $Id : \mathbb{R}^2 \to \mathbb{R}^2$.

Then, let $g = \Psi \circ h_1$. It follows that $g(\xi_0) = \Psi(h_1(\xi_0)) = \Psi(\eta_0)$. Since $\Psi^i(\eta_0) \in \Upsilon_0$ for $i = 1, \cdots, s-1$, we have $\Psi^i(\eta_0) \notin W$ for $i = 1, \cdots, s-1$. Hence, $g(\Psi^i(\eta_0)) = \Psi(\Psi^i(\eta_0)) = \Psi^{i+1}(\eta_0)$. It follows that

$$g^s(\xi_0) = g^{s-1}(g(\xi_0)) = g^{s-1}(\Psi(\eta_0)) = \Psi^s(\eta_0) = \xi_0.$$

Hence, ξ_0 is a periodic point of g. On the other hand, since $\Psi(W) \cap W = \emptyset$, we have

$$g(W) \cap W = \Psi(h_1(W)) \cap W \subset \Psi(W) \cap W = \emptyset,$$

which implies that ξ_0 cannot be a fixed point of g. Hence, the minimal period n of the periodic point ξ_0 is larger than 1 (i.e., $n \ge 2$).

Now, for the above map g, using the following Brown Lemma (see [15]), we have a simple closed curve J such that $\text{Ind}[V_1, J] = 1$, where V_1 is the vector field defined by $V_1(z) = g(z) - z$.

Brown Lemma. *If g is a direction-preserving map from $\mathbb{R}^2$ to itself and has a n-periodic point with minimal period $n \ge 2$, then there is a simple closed curve $J \subset \mathbb{R}^2$, such that the index of the vector field $V_1(z) = g(z) - z$ along J is equal to 1 (i.e., $\text{Ind}[V_1, J] = 1$).*

Now, consider the family of vector fields

$$V_\lambda(z) = \Psi(h_\lambda(z)) - z \quad (0 \le \lambda \le 1), \qquad z \in \mathbb{R}^2.$$

We want to prove

$$V_\lambda(z) \ne 0 \quad (0 \le \lambda \le 1), \qquad z \in J. \tag{4.29}$$

In fact, let $z \in J$. If $z \notin W$, we have

$$V_\lambda(z) = \Psi(h_\lambda(z)) - z = \Psi(z) - z = g(z) - z = V_1(z) \ne 0, \qquad z \in J,$$

where we use the fact that the definition of the index $\text{Ind}[V_1, J](= 1)$ implies that the vector field V_1 along J is nonsingular (i.e., $V_1(z) \ne 0$ for $z \in J$).

If $z \in W$, we have
$$h_\lambda(z) \in W, \qquad \Psi(h_\lambda(z)) \in \Psi(W),$$
which together with $\Psi(W) \cap W = \emptyset$ implies
$$V_\lambda(z) = \Psi(h_\lambda(z)) - z \neq 0.$$

Hence, (4.29) is proved. Therefore, V_0 is homotopic to V_1 and, therefore, $\text{Ind}[V_0, J] = \text{Ind}[V_1, J] = 1$. It follows that the map Ψ has at least a fixed point ξ_0 in the region D bounded by J. Since $\Omega(\Sigma)$ contains no fixed point, we assert that $\xi_0 \notin \Omega(\Sigma)$.

It can be seen that if the fixed points of Ψ are isolated, then there is a fixed point of Ψ having positive Poincaré index.

The proof of Theorem 4.9 is thus completed. $\square$

4.5 Invariant Continuum

Assume the smooth map
$$f: \quad \mathbb{R}^n \quad \to \quad \mathbb{R}^n$$
has a fixed point. Without loss of generality, assume the origin o is a fixed point. Then f can be written in the form
$$f(x) = Ax + N(x), \qquad \text{when } |x| \ll 1,$$
where $A = f'(o)$ and $N(x) = \mathcal{O}(|x|^2)$.

It is well-known in literature that if o is a hyperbolic fixed point,[2] f has a positively invariant S^k-manifold and a negatively invariant manifold U^m across the fixed point o, where the dimensions k and m satisfy the condition $k + m = n$.

In the following theorem, we generalize this result to a continuous map, concerning the positively invariant continuum[3] and the negatively invariant continuum across a fixed point (see [38]).

Theorem 4.10 *If z_0 is a fixed point of Ψ, then for any constant $r > |z_0|$ there is a simply connected compact continuum K_r in the closed disk D_r centered at the origin 0 with radius r, such that K_r connects z_0 to the circle*

---
[2] That is, all the real parts of characteristic roots of A are non-vanishing.
[3] In what follows, the closed connected set is called the *continuum* for brevity.

$C_r = \partial D_r$ with the property that

$$\Psi(K_r) \subset K_r \qquad (\text{or } \Psi^{-1}(K_r) \subset K_r).$$

Proof. Let $r > |z_0|$. Denote by B_r the open disk bounded by the circle C_r. It is obvious that $z_0 \in B_r$. Let $H_r^0 = B_r$.

Since z_0 is a fixed point of Ψ, we have

$$z_0 \in H_r^0 \cap \Psi(H_r^0). \qquad (4.30)$$

If $\Psi(H_r^0) \cap C_r = \emptyset$, then we have

$$\Psi(H_r^0) \subset H_r^0.$$

In this case, letting $K_r = \overline{H_r^0}$ (the closure of H_r^0), then we have

$$\Psi(K_r) = \Psi(\overline{H_r^0}) = \overline{\Psi(H_r^0)} \subset \overline{H_r^0} = K_r.$$

It means that K_r is positively invariant. Therefore, for any point $p \in K_r$, the orbit

$$\Sigma_p = \{\, \Psi^k(p), \quad k \in \mathbb{Z}^1 \,\}$$

is positively bounded in K_r.

If $\Psi(H_r^0) \cap C_r \neq \emptyset$, then (4.30) implies that the open set $H_r^0 \cap \Psi(H_r^0)$ has a component (a maximal connected open subset set)

$$H_r^1 = \sigma_{z_0}[H_r^0 \cap \Psi(H_r^0)],$$

containing z_0, such that $\overline{H_r^1} \cap C_r \neq \emptyset$. It follows that

$$H_r^1 \subset H_r^0, \quad H_r^1 \subset \Psi(H_r^0), \quad \Psi^{-1}(H_r^1) \subset H_r^0.$$

It can be seen that the closure $\overline{H_r^1} \subset B_r$ is a simply connected compact set connecting z_0 to C_r. Not that

$$z_0 \in H_r^1 \cap \Psi(H_r^1). \qquad (4.31)$$

If $\Psi(H_r^1) \cap C_r = \emptyset$, then we have

$$\Psi(H_r^1) \subset H_r^1.$$

In this case, $K_r = \overline{H_r^1}$ is a positively invariant compact set connecting z_0 to the circle C_r.

If $\Psi(H_r^1) \cap C_r \neq \emptyset$, then (4.31) implies that the open set $H_r^1 \cap \Psi(H_r^1)$ has a component

$$H_r^2 = \sigma_{z_0}[H_r^1 \cap \Psi(H_r^1)]$$

containing z_0, such that $\overline{H_r^2} \cap C_r \neq \emptyset$. It follows that

$$H_r^2 \subset H_r^1, \quad H_r^2 \subset \Psi(H_r^1), \quad \Psi^{-1}(H_r^2) \subset H_r^1.$$

It can be seen that the closure $\overline{H_r^2} \subset B_r$ is a simply connected compact set connecting z_0 to C_r. Not that

$$z_0 \in H_r^2 \cap \Psi(H_r^2). \tag{4.32}$$

In this manner, we arrive at the following conclusion:

(1) In the m-th step, we get a simply connected open set H_r^m, satisfying

$$\overline{H_r^m} \cap C_r \neq \emptyset, \quad \Psi(H_r^m) \subset H_r^m \subset B_r.$$

Then $K_r = \overline{H_r^m}$ is a positively invariant compact set connecting z_0 to the circle C_r;

(2) There is a sequence of simply connected open sets

$$B_r = H_r^0 \supset H_r^1 \supset \cdots \supset H_r^k \supset \cdots\cdots ,$$

where each open set H_r^k connects z_0 to the circle C_r. It follows from $H_r^{k+1} \subset \Psi(H_r^k)$ that

$$\Psi^{-1}(H_r^{k+1}) \subset H_r^k.$$

Let

$$K_r = \bigcap_{k=0}^{\infty} \overline{H_r^k}.$$

Then K_r is a connected compact set connecting z_0 to the circle C_r satisfying

$$\Psi^{-1}(K_r) = \bigcap_{k=0}^{\infty} \Psi^{-1}\left(\overline{H_r^{k+1}}\right) \subset \bigcap_{k=0}^{\infty} \overline{H_r^k} = K_r.$$

Therefore, K_r is a negatively invariant compact set connecting z_0 to the circle C_r.

Theorem 4.10 is thus proved. □

4.5. Invariant Continuum

It is apparent that

$$K_{r_2} \supset K_{r_1}, \quad \text{if } r_2 > r_1 > |z_0|.$$

The above set K_r is called the *stable continuum* of the fixed point z_0 if it is positively invariant under Ψ and the *unstable continuum* of the fixed point z_0 if it is negatively invariant under Ψ.

It is noted that the above proof of Theorem 4.10 is also valid in the higher dimensional phase-space.

In general, we have nothing further to say about the above invariant continuum. However, when it is a Jordan closed curve J, many interesting results can be derived as mentioned in the KAM theorem. For example, if f is the Poincaré map of the periodic differential equations (4.27), then we have the conclusions:

(1) $\mathbb{T}^2 = J \times \mathbb{S}^1$ is an invariant torus of (4.27);
(2) The solution of (4.27) starting from the inside of $\mathbb{T}^2$ is bounded;
(3) The solution of (4.27) on the invariant torus $\mathbb{T}^2$ is almost periodic if the rotation number of f on J is irrational (see [97]).

Chapter 5

Dynamical Systems

5.1 The Originality

5.1.1 *Fundamental Properties*

The so-called *theory of dynamical systems* is indeed originated from the *qualitative theory of differential equations*, which is concerned with the basic geometrical properties of autonomous differential equations as shown below.

Let $x = \varphi(t,p)$ be the (unique) solution of (3.1) satisfying the initial condition $\varphi(0,p) = p$. Then $x = \varphi(t,p)$ has the following properties:

(P_1) $\varphi(0,p) = p, \quad p \in \mathbb{R}^n$;
(P_2) $\varphi(t_2, \varphi(t_1,p)) = \varphi(t_2 + t_1, p), \quad t_1, t_2 \in \mathbb{R}^1$;
(P_3) $x = \varphi(t,p)$ is continuous in $(t,p) \in \mathbb{R}^1 \times \mathbb{R}^n$.

Firstly, the property (P_1) is a direct consequence of the definition of $x = \varphi(t,p)$.

Nextly, since $x = \varphi(t, \varphi(t_1,p))$ is a solution of (3.1) satisfying the initial condition $x(0) = \varphi(t_1,p)$ and $x = \varphi(t + t_1, p)$ is the solution of (3.1) satisfying the initial condition $x(0) = \varphi(t_1,p)$, the uniqueness theorem of solution implies

$$\varphi(t, \varphi(t_1, p)) = \varphi(t + t_1, p), \quad t \in \mathbb{R},$$

which is just (P_2) if $t = t_2$.

Finally, it is noted that the uniqueness of solution implies the continuity of $x = \varphi(t,p)$ with respect to the initial condition. Hence, (P_3) is valid. □

In ancient time, the theory of differential equations merely considered the solution $x = \varphi(t,p)$ of (3.1) as a function of t, where the initial point p is understood as a parameter.

5.1. The Originality

It seems likely that H. Poincaré was the first person to consider the solution $x = \varphi(t,p)$ with a different point of view. Indeed, the so-called Poincaré map

$$\varphi^t : \quad \mathbb{R}^n \to \mathbb{R}^n ; \quad p \mapsto \varphi(t,p)$$

is defined for the given $t \in \mathbb{R}^1$, where p acts the role of a variable and t is understood as a parameter.

With this idea, G. Birkhoff used the properties (P_1), (P_2) and (P_3) as assumptions to establish the theory of dynamical systems. That is, the family $\{\varphi^t\}_{t \in \mathbb{R}}$ of homeomorphisms of $\mathbb{R}^n$ is defined as a *dynamical system*, if it satisfies

(1) : The initial condition:

$$\varphi^0(p) = p \quad (p \in \mathbb{R}^n).$$

(2) : The group[1] condition:

$$\varphi^{t_2} \circ \varphi^{t_1} = \varphi^{t_2+t_1}, \quad \forall\, t_1, t_2 \in \mathbb{R}^1.$$

(3) : The continuity condition:

$$\varphi^t(p) \text{ is continuous in } (t,p) \in \mathbb{R} \times \mathbb{R}^n.$$

In other words, a dynamical system is a commutative continuous group of one-parameter containing the unit-element φ^0. The family $\{\varphi^t\}$ of the Poincaré mappings of differential equation (3.1) is a natural example of dynamical systems. The new topic has greatly inspired the development of the qualitative theory of differential equations.

5.1.2 Continuous Dynamical Systems

Moreover, the definition of dynamical systems can be set up in a more general form.

Let $\mathcal{R}$ be a complete metric space, and let

$$f^t : \quad \mathcal{R} \to \mathcal{R} \quad (\text{where } t \in \mathbb{R}^1 \text{ is a parameter})$$

be a family of homeomorphisms of $\mathcal{R}$. The parameter t will be called the *time* for convenience. Assume $\{f^t\}$ satisfies the following conditions.

---
[1] Where the operator $\circ$ denotes the composition of mappings.

(C_1) : The initial condition:
$$f^0(p) = p \qquad (\forall\, p \in \mathcal{R}).$$

(C_2) : The group condition:
$$f^{t_2} \circ f^{t_1} = f^{t_2+t_1} \qquad (\forall\, t_1, t_2 \in \mathbb{R}^1).$$

(C_3) : The continuity condition:
$$f^t(p) \text{ is continuous in } (t,p) \in \mathbb{R}^1 \times \mathcal{R}.$$

It means that the family $\{f^t\}$ is a continuous commutative group of one-parameter containing the unit-element f^0. This group $\{f^t\}$ is called an *abstract dynamical system* or a *topological dynamical system*. Since t is a continuous parameter t, $\{f^t\}$ is called a *continuous* dynamical system. Sometimes, f^t is also called a *flow*. It seems as if the dynamical system has the mechanical meaning of fluid.

The family of Poincaré mappings $\{\varphi^t\}$ of (3.1) is a special example of the abstract dynamical system. In fact, since $\mathcal{R}$ is a metric space, the point p in $\mathcal{R}$ may have certain generalized meaning other than a point in Euclidean space. For example, suppose $\mathcal{R}$ is the space of real continuous functions on the interval $[0,1]$. In this case, the motion of the dynamical system may be the solution of partial differential equation. Hence, the abstract dynamical system extends essentially the qualitative theory of ordinary differential equations.

Similar to the qualitative theory of differential equations, the central topic of dynamical systems is concerned with the behavior of motion. It is noted that the behavior of motion is completely determined by the closure of the orbit.

For example, let $x = \varphi(t,p)$ be a motion of the system (3.1) and let $\widehat{\varGamma}_p$ be the closure of its orbit. It is noted that the vector field $V(x)$ can be smoothly changed to a different vector field $U(x)$, such that

$$\begin{cases} U(x) = V(x), & \forall\, x \in \widehat{\varGamma}_p; \\ U(x) \neq V(x), & \forall\, x \notin \widehat{\varGamma}_p. \end{cases}$$

It follows that $x = \varphi(t,p)$ also satisfies the autonomous system $\dot{x} = U(x)$. In this sense, the behavior of the solution $x = \varphi(t,p)$ of (3.1) is determined by the vector field on $\widehat{\varGamma}_p$ and is independent of the vector field outside $\widehat{\varGamma}_p$.

Analogously, the behavior of the motion $z = f^t(p)$ of a dynamical system depends upon merely the conditions on $\widehat{\Gamma}_p$.

5.1.3 Discrete Dynamical System

Now, consider a T-periodic system of differential equations

$$\frac{dx}{dt} = W(t,x), \qquad (t,x) \in \mathbb{S}^1 \times \mathbb{R}^n, \tag{5.1}$$

where $\mathbb{S}^1 = \mathbb{R}^1/\mathrm{mod}(T)$, the vector-field $W(t,x)$ is continuous in (t,x) and satisfies a local Lipschitz condition with respect to the variable x. In general, the non-autonomous system (5.1) does not define a (continuous) dynamical system in the phase-space $\mathbb{R}^n$.

If we set

$$y = \begin{pmatrix} s \\ x \end{pmatrix} \qquad \text{and} \qquad U(y) = \begin{pmatrix} 1 \\ W(s,x) \end{pmatrix}, \tag{5.2}$$

then (5.1) is equivalent to an autonomous system

$$\frac{dy}{dt} = U(y), \qquad y \in \mathbb{R}^{n+1} \tag{5.3}$$

which defines a (continuous) dynamical system in $\mathbb{R}^{n+1}$. But the expense is that the order of differential equation is raised from n to $n+1$. Moreover, all the orbits of (5.3) are unbounded in the phase space $\mathbb{R}^{n+1}$ since $s = t$ is the first component of y for $t \in \mathbb{R}^1$.

However, we have another choice. Denote by $x = \psi(t,p)$ the solution of (5.1) satisfying the initial condition $x(0) = p$. Then $x = \psi(t,p)$ is uniquely determined in some interval $\alpha < t < \beta$, and is continuous in (t,p). For simplicity, assume the maximal existence-interval of the solution $x = \psi(t,p)$ is the whole t-axis $\mathbb{R}$. Then we have the Poincaré map

$$\Psi : \quad \mathbb{R}^n \quad \to \quad \mathbb{R}^n; \qquad p \quad \mapsto \quad \psi(T,p).$$

It can be seen that the family $\{\Psi^i\}_{i \in \mathbb{Z}}$ satisfies the following conditions
  (C_1) $\Psi^0 = id$;
  (C_2) $\Psi^k \circ \Psi^j = \Psi^{j+k} = \Psi^j \circ \Psi^k \qquad (k, j \in \mathbb{Z})$.

Hence, the family $\{\Psi^i\}$, defined by the Poincar'e map of periodic system (5.1), is a commutative group of one-parameter $i \in \mathbb{Z}^1$ which contains the unit-element Ψ^0. Since the parameter $i \in \mathbb{Z}^1$ is discrete, the group $\{\Psi^i\}$ is called a *discrete dynamical system*.

Indeed, the discrete dynamical system $\{\Psi^i\}$ expresses the dynamical behavior of solutions of (5.1). For examples,

(1) if $\{\Psi^i(p)\}$ is a periodic orbit with minimal period $m > 0$ (i.e., $\Psi^m(p) = p$, with $\Psi^j(p) \neq p$, $0 < j < m$), then $x = \psi(t,p)$ is a periodic solution of (5.1) with minimal period mT;

(2) if Γ is an invariant closed curve of Ψ, then $\mathbb{T}^2 = \Gamma \times \mathbb{S}^1$ is an invariant torus of (5.1).

Analogous to the continuous dynamical system, the discrete dynamical system can be defined in an abstract manner. Let f be a homeomorphism of the complete metric space $\mathcal{R}$. Then $\{f^i\}$ satisfies the conditions which look like (C_1) and (C_2). It follows that $\{f^i\}$ is a commutative group containing the unit element f^0, which is then called a (*discrete*) dynamical system in $\mathcal{R}$.

5.1.4 *Method of Suspension*

On the one hand, given a continuous dynamical system f^t in a complete metric space $\mathcal{R}$, then the time-T map $g = f^T$ is a homeomorphism of the space $\mathcal{R}$, where T is a given positive constant. It follows that the series $\{g^i\}$ is a discrete dynamical system in $\mathcal{R}$. Indeed, the discrete dynamical system $\{g^i\}$ is an interpolation of the continuous dynamical system f^t for $t = iT$ ($i \in \mathbb{Z}$).

On the other hand, given a discrete dynamical system $\{g^i\}$ in the space $\mathcal{R}$, it is usually impossible to find a continuous dynamical system $\{f^t\}$ in $\mathcal{R}$, such that $\{g^i\}$ is an interpolation of $\{f^t\}$.

For example, consider the discrete dynamical system $\{g^i\}$ on the plane $\mathbb{R}^2$, defined by the orientation-preserving homeomorphism

$$g: \quad (x,y) \mapsto (-2x, -2^{-1}y), \quad (x,y) \in \mathbb{R}^2.$$

It follows that

$$\lim_{n \to +\infty} g^n(x_0, y_0) = \lim_{n \to +\infty} ((-2)^n x, (-2)^{-n} y) = (\pm\infty, 0), \quad (5.4)$$

whenever $x_0 \neq 0$. Now, suppose that the above discrete dynamical system $\{g^i\}$ is an interpolation of a continuous dynamical system $\{g^t\}$. Let the orbit of $\{g^t\}$ passing through the initial point $\xi = (x_0, y_0)$ be denoted by Γ_ξ. It can be seen from (5.4) that the positive semi-orbit Γ_ξ^+ continuously approaches to the limits $(+\infty, +0)$ and $(-\infty, -0)$ indefinitely. It follows that the semi-orbit Γ_ξ^+ will intersect itself indefinitely. It is in conflict with the property that a non-periodic orbit of the continuous dynamical system can not intersect itself. Therefore, the above planar discrete dynamical system $\{g^i\}$ can not be interpolated in a planar continuous dynamical system.

In what follows, we will show by the *method of suspension* that a discrete dynamical system $\{g^i\}$ in $\mathcal{R}$ can be interpolated in a continuous dynamical system which is defined in higher dimensional phase space.

It is obvious that

$$\mathcal{R}^* := \{(s, x) : \quad 0 \leq s \leq 1, \ x \in \mathcal{R}, \quad \text{with } (1, x) \cong (0, g(x))\}$$

is a complete metric space, which is called the *suspension* of $\mathcal{R}$.

Then, define

$$G^t : \quad \mathcal{R}^* \to \mathcal{R}^*; \quad (s, x) \mapsto (\tau, g^k(x)),$$

where $k = [t+s]$ (i.e., the integral part of $t+s$), and $\tau = t+s-k \in [0, 1)$. It can be easily verified that G^t is a continuous dynamical system on $\mathcal{R}^*$, such that $G^i(0, x) = (0, g^i(x))$. Hence, the discrete dynamical system $\{g^i\}$ in $\mathcal{R}$ is an interpolation of the continuous dynamical system G^t in the suspension space $\mathcal{R}^*$.

For convenience, G^t is called the *suspension flow* of the map g.

Finally, we take some analogue for the complexity between a continuous dynamical system and a discrete dynamical system. Suppose that there are an ant and a flea on a plane. The ant moves on a continuous path similar to the orbit of a planar flow, and the flea jumps on a path of disjoint points similar to an orbit of a planar discrete dynamical system. Really, it is more complicated to predict the jump of a flea than the move of an ant.

5.2 Recurrence

The character of a dynamical system is heavily dependent on the recurrence property of motion. There are different types of recurrence.

5.2.1 *Poisson's Recurrence*

Assume f^t is a continuous dynamical system in $\mathcal{R}$. Then, $x = f^t(p)$ is the motion starting from p. Denote by Γ_p the orbit of $x = f^t(p)$. Then, the ω-limit set Ω_p, the α-limit set A_p, and the invariant set of f^t can be defined in a similar manner as that of the autonomous systems of differential equations in $\mathbb{R}^n$. It can be shown that Ω_p (A_p) is a connected compact invariant set if the positive semi-orbit Γ_p^+ (the negative semi-orbit Γ_p^-) is contained in a compact set in $\mathcal{R}$. In this case, Γ_p^+ (or Γ_p^-) is called L^+ (or L^-)-*stable* (i.e., positively (or negatively) stable in the Langrange sense). A motion is called L-stable if it is both L^+ and L^--stable.

A motion $x = f^t(p)$ and its orbit Γ_p are called *positively (or negatively) recurrent* in the Poisson's sense if $\Gamma_p \cap \Omega_p$ (or $\Gamma_p \cap A_p$) $\neq \emptyset$. For simplicity, a *positively (or negatively) recurrent* motion in the Poisson's sense is said to be P^+ (or P^-)-recurrent. A motion is called P-recurrent if it is both P^+ and P^--recurrent.

Since $\Gamma_p \cap \Omega_p$ (or $\Gamma_p \cap A_p$) $\neq \emptyset$ implies $p \in \Gamma_p \subset \Omega_p$ (or A_p), we conclude that if Γ_p is P^+ (or P^-)-recurrent, then for given constant $\varepsilon > 0$, there is a sequence t_k tending to $+\infty$ (or $-\infty$), such that

$$\text{dist}\,[f^{t_k}(p), p] < \varepsilon.$$

The sequence t_k depends, in general, on the constant ε and the initial point p. Let

$$\Delta t_k = |t_{k+1} - t_k|$$

be the *time periods* of recurrence. Generally speaking, the series Δt_k may be bounded or not.

5.2.2 Birkhoff's Recurrence

In the sequel, we are concerned with the concepts of Birkhoff's recurrence.

(1) Assume the motion $x = f^t(p)$ is P^+ (or P^-)-recurrent. For given $\varepsilon > 0$, let Δt_k be the time periods of recurrence as defined above. If there is a constant $L = L(\varepsilon, p)$ such that

$$0 < \Delta t_k < L, \qquad \text{for all } t_k,$$

then $x = f^t(p)$ is called a B^+ (or B^-)-*recurrent motion* (i.e., a *positively (or negatively)* recurrent motion in the Birkhoff's sense). A motion is called B-recurrent if it is both B^+ and B^--recurrent.

Example 5.1 It is not hard to verify the following conclusions.

a) A periodic (or an equilibrium) motion of any given dynamical system is P-recurrent as well as B-recurrent.

b) Each motion of the system

$$\dot{x} = 1, \qquad \dot{y} = \lambda \qquad (\lambda \text{ is an irrational number})$$

on the torus $\mathbb{T}^2$ is P-recurrent as well as B-recurrent.

c) Consider the system

$$\begin{cases} \dot{x} = \sin^2 \dfrac{x}{2} + \sin^2 \dfrac{y}{2}, \\ \dot{y} = \lambda \left(\sin^2 \dfrac{x}{2} + \sin^2 \dfrac{y}{2} \right), \end{cases} \qquad (\lambda \text{ is an irrational number}) \qquad (5.5)$$

on the torus $\mathbb{T}^2 = \mathbb{R}^2/\mathrm{mod}(2\pi\mathbb{Z}^2)$. It is obvious that $\mathfrak{o} = (0,0)$ is a unique singular point of (5.5). Let $p = (0, \eta) \in \mathbb{T}^2$ with $0 < \eta < 2\pi$. It can be seen that both the positive semi-orbit Γ_p^+ and the negative semi-orbit Γ_p^- are dense in the torus $\mathbb{T}^2$. Therefore, when $t \to \pm\infty$, the motion $z = \varphi(t, p)$ will come arbitrarily near to $\mathfrak{o}$. When the motion $z = f^t(p)$ comes nearer and nearer to the singular point $\mathfrak{o}$, its velocity will become slower and slower. It follows that the time periods Δt_k of recurrence becomes unbounded. Hence, the motion $z = f^t(p)$ is P-recurrent, but it is neither B^+ nor B^--recurrent.

Let $q = (2\pi, 2\pi\lambda)$. Since the motion $z = \varphi(t, q)$ tends negatively to the singular point $\mathfrak{o}$, it is not P^--recurrent. However, since the positive semi-orbit Γ_q is dense in $\mathbb{T}^2$, it is P^+-recurrent in $\mathbb{T}^2$, with unbounded time periods Δt_k of recurrence. Hence, the motion $z = \varphi(t, q)$ is not B^+-recurrent.

Let $r = (-2\pi, -2\pi\lambda)$. In a similar manner, it can be shown that the motion $z = \varphi(t, r)$ is P^--recurrent, but not B^--recurrent.

(2) Birkhoff also defined the regional recurrence of dynamical systems as follows.

A dynamical system f^t in $\mathcal{R}$ has the property of *regional recurrence* if for any region $G \subset \mathcal{R}$ and any $T > 0$ there can be found a value $t > T$ such that $f^t(G) \cap G \ne \emptyset$.

Applying the homeomorphism f^{-t} to the above inequality, we have also $G \cap f^{-t}(G) \ne \emptyset$. It means that the definition of regional recurrence refers simultaneously to positive and negative values of t.

We shall call the motion $z = f^t(p)$ *wandering* if there is a neighborhood $N(p)$ of p and a positive number T such that

$$f^t(N(p)) \cap N(p) = \emptyset, \qquad \forall\, t \ge T.$$

Applying f^{-t} to this equality, we obtain $f^{-t}(N(p)) \cap N(p) = \emptyset$. Hence, the definition of wandering motion is symmetrical with respect to positive and negative values of t. For convenience, a point p is called *wandering* if the motion $z = f^t(p)$ is wandering.

The set W of wandering points is invariant since each point of the orbit Γ_p is wandering if p is wandering.

Furthermore, the set W of wandering points is open since, by definition, along with the point p all points of the neighborhood $N(p)$ are also wandering.

Consider the set of non-wandering points with respect to $\mathcal{R}$; that is

$$M_1 = \mathcal{R} \setminus W,$$

which is a closed invariant set of f^t. It may be empty or not.

Theorem 5.1 *If a dynamical system f^t in $\mathcal{R}$ has at least one motion $z = f^t(p)$ which is L^+ (or L^-) stable, then $M_1 \neq \emptyset$.*

Proof. Let $z = f^t(p)$ be L^+-stable. Then Ω_p is a compact and invariant (non-empty) set. Therefore, regarding Ω_p as the space $\mathcal{R}$ of motions, we can assume $\mathcal{R}$ is compact without loss of generality.

Now, assume that $M_1 = \mathcal{R} \setminus W = \emptyset$.

Then for every point $p \in \mathcal{R}$ there can be found a neighbohood $N(p)$ satisfying $f^t(N(p)) \cap N(p) = \emptyset$ for $t > T$. Because $\mathcal{R}$ is compact, we can choose from these neighborhoods a finite number $N(p_1), N(p_2), \cdots, N(p_m)$, such that

$$\mathcal{R} = \bigcup_{1 \leq j \leq m} N(p_j);$$

let these neighborhoods correspond to the numbers $T_1, T_2, \cdots, T_m$, and let $T^* = T_1 + T_2 + \cdots + T_m$. It follows that for any point $p \in \mathcal{R}$ the motion $z = f^t(p)$ will leave $\mathcal{R}$ for $t > T^*$. This is a contradiction. Therefore, the proof of the theorem is complete. □

It can be seen that an ω (or α)-limit point q is non-wandering (i.e., $q \in M_1$). But the converse is not true.

Example 5.2 Consider the planar system

$$\dot{x} = -y[x^2 + (y-1)^2], \qquad \dot{y} = x[x^2 + (y-1)^2]. \tag{5.6}$$

It can be seen that (5.6) has a first integral

$$x^2 + y^2 = \sigma^2 \qquad (\sigma \geq 0),$$

which determines a circle C_σ for $\sigma > 0$.

It is obvious that $q_0 = (0,0)$ and $q_1 = (0,1)$ are singular points, and C_σ is a closed orbit if $0 < \sigma \neq 1$. It can be seen that every point $z \in \mathbb{R}^2$ is non-wandering. On the other hand, for a given point $\eta \in (C_1 \setminus q_1)$, we have $\Omega_\eta = A_\eta = q_1$. Therefore, the orbit Γ_η is neither P^+-recurrent nor P^--recurrent although it is non-wandering.

In the theory, the motion $z = f^t(p)$ is *observable* whenever it returns to a neighborhood of the initial point p for a series of times t_j with $t_j \to +\infty$ (or $-\infty$). The above example shows that a non-wandering motion may be non-observable. The property of $P^\pm$-recurrence is really necessary for the observable motions.

5.3 Quasi-Minimal Set

Let Γ_p be a P^+ (or P^-)-recurrent orbit. The closure $Q = \widehat{\Gamma}_p$ is called a *quasi-minimal set* if it is compact. Indeed, the quasi-minimal set is deeply involved with the property of $P^\pm$-recurrence.

For examples,

(1) A singular point is a quasi-minimal set, and a closed orbit is a quasi-minimal set too. The quasi-minimal set is said to be trivial if it is a singular point or a closed orbit.

(2) The torus $\mathbb{T}^2$ is a (non-trivial) quasi-minimal set of the system (5.5).

Theorem 5.2 *Assume f^t is a dynamical system in a complete metric space $\mathcal{R}$ with a quasi-minimal set Q. Let*

$$D = \{z \in Q : \text{ the orbit } \Gamma_z \text{ is dense in } Q\}.$$

Then D is a residual set in Q.[2]

Proof. Consider the sets

$$F^+ = \{u \in Q : \text{ the positive semi-orbit } \Gamma_u^+ \text{ is not dense in } Q\},$$

and

$$F^- = \{v \in Q : \text{ the negative semi-orbit } \Gamma_v^- \text{ is not dense in } Q\}.$$

Assume

$$U_1, U_2, \cdots, U_k, \cdots\cdots$$

---

[2] That is, A residual set in Q is the intersection of countable open dense sets in Q.

is a countable base of Q. Let T_j be a sequence of positive constants tending to $+\infty$. It follows that if an orbit Γ_p is not dense in Q, then there are some open set U_m in the countable base and some constant T_n in the sequence, such that $f^t(p) \cap U_m = \emptyset$ for $t \geq T_n$ (or $t \leq -T_n$). On the other hand, let

$$F^+_{k,n} = \{ y \in Q : \quad f^t(y) \cap U_k = \emptyset, \quad \forall \, t \geq T_n \},$$

and

$$F^-_{k,n} = \{ y \in Q : \quad f^t(y) \cap U_k = \emptyset, \quad \forall \, t \leq -T_n \}.$$

Then, both $F^+_{k,n}$ and $F^-_{k,n}$ are closed sets in Q, and $(Q \setminus \{F^+_{k,n} \cup F^-_{k,n}\})$ is an open set in Q.

It follows from

$$F^+ = \bigcup_{1 \leq k,\, n < \infty} F^+_{k,n} \quad \text{and} \quad F^- = \bigcup_{1 \leq k,\, n < \infty} F^-_{k,n}$$

that

$$D = Q \setminus (F^+ \cup F^-) = \bigcap_{1 \leq k,\, n < \infty} (Q \setminus \{F^+_{k,\, n} \cup F^-_{k,\, n}\}). \tag{5.7}$$

Hence, we have

$$D \subset (Q \setminus \{F^+_{k,\, n} \cup F^-_{k,\, n}\}),$$

which implies that the open set $(Q \setminus \{F^+_{k,\, n} \cup F^-_{k,\, n}\})$ is dense in Q. Hence, the theorem is proved by (5.7). □

Corollary 5.1 *If Q is a non-trivial quasi-minimal set, then the set of dense orbits in Q is uncountable.*

Proof. Assume the contrary. Then there is a nontrivial quasi-minimal set Q, such that the set of dense orbits in Q is countable. Denote these dense orbits by Γ_k ($k \in I \subset \mathbb{Z}$) passing through the initial point p_k. On the other hand, since

$$D = \bigcup_{k \in I} \Gamma_k = \bigcup_{k \in I} \bigcup_{j \in \mathbb{Z}^+} \Gamma_k[-j, j],$$

where $\Gamma_k[-j, j]$ is the arc of Γ_k from the point $f^{-j}(p_k)$ to the point $f^j(p_k)$. It follows that

$$(Q \setminus D) = \bigcap_{k \in I,\, j \in \mathbb{Z}^+} (Q \setminus \Gamma_k[-j, j]). \tag{5.8}$$

When Q is not trivial, $(Q\setminus \Gamma_k[-j,j])$ is a dense open set in Q. It follows from (5.8) that $(Q\setminus D)$ is a residual set in Q. This contradicts the fact that D is a residual set in Q.

The proof of Corollary 5.1 is thus completed. □

Finally, we remark that the set of both positively and negatively dense orbits in Q is also a residual set in Q. The proof is apparent in the above arguments.

5.4 Minimal Set

A compact invariant set M of f^t is called a *minimal set* if every orbit of f^t in M is dense in M.

Or equivalently, a compact invariant set M of f^t is minimal if f^t has no closed invariant set which is a proper (non-empty) subset of M.

5.4.1 *Property of Minimal Set*

A minimal set is called *trivial* if it is a singular point or a closed orbit. It follows from the Poincaré-Bendixson theorem that the minimal set of any planar (continuous) dynamical system is trivial. In other words, there is no non-trivial minimal set in $\mathbb{R}^2$.

Theorem 5.3 *If Γ_p^+ (or Γ_p^-) is a B^+ (or B^-)-recurrent semi-orbit of f^t, then the closure $\widehat{\Gamma}_p$ is a minimal set of f^t.*

Proof. It is noticed that $\widehat{\Gamma}_p^+$ and $\widehat{\Gamma}_p^-$ are closed sets. Moreover, since Γ_p^+ (or Γ_p^-) is B^+ (or B^-)-recurrent, then for a constant $\varepsilon > 0$ there is a constant $L = L(\varepsilon, p) > 0$, such that in any interval $(s, s+L)$ there is a time $\tau \in (s, s+L)$ satisfying

$$\rho[f^\tau(p), p] < \varepsilon. \tag{5.9}$$

On the other hand, it is clear that the set

$$K = \{z \in \mathcal{R} : \; z = f^t(\xi), \text{ for } \rho[\xi, p] \leq \varepsilon, \; 0 \leq t \leq L\}$$

is compact. Moreover, it follows from (5.9) that

$$\Gamma_p^+ \text{ (or } \Gamma_p^-) \subset K,$$

which implies
$$\widehat{\Gamma}_p^+ \text{ (or } \widehat{\Gamma}_p^-) \subset K.$$

Hence, we have
$$\Omega_p \text{ (or } A_p) \subset \widehat{\Gamma}_p^+ \text{ (or } \widehat{\Gamma}_p^-) \subset K.$$

Since Γ_p^+ (or Γ_p^-) is B^+ (or B^-)-recurrent, we have
$$\Gamma_p \subset \Omega_p \text{ (or } A_p) \subset \widehat{\Gamma}_p^+ \text{ (or } \widehat{\Gamma}_p^-) \subset K,$$

which implies
$$\widehat{\Gamma}_p \subset K.$$

We have thus proved that $\widehat{\Gamma}_p$ is a compact invariant set.

It is suffices to prove that any compact proper (non-empty) subset in $\widehat{\Gamma}_p$ is not invariant for f^t.

In fact, assume the contrary. Then f^t has a compact invariant set $N \subset \widehat{\Gamma}_p$, such that $\Upsilon = (\widehat{\Gamma}_p \setminus N) \neq \emptyset$.

Let $p \in N$. Since N is invariant and compact, we have $\widehat{\Gamma}_p \subset N$, which is in conflict with $\Upsilon \neq \emptyset$.

Hence, we have $p \notin N$, such that the distance
$$d = \rho[p, N] > 0.$$

Then, consider the σ-neighborhood $B_\sigma(p)$ centered at p with radius $\sigma = d/2$. Since $z = f^t(p)$ is B^+ (or B^-)-recurrent, there is a sequence $t_k \to \infty$ (or $\to -\infty$), such that
$$f^{t_k}(p) \subset B_\sigma(p), \qquad 0 < |t_{k+1} - t_k| < L_0, \tag{5.10}$$

where $L_0 > 0$ is a constant. On the other hand, let $q \in N$. Since N is invariant, we have
$$f^t(q) \in N, \qquad \text{for all } t \in \mathbb{R}^1.$$

It follows from the continuity of motion for initial conditions that there is a sufficiently small constant $\delta > 0$, such that for any point $\xi \in B_\delta(p)$ we have
$$\rho[f^t(\xi), f^t(q)] < \sigma, \qquad 0 \leq |t| \leq 2L_0.$$

It follows from $f^t(q) \in N$ that

$$\rho[f^t(\xi), N] < \sigma, \qquad 0 \leq |t| \leq 2L_0. \tag{5.11}$$

Since $q \in N \subset \widehat{\Gamma}_p$, there is a sufficient large $s > 0$, such that $\xi = f^s(p) \in B_\delta(p)$. It follows from (5.11) that

$$\rho[f^{t+s}(p), N] < \sigma, \qquad 0 \leq |t| \leq 2L_0,$$

which implies that the motion $z = f^t(p)$ will not return to the neighborhood $B_\sigma(p)$ during the time interval $[s, s+2L_0]$. However, this is in conflict with (5.10). The proof of theorem is thus completed. □

Theorem 5.4 *If M is a minimal set of f^t, then every orbit Γ_p in M is B-recurrent.*

Proof. Since M is minimal, every orbit Γ_p in M is dense in M; that is, $\widehat{\Gamma}_p = M$. Hence, Γ_p is either P^+-recurrent or P^--recurrent.

First, we prove that Γ_p is P-recurrent.

In fact, if Γ_p is not P-recurrent, then it is either not P^+-recurrent or not P^--recurrent. It follows that either the compact invariant set Ω_p or A_p is a compact invariant proper subset of M. This is in conflict with the fact that M is minimal. Therefore, Γ_p is P-recurrent.

Next, we claim that Γ_p is B^+-recurrent.

Assume the contrary. Then we have a constant $\delta_0 > 0$ and a time sequence

$$0 < t_1 < t_2 < \cdots < t_k < \cdots,$$

such that

$$\rho[f^{t_j}(p), p] = \delta_0, \qquad \forall\, t_j$$

and

$$\rho[f^t(p), p] \geq \delta_0, \qquad \text{for } t \in (t_j, s_j), \tag{5.12}$$

where s_j is a constant satisfying

$$t_j < s_j < t_{j+1}, \qquad \lim_{j \to \infty}(s_j - t_j) = +\infty.$$

Then

$$\tau_j = \frac{1}{2}(t_j + s_j) \in (t_j, s_j).$$

It follows from (5.12) that
$$\rho[f^{\tau_j}(p), p] \geq \delta_0.$$

Since $f^{\tau_j}(p)$ is a sequence in the compact set M, it has a convergent subsequence. Without loss of generality, assume
$$\lim_{\tau_j \to \infty} f^{\tau_j}(p) = q \in M.$$

Moreover, it can be seen that
$$\rho[\Gamma_q^-, p] \geq \delta_0. \tag{5.13}$$

In fact, assume (5.13) is false. Then there is a constant $s > 0$, such that $f^{-s}(q) \in B_{\delta_0}(p)$. On the other hand, in the finite interval $[-s, 0]$, using the continuity of solution for initial conditions, we conclude that there is a sufficiently small neighborhood $B_\varepsilon(q)$ at q, satisfying
$$f^{-s}(B_\varepsilon(q)) \subset B_{\delta_0}(p).$$

It is noticed that if $\tau_j \gg 1$ we have $f^{\tau_j}(p) \in B_\varepsilon(q)$. It follows that
$$f^{\tau_j - s}(p) \subset B_{\delta_0}(p). \tag{5.14}$$

Since $s > 0$ is a constant, we have $t_j < \tau_j - s < s_j$ for sufficiently large j. Then (5.12) and (5.14) are in conflict with each other. This contradiction proves (5.13).

It follows from (5.13) that $\rho[A_q, p] \geq \delta_0$. Therefore, there is a compact invariant set A_q, which is proper in M.

This contradiction proves that Γ_p is B^+-recurrent. In a similar manner, we can prove that Γ_p is B^--recurrent. Therefore, Γ_p is B-recurrent.

Theorem 5.4 is thus proved. □

The following corollary is a direct consequence of Theorems 5.3 - 5.4, which is a simple geometrical method to find B-recurrent motions.

Corollary 5.2 *Every orbit Γ_p in a minimal set is B-recurrent.*

5.4.2 Generalized Poincaré-Bendixon Theorem

Analogous to the Poincaré-Bendixson theorem on the existence of closed orbit (i.e., trivial minimal set) for planar systems, the following *generalized Poincaré-Bendixon theorem* is concerning with the existence of minimal set for general dynamical systems.

Theorem 5.5 *If a dynamical system f^t has a semi-orbit Γ_p^+ (or Γ_p^-) in a compact set, then it has at least a minimal set in Ω_p (or A_p).*

Proof. It follows that $N_1 = \Omega_p$ or (A_p) is a compact invariant set of f^t. We shall prove the theorem by the following process.

If N_1 is minimal, then the theorem is proved.

If N_1 is not minimal, then f^t has at least a compact invariant proper (non-empty) subset N_2 in N_1.

If N_2 is minimal, then the theorem is proved.

If N_2 is not minimal, then f^t has at least a compact invariant proper (non-empty) subset N_3 in N_2.

Continuing this process, in the finite m-th step, we obtain a minimal set N_m; otherwise, we obtain a chain of compact invariant sets

$$N_1 \supset N_2 \supset \cdots \supset N_k \supset \cdots\cdots.$$

It follows that

$$N_\omega = \bigcap_{1 \leq k < \infty} N_k,$$

is a compact invariant (non-empty) set, which is invariant for f^t.

Then this process can be continued to the second kind of transfinite numbers by the method of transfinite induction:

Assume α is a number of first kind and N_α is already defined. If N_α is minimal, then the theorem is proved and the process is finished. If N_α is not minimal, then the process defines $N_{\alpha+1}$ as above.

Assume β is a transfinite number of second kind and all N_α ($\alpha < \beta$) are already defined, then define

$$N_\beta = \bigcap_{\alpha < \beta} N_\alpha$$

which is a compact invariant set of f^t. If N_β is minimal, then the theorem is proved and the process is finished. If N_β is not minimal, then the process defines the set $N_{\beta+1}$ as above.

In this manner, we obtain a transfinite sequence of compact invariant sequence of f^t

$$N_1 \supset N_2 \supset \cdots \supset N_k \supset \cdots \supset N_\omega \supset \cdots \supset N_\alpha \supset \cdots. \tag{5.15}$$

complement of $\{I_n : n \in \mathbb{Z}\}$ is a nowhere dense closed set Ω on $S(2\pi + a)$. Consequently, Ω is homeomorphic to the Cantor set.

Denote by Ω_1 the subset of Ω that are the endpoints of I_n ($n \in \mathbb{Z}$). Then $\Omega_2 = (\Omega \setminus \Omega_1)$ is the subset of Ω that are not the endpoints of I_n ($n \in \mathbb{Z}$).

Since the mutual arrangement of the intervals I_n ($n \in \mathbb{Z}$) is similar to the mutual arrangement of x_n ($n \in \mathbb{Z}$), there exists an orientation-preserving continuous mapping

$$h: \quad S(2\pi + a) \quad \to \quad S(2\pi)$$

that carries the close interval $\hat{I}_n$ into x_n for all $n \in \mathbb{Z}$, and is one-to-one on the set Ω_2.

We begin the construction of the desired diffeomorphism

$$f: \quad S(2\pi + a) \quad \to \quad S(2\pi + a),$$

satisfying the commutative diagram

$$\begin{array}{ccc} S(2\pi + a) & \xrightarrow{h} & S(2\pi) \\ f \downarrow & & \downarrow R \\ S(2\pi + a) & \xrightarrow{h} & S(2\pi). \end{array}$$

Let us first construct the derivative of f:

$$F = f': \quad S(2\pi + a) \quad \to \quad \mathbb{R}^1.$$

Assume first $F\mid_\Omega = 1$. For $x \in I_n = (\alpha_n, \beta_n)$, let

$$F(x) = 1 + \frac{(x - \alpha_n)(\beta_n - x)}{a_n^2} \kappa_n, \quad \text{where } \kappa_n = 6\left(\frac{a_{n+1}}{a_n} - 1\right).$$

It can be verified directly that

$$\int_{I_n} F\, dx = a_{n+1}. \tag{5.16}$$

Let us compute the ratio

$$\frac{a_{n+1}}{a_n} = \begin{cases} \dfrac{n+2}{n+1} & \text{for } n \geq 0, \\[2mm] \dfrac{|n|+3}{|n|+1} & \text{for } n \leq -1, \end{cases}$$

which implies

$$\kappa_n = \begin{cases} -\dfrac{12}{n+4} & \text{for } n \geq 0, \\ \dfrac{12}{|n|+1} & \text{for } n \leq -1. \end{cases}$$

Since $F(x)$ takes a maximal value on the interval $I_n = (\alpha_n, \beta_n)$ at $x = \frac{1}{2}(\alpha_n + \beta_n)$, we get the inequalities

$$\begin{aligned} \frac{1}{4} \leq 1 - \frac{3}{n+4} \leq F(x) \leq 1, & \quad n \geq 0, \\ 1 \leq F(x) \leq 1 + \frac{3}{|n|+1} \leq \frac{5}{2}, & \quad n \leq -1, \end{aligned} \tag{5.17}$$

where $x \in I_n$. Obviously, $F(x)$ is continuous on the intervals I_n, $n \in \mathbb{Z}$. It follows from (5.17) that $F(x)$ is continuous at the points of Ω. Therefore, $F(x)$ is continuous on $S(2\pi + a)$.

Denote by

$$P: \quad \mathbb{R}^1 \quad \to \quad S(2\pi + a)$$

the projection from the covering space $\mathbb{R}^1$ to the circle $S(2\pi+a)$. It can be assumed without loss of generality that $P(0) = \alpha_0$ is the left-hand endpoint of the interval $I_0 = (\alpha_0, \beta_0)$. Let $\tilde{F}: \mathbb{R}^1 \to \mathbb{R}^1$ be a lift mapping of F; that is, $\tilde{F} = P^{-1} \circ F \circ P$.

Denote by $\nu(X)$ the Lebesgue measure of a subset $X \subset S(\lambda)$. Let $\tilde{\alpha}_1 \in [0, 2\pi + a] \subset \mathbb{R}^1$ be a point such that $P(\tilde{\alpha}_1) = \alpha_1$ is the left-hand endpoint of the interval $I_1 = (\alpha_1, \beta_1)$. We define the mapping $\tilde{f}: \mathbb{R}^1 \to \mathbb{R}^1$ by

$$\tilde{f}(x) = \tilde{\alpha}_1 + \int_0^x \tilde{F}(x)\, dx.$$

It is noted that $D\tilde{f}(x) = \tilde{F}(x) \neq 0$, and

$$\begin{aligned} \tilde{f}(x + 2\pi + a) &= \tilde{\alpha}_1 + \int_0^x \tilde{F}(x)\, dx + \int_x^{x+2\pi+a} \tilde{F}(x)\, dx \\ &= \tilde{f}(x) + \int_{S(2\pi+a)} \tilde{F}(x)\, dx = \tilde{f}(x) + \int_\Omega f\, dx + \int_{\cup I_n} F\, dx \\ &= \tilde{f}(x) + \nu(\Omega) + \sum_{-\infty}^{\infty} a_n = \tilde{f}(x) + 2\pi + a. \end{aligned}$$

Consequently, the mapping $\tilde{f}$ is a lift for some monotonically increasing mapping $f : S(2\pi + a) \to S(2\pi + a)$, such that $Df = F$. Since F is continuous on $S(2\pi+a)$, f is a C^1-diffeomorphism of the circle $S(2\pi+a)$.

The diffeomorphism f is the desired Denjoy diffeomorphism. Let us consider its properties.

It follows from the construction of the mapping $h : S(2\pi+a) \to S(2\pi)$ that for any interval $[\alpha, \beta] \subset S(2\pi + a)$ with endpoints $\alpha, \beta \in \Omega_2$

$$\nu([\alpha, \beta] \cap \Omega) = \nu([h(\alpha), h(\beta)]). \qquad (5.18)$$

Claim that $f(\Omega) = \Omega$ and $R \circ h = h \circ f$.

In fact, let $x \in \Omega_2$ and let $\mathcal{J}$ be the index set such that $I_k \subset [\alpha_0, x]$ if and only if $k \in \mathcal{J}$. Then $x_k \in [h(\alpha_0), h(\beta_0)]$ for $k \in \mathcal{J}$. By (5.16) and (5.18) and the fact that R is a rotation,

$$\nu([\alpha, f(x)]) = \int_{[\alpha_0, x]} F\, dx = \int_{\cup I_k} F\, dx + \int_{\Omega \cap [\alpha_0, x]} F\, dx$$
$$= \sum_{k \in \mathcal{J}} a_{k+1} + \nu([R \circ h(\alpha_0), R \circ h(x)]).$$

Since $x \in \Omega_2$, it follows that $h(x) \notin \mathcal{O}(x_0)$ and $R \circ h(x) \notin \mathcal{O}(x_0)$, and hence $h^{-1} \circ R \circ h(x) \in \Omega_2$.

It is clear that $x_k \in [R \circ h(\alpha_0), R \circ h(x)] = [x_1, R \circ h(x)]$ if and only if $k - 1 \in \mathcal{J}$. By the construction of the mapping h, this means that $I_k \subset [\alpha_1, h^{-1} \circ R \circ h(x)]$ if and only if $k - 1 \in \mathcal{J}$. Consequently, we get

$$\nu([\alpha_1, h^{-1} \circ R \circ h(x)]) = \sum_{k \in \mathcal{J}} a_{k+1} + \nu(\Omega \cap [\alpha_1, h^{-1} \circ R \circ h(x)])$$
$$= \sum_{k \in \mathcal{J}} a_{k+1} + \nu([R \circ h(\alpha_0), R \circ h(x)]).$$

This implies that $\nu([\alpha_1, h^{-1} \circ R \circ h(x)]) = \nu([\alpha_1, f(x)])$, so that

$$f(x) = h^{-1} \circ R \circ h(x) \in \Omega_2.$$

We have thus proved the inclusion: $f(\Omega_2) \subset \Omega_2$. Similarly, we can prove the inclusion: $f^{-1}(\Omega_2) \subset \Omega_2$. Thus, $f(\Omega_2) = \Omega_2$, and the relation $h \circ f = R \circ h$ holds on the set Ω_2. Since Ω_2 is dense in Ω and f, f^{-1} are continuous, we conclude that $f(\Omega) = f^{-1}(\Omega) = \Omega$, and the relation $h \circ f = R \circ h$ holds on Ω. This and the monotonicity of f imply that $f(I_n) = I_{n+1}$ for all $n \in \mathbb{Z}$. Since $h(I_n) = x_n$ for $n \in \mathbb{Z}$, the relation $h \circ f = R \circ h$ holds on the whole circle $S(2\pi + a)$.

If $x \in \Omega$, then it follows from the monotonicity of h, the equality $h(\Omega) = S(2\pi)$, and the denseness of each orbit of the rotation R that the limit set of of x coincide with Ω.

If $x \notin \Omega$, then the orbit $\mathcal{O}(x)$ intersects each interval at precisely one point. Since the length of I_n tends to zero as $|n| \to \infty$, the limit set of $\mathcal{O}(x)$ also coincides with Ω. Consequently, the limit set of the Denjoy diffeomorphism f is the Cantor set Ω.

The suspension flow φ^t of the above Denjoy diffeomorphism f gives the desired Denjoy flow on torus. It follows from the foregoing that the limit set $\mathcal{D}$ of φ^t consists of the trajectories passing through points of Ω, is locally homeomorphic to the topological product of a compact interval and the Cantor set.

It is noticed that each orbit of the rotation R is dense, the relation $h \circ f = R \circ h$ holds, and h is monotonic. Therefore, any orbit in Ω with respect to f is dense in Ω. Therefore, each orbit of φ^t in $\mathcal{D}$ is dense in $\mathcal{D}$. We have thus obtained the Denjoy's Cantorus $\mathcal{D}$.

Remark 1. The above Denjoy's flow is a C^1-flow on the torus. In fact, a C^2-flow on the torus does not have irregular minimal set (see [21]).

Remark 2. For the Denjoy flow φ^t, we have

$$\varphi^m([\alpha_k, \beta_k]) = [\alpha_0, \beta_0], \qquad \text{for some integer } m > 0,$$

where

$$\rho([\alpha_k, \beta_k]) \ll 1 \qquad (\text{as } k < 0, \quad |k| \gg 1).$$

It follows that the Denjoy's flow φ^t is Liapunov unstable on the torus.

5.5 Almost Periodic Motion

A motion $z = f^t(p)$ is called *almost periodic* if for any $\varepsilon > 0$ there exists a number $L = L(\varepsilon) > 0$, such that the *set of ε - displacements*

$$E(\varepsilon) = \{ \tau : \quad \rho[f^{t+\tau}(p), f^t(p)] < \varepsilon, \quad \forall\, t \in \mathbb{R}^1 \}$$

is relatively dense with respect to the constant L. It means that any interval of length L in $\mathbb{R}^1$ contains at least a point of τ in $E(\varepsilon)$.

It is clear that a periodic motion is an almost periodic motion. In fact, the almost periodic motion is a generalization of periodic motion. Similarly, the B-recurrent motion is also a generalization of periodic motion. On the other hand, an almost periodic motion is apparently B-recurrent. But it is not clear that a B-recurrent motion is almost periodic or not.

In what follows, we shall explore the relation between the B-recurrent motion and the almost periodic motion.

5.5.1 Set of ε-Displacements

Assume the motion $z = f^t(p)$ is B-recurrent. For a given $s \in \mathbb{R}^1$, let $q = f^s(p)$. It is clear that the motion $z = f^t(q)$ is also B-recurrent. It follows from the B-recurrence that for a given $\varepsilon > 0$ the set

$$E_q(\varepsilon) = \{\tau : \quad \rho[f^\tau(q), q] < \varepsilon\} \tag{5.19}$$

is relatively dense with respect to a constant $L_s = L_s(\varepsilon)$, which depends upon not only the constant ε but also the number s (i.e., the point $q \in \Gamma_p$). In general, if $s \in \mathbb{R}^1$ varies, L_s may be bounded or not.

It can be seen that if L_s is bounded for $s \in \mathbb{R}^1$, then the motion $z = f^t(p)$ is almost periodic. In this case, we have a constant $L(\varepsilon) > 0$, such that the set (5.19) of ε-displacements is relatively dense with respect to the constant $L(\varepsilon)$. In fact, $E_q(\varepsilon)$ is the set of ε-displacements.

It is noticed that if $z = f^t(p)$ is an almost periodic motion with the set $E(\varepsilon)$ of ε-displacements, then for any given $q \in \Gamma_p$, the motion $z = f^t(q)$ is also almost periodic with same $E(\varepsilon)$ as the set of ε-displacements.

5.5.2 Examples of Almost Periodic Motion

Example 5.3 Assume $z = f^t(p_0)$ is a periodic motion with period $T > 0$. Then, for any integer k, we have

$$\rho[f^{t+kT}(p_0), f^t(p_0)] = 0, \quad \text{for all } t \in \mathbb{R}^1,$$

which implies

$$\rho[f^{t+kT}(p_0), f^t(p_0)] < \varepsilon, \quad \text{for all } t \in \mathbb{R}^1,$$

for any given constant $\varepsilon > 0$. It follows that

$$kT \in E^*(\varepsilon) := \{\tau : \quad \rho[f^{t+\tau}(p_0), f^t(p_0)] < \varepsilon.$$

Therefore, $E^*(\varepsilon)$ is a relatively dense set with respect to a constant T_1 (whenever $T_1 > T$). It follows that the periodic motion $z = f^t(p_0)$ is almost periodic.

Example 5.4 Consider the system
$$\dot{x} = 1, \quad \dot{y} = \lambda, \quad (x, y) \in \mathbb{T}^2,$$
where $\lambda > 0$ is a constant. Then it has the motion
$$z = f^t(p_0) = (t + x_0, \lambda t + y_0)$$
starting from the initial point $p_0 = (x_0, y_0) \in \mathbb{T}^2$. It is known that
(1) $z = f^t(p_0)$ is a periodic motion on $\mathbb{T}^2$ if λ is a rational number;
(2) $z = f^t(p_0)$ is not a periodic motion on $\mathbb{T}^2$ if λ is an irrational number.

Now, assume λ is an irrational number. We claim that the set
$$E_1(\varepsilon) := \{\tau : \ \rho[f^{t+\tau}(p_0), f^t(p_0)] = |\tau| + |\lambda\tau| < \varepsilon, \quad (\mathrm{mod}\, 2\pi)\}$$
is relatively dense on $\mathbb{R}^1$.

In fact, if $\tau = 2k\pi$ ($k \in \mathbb{Z}^1$), we have
$$|\tau| + |\lambda\tau| = |2k\pi\lambda| \quad (\mathrm{mod}\, 2\pi). \tag{5.20}$$

When λ is an irrational number, the number set (5.20) is dense in the neighborhood at $0 \in \mathbb{S}^1$. Hence, the number set
$$E^*(\varepsilon) = \{\tau \in \mathbb{R}^1 : \ \text{where } \tau = 2k\pi \text{ satisfy } |2k\pi\lambda| < \varepsilon \quad (\mathrm{mod}\, 2\pi)\}$$
is relatively dense in $\mathbb{R}^1$. It follows from $E^*(\varepsilon) \subset E_1(\varepsilon)$ that $E_1(\varepsilon)$ is relatively dense in $\mathbb{R}^1$. We have thus proved that $z = f^t(p_0)$ is an almost periodic motion on $\mathbb{T}^2$.

Now, it is a position to resume the types of recurrence as follows.

Assume P is the set of periodic motions, A is the set of almost periodic motions, B is the set of B-recurrent motions, Q is the set of P-recurrent motions, and G is the set of regional recurrent motions. Then we have the following inclusions:
$$P \subset A \subset B \subset Q \subset G. \tag{5.21}$$

It can be seen that Example 5.4 shows that the inclusion $P \subset A$ is proper; Example 5.2 shows that the inclusion $Q \subset G$ is proper; Example 3.10 shows that the inclusion $B \subset Q$ is proper.

In history, Birkhoff was the first person to discover the B-recurrent motions, such that an almost periodic motions is a B-recurrent motion. He also construct a counter example of differential equations to show that a B-recurrent motion is not an almost periodic motion, so that the inclusion

$A \subset B$ is also proper. Hence, B is a new kind of recurrent motions. However, the counter example of differential equations constructed by Birkhoff is discontinuous. In order to demonstrate the naturalness of B-recurrent motions, Birkhoff wanted to construct the B-recurrent motions defined by analytic differential equations, and conjectured the existence of such examples. The conjecture attracted the attention of mathematical circle. For examples, Franks (1928), V. Nemytskii (1949) and L. Cesari (1987) repeated to notice this problem in their works.

We shall give later a positive response to the Birkhoff's conjecture on the existence of analytic B-recurrent but non-almost periodic motion. For this aim, we have to study furthermore some properties of almost periodic motions.

5.5.3 Liapunov Stability of Almost Periodic Motion

Assume Λ is an invariant set of f^t in $\mathcal{R}$. For a given $p \in \Lambda$, consider the motion $z = f^t(p)$. If for any given $\varepsilon > 0$ there is a $\delta = \delta(\varepsilon, p) > 0$, such that we have

$$\rho[f^t(x), f^t(p)] < \varepsilon, \qquad \forall\, t \geq 0,$$

whenever $\rho[x, p] < \delta$ for $x \in \Lambda$, then the motion $z = f^t(p)$ is said to be *positively Liapunov stable* with respect to Λ.

If, for any $p \in \Lambda$, $z = f^t(p)$ is positively Liapunov stable with respect to Λ, f^t is said to be positively Liapunov stable in Λ. If the above δ is independent of $p \in \Lambda$, then f^t is said to be positively *uniformly* Liapunov stable in Λ.

In a similar manner, we can define the motion $z = f^t(p)$ being negatively Liapunov stable with respect to Λ, and so on.

Theorem 5.6 *Assume f^t is positively (or negatively) Liapunov stable in the invariant set Λ which is compact. Then f^t is positively (or negatively) uniformly Liapunov stable in Λ.*

Proof. Consider first the case of positively Liapunov stable motions.

Let $\varepsilon > 0$ be a given constant. Then, for any $p \in \Lambda$, there is a $\sigma = \sigma(p, \varepsilon) > 0$, such that for any $x \in \Lambda$ we have

$$\rho[f^t(x), f^t(p)] < \frac{\varepsilon}{2}, \qquad \forall\, t \geq 0,$$

whenever $\rho[x, p] < \sigma$. It follows from the compactness of Λ that there is a finite number of points $p_1, \cdots, p_m$, such that

$$\Lambda \subset \bigcup_{1 \leq j \leq m} B_{\frac{\sigma_j}{2}}(p_j), \qquad \text{where } \sigma_j = \sigma(p_j, \varepsilon).$$

Let

$$\delta(\varepsilon) = \min\left\{\frac{\sigma_1}{2}, \cdots, \frac{\sigma_m}{2}\right\}.$$

Then, for any given point $p \in \Lambda$, there is some p_j, such that

$$\rho[p, p_j] < \frac{\sigma_j}{2}.$$

It follows from the definition of σ_j that

$$\rho[f^t(p), f^t(p_j)] < \frac{\varepsilon}{2}, \qquad \forall \, t \geq 0.$$

On the other hand, if $\rho[x, p] < \delta$ for $x \in \Lambda$, we have

$$\rho[x, p_j] \leq \rho[x, p] + \rho[p, p_j] < \sigma_j,$$

which implies

$$\rho[f^t(x), f^t(p_j)] < \frac{\varepsilon}{2}, \qquad \forall \, t \geq 0.$$

Therefore, we have

$$\rho[f^t(x), f^t(p)] \leq \rho[f^t(x), f^t(p_j)] + \rho[f^t(p_j), f^t(p)] < \varepsilon, \qquad \forall \, t \geq 0,$$

whenever $\rho[x, p] < \delta$ for $x \in \Lambda$. Note that $\delta = \delta(\varepsilon)$ is independent of p.
We have thus proved the theorem. □

Theorem 5.7 *An almost periodic motion $z = f^t(p)$ is both positively and negatively uniformly Liapunov stable with respect to its orbit Γ_p.*

Proof. Let $\varepsilon >$ be a given constant. Since $z = f^t(p)$ is almost periodic, there is a constant $L > 0$, such that the set of $(\frac{\varepsilon}{3})$-displacements

$$E\left(\frac{\varepsilon}{3}\right) = \{\tau : \quad \rho[f^{t+\tau}(p), f^t(p)] < \frac{\varepsilon}{3}, \quad \forall \, t \in \mathbb{R}^1\}$$

is relatively dense with respect to L.

On the other hand, Since $z = f^t(p)$ is B-recurrent, the closure $\widehat{\Gamma}_p$ of its orbit is a compact minimal set. It follows that on the finite interval

$0 \leq t \leq L$, $z = f^t(x)$ is uniformly continuous for the initial point $x \in \widehat{\varGamma}_p$. Hence, for the constant $(\frac{\varepsilon}{3})$, there is a $\delta > 0$, such that

$$\rho[f^t(x), f^t(y)] < \frac{\varepsilon}{3}, \qquad 0 \leq t \leq L,$$

whenever $\rho[x,y] < \delta$ for $x, y \in \widehat{\varGamma}_p$.

Now, it suffices to prove that

$$\rho[f^t(u), f^t(v)] < \varepsilon, \qquad t \in \mathbb{R}^1, \tag{5.22}$$

whenever $\rho[u,v] < \delta$ for $u, v \in \varGamma_p$.

In fact, for an arbitrarily fixed t, choose $\tau \in E(\frac{\varepsilon}{3})$ such that

$$-t \leq \tau \leq -t + L \qquad (i.e., \ 0 \leq t + \tau \leq L).$$

Since $u, v \in \varGamma_p$, we have $u = f^{t_1}(p)$ and $v = t^{t_2}(p)$. Using the property of $E(\frac{\varepsilon}{3})$, we obtain

$$\rho[f^{t_1+t+\tau}(p), f^{t_1+t}(p)] < \frac{\varepsilon}{3}, \qquad \forall \ t_1 + t \in \mathbb{R}^1$$

and

$$\rho[f^{t_2+t+\tau}(p), f^{t_2+t}(p)] < \frac{\varepsilon}{3}, \qquad \forall \ t_2 + t \in \mathbb{R}^1,$$

whenever $\tau \in E(\frac{\varepsilon}{3})$; that is,

$$\rho[f^{t+\tau}(u), f^t(u)] < \frac{\varepsilon}{3}, \qquad \forall \ t \in \mathbb{R}^1 \tag{5.23}$$

and

$$\rho[f^{t+\tau}(v), f^t(v)] < \frac{\varepsilon}{3}, \qquad \forall \ t \in \mathbb{R}^1, \tag{5.24}$$

whenever $\tau \in E(\frac{\varepsilon}{3})$.

Finally, using the property of the above δ, we obtain

$$\rho[f^{t+\tau}(u), f^{t+\tau}(v)] < \frac{\varepsilon}{3}, \qquad 0 \leq t + \tau \leq L, \tag{5.25}$$

whenever $\rho[u,v] < \delta$ for $u, v \in \varGamma_p$.

It follows from (5.23), (5.24) and (5.25) that

$$\rho[f^t(u), f^t(v)] \leq \rho[f^t(u), f^{t+\tau}(u)] + \rho[f^{t+\tau}(u), f^{t+\tau}(v)]$$

$$+ \rho[f^{t+\tau}(v), f^t(v)] < \varepsilon, \qquad \forall \ t \in \mathbb{R}^1,$$

whenever $\rho[u,v] < \delta$ for $u, v \in \varGamma_p$. We have thus proved (5.22).

The proof of the theorem is complete. □

Theorem 5.8 *Assume $z = f^t(p)$ is an almost periodic motion. Then for any sequence $\{p_j\} \subset \Gamma_p$ the corresponding sequence of motions $\{f^t(p_j)\}$ ($t \in \mathbb{R}^1$) has a uniformly convergent subsequence $\{f^t(p_{j'})\}$, such that its limit*

$$z = f^t(q) = \lim_{j' \to \infty} f^t(p_{j'}) \qquad (\text{where } q = \lim_{j' \to \infty} p_{j'})$$

is an almost periodic motion.

Proof. Let $p_j = f^{t_j}(p)$. Since the almost periodic motion $z = f^t(p)$ is also B-recurrent, the closure $\widehat{\Gamma}_p$ of its orbit is compact. It follows that the sequence $\{p_j\} \subset \widehat{\Gamma}_p$ has a convergent subsequence $\{p_{j'}\}$. Let

$$q = \lim_{j' \to \infty} p_{j'} = \lim_{t_{j'} \to \infty} f^{t_{j'}}(p) \in \widehat{\Gamma}_p.$$

According to Theorem 5.7, $z = f^t(p)$ is both positively and negatively uniformly Liapunov stable with respect to its orbit Γ_p. Hence, for any $\varepsilon > 0$, there is a $\delta = \delta(\varepsilon) > 0$, such that

$$\rho[f^{t+s_1}(p), f^{t+s_2}(p)] < \varepsilon, \qquad \forall\, t \in \mathbb{R}^1, \tag{5.26}$$

whenever $\rho[f^{s_1}(p), f^{s_2}(p)] < \delta$.

On the other hand, since $\{f^{j'}(p)\}$ is convergent, the Cauchy principle of convergence implies that there is a sufficiently large integer $N > 0$, such that

$$\rho[f^{t_{j'}}(p), f^{t_{j'}+m}(p)] < \delta, \qquad \text{if } j' \geq N \text{ and } m \geq 0.$$

Letting $s_1 = t_{j'}$ and $s_2 = t_{j'+m}$ in (5.26), we have

$$\rho[f^t(p_{j'}), f^t(p_{j'+m})] = \rho[f^{t_{j'}+t}(p), f^{t_{j'+m}+t}(p)] < \varepsilon, \qquad t \in \mathbb{R}^1,$$

which implies that, in the interval $-\infty < t < \infty$, the sequence $\{f^t(p_{j'})\}$ is uniformly convergent to $f^t(q)$.

Claim: $z = f^t(q)$ is an almost periodic motion.

In fact, since $z = f^t(p)$ is almost periodic, then for any given $\varepsilon > 0$ there is a $L > 0$, such that the set of ε-displacements

$$E(\varepsilon) = \{\tau : \quad \rho[f^{t+\tau}(p), f^t(p)] < \varepsilon \quad (\forall\, t \in \mathbb{R}^1)\}$$

is relatively dense in $\mathbb{R}^1$ with respect to L. It follows that

$$\rho[f^{t+\tau}(p_{j'}), f^t(p_{j'})] = \rho[f^{t+j'+\tau}(p), f^{t+j'}(p)] < \varepsilon, \qquad \forall\, t \in \mathbb{R}^1.$$

Then, letting $j' \to \infty$ for any fixed t, we have

$$\rho[f^{t+\tau}(q), f^t(q)] \leq \varepsilon.$$

Since t is arbitrarily fixed, we have

$$\rho[f^{t+\tau}(q), f^t(q)] \leq \varepsilon, \qquad \forall\, t \in \mathbb{R}^1$$

whenever $\tau \in E(\varepsilon)$. Hence, $z = f^t(q)$ is almost periodic. □

Theorem 5.9 *An almost periodic motion $z = f^t(p)$ is both positively and negatively uniformly Liapunov stable with respect to the closure $\widehat{\Gamma}_p$ of its orbit.*

Proof. According to Theorem 5.7, the motion $z = f^t(p)$ is both positively and negatively uniformly Liapunov stable with respect to its orbit Γ_p. Hence, for given $\varepsilon > 0$, there is a $\delta = \delta(\frac{\varepsilon}{2}) > 0$, such that

$$\rho[f^t(x), f^t(p)] < \frac{\varepsilon}{2}, \qquad \forall\, t \in \mathbb{R}^1,$$

whenever $x = f^s(p)$ satisfying $\rho[x, p] < \delta$.

Assume $q \in \widehat{\Gamma}_p$ with $\rho[q, p] < \frac{\delta}{2}$, and let $p_j = f^{t_j}(p) \to q$. Without destroying the generality, let $\rho[p_j, q] < \frac{\delta}{2}$. Hence, we have $\rho[p_j, p] < \delta$. It follows that

$$\rho[f^t(p_j), f^t(p)] < \frac{\varepsilon}{2}, \qquad \forall\, t \in \mathbb{R}^1. \tag{5.27}$$

On the other hand, according to Theorem 5.8, $\{f^t(p_j)\}$ has a uniformly convergent subsequence. Without loss of generality, assume the sequence $\{f^t(p_j)\}$ is uniformly convergent; that is,

$$\rho[f^t(p_j), f^t(q)] < \frac{\varepsilon}{2}, \qquad \forall\, t \in \mathbb{R}^1,$$

which together with (5.27) implies

$$\rho[f^t(q), f^t(p)] < \varepsilon, \qquad \forall\, t \in \mathbb{R}^1,$$

whenever $\rho[q, p] < \frac{\delta}{2}$ for $q \in \widehat{\Gamma}_p$.

We have thus proved the theorem. □

Theorem 5.10 *Assume the motion $z = f^t(p)$ is B-recurrent. If the motion is positively (or negatively) Liapunov stable with respect to the orbit Γ_p, then $z = f^t(p)$ is an almost periodic motion.*

Proof. Assume $z = f^t(p)$ is positively Liapunov stable with respect to the orbit Γ_p. When $z = f^t(p)$ is negatively Liapunov stable with respect to the orbit Γ_p, the proof is similar.

For a given $\varepsilon > 0$, there is a $\delta = \delta(\frac{\varepsilon}{2}) > 0$, such that

$$\rho[f^t(x), f^t(p)] < \frac{\varepsilon}{2}, \quad \forall\, t \in \mathbb{R}^1,$$

whenever $x \in \Gamma_p$ satisfying $\rho[x, p] < \delta$. On the other hand, since $z = f^t(p)$ is B-recurrent, the set

$$E^* = \{\tau \in \mathbb{R}^1 : \rho[f^\tau(p), p] < \frac{\delta}{2}\}$$

is relatively dense in $\mathbb{R}^1$.

Claim: E^* is a set of ε-displacements for the motion $z = f^t(p)$.

In fact, choose a constant $\tau \in E^*$ ($\tau \neq 0$). In the interval $|t| \leq |\tau|$, using the continuity of motion with respect to the initial point, we have a constant $\sigma > 0$ such that $\rho[f^\tau(q), f^\tau(p)] < \frac{\delta}{2}$ whenever $\rho[q, p] < \sigma$.

For any given time t, it follows from the B-recurrence that there is a $t_1 < t$ such that

$$\rho[f^{t_1}(p), p] < \min\{\sigma, \delta\}. \tag{5.28}$$

According to the definition of σ, we have

$$\rho[f^{t_1+\tau}(p), f^\tau(p)] < \frac{\delta}{2}.$$

On the other hand, it follows from $\tau \in E^*$ that

$$\rho[f^{t_1+\tau}(p), p] \leq \rho[f^{t_1+\tau}(p), f^\tau(p)] + \rho[f^\tau(p), p] < \delta. \tag{5.29}$$

It is noticed that $t - t_1 \geq 0$. Since $z = f^t(p)$ is positively Liapunov stable with respect to the orbit Γ_p, it follows from (5.28) and (5.29) respectively that

$$\rho[f^t(p), f^{t-t_1}(p)] < \frac{\varepsilon}{2} \quad \text{and} \quad \rho[f^{t+\tau}(p), f^{t-t_1}(p)] < \frac{\varepsilon}{2},$$

which yields

$$\rho[f^{t+\tau}(p), f^t(p)] < \varepsilon, \quad \forall\, t \in \mathbb{R}^1,$$

with $\tau \in E^*$. This proves that E^* is a set of ε-displacements for the motion $z = f^t(p)$. Hence, $z = f^t(p)$ is an almost periodic motion.

The proof of the theorem is thus completed. □

Compared with Theorem 5.10, it is interesting to note the following theorem [100], where the property of recurrence is not assumed as a prior condition.

Theorem 5.11 *Assume the motion $z = f^t(p)$ is positively (or negatively) uniformly Liapunov stable with respect to the orbit Γ_p and is negatively (or positively) Lagrange bounded,[4] then it is almost periodic.*

Proof. On account of Theorem 5.10 it suffices to prove that the motion $z = f^t(p)$ is B-recurrent under the conditions of the present theorem.

If we assumed the contrary, we would find a minimal set M in the compact set A_p contained properly in the closure $\widehat{\Gamma}_p$ while, evidently, the point p does not belong to M and therefore $\rho[p, M] = \alpha > 0$.

We will show that every point $q \in A_p$ is positively Liapunov stable with respect to Γ_p. Given $\varepsilon > 0$ we determine the number $\delta = \delta(\varepsilon) > 0$ from the positive Liapunov stability of $f^t(p)$. Because of the way q was defined there is a sequence $\{p_n\}$, $p_n = f^{-t_n}(p)$, $t_n \to \infty$, and $p_n \to q$. We define N such that for $n \geq N$, $\rho[p_n, q] < \frac{\delta}{2}$, and hence $\rho[p_n, p_{n+m}] < \delta$, $m = 1, 2, \cdots$. But then the inequality

$$\rho[f^t(p_n), f^t(p_{n+m})] < \varepsilon,$$

for any $t > 0$, follows from the uniform positive Liapunov stability.

We fix t and n and let m become infinite. It follows that

$$\rho[f^t(p_n), f^t(q)] \leq \varepsilon,$$

for $t > 0$ and $\rho[p_n.q] < \frac{\delta}{2}$.

Now, let $q \in M \subset A_p$. We put $\varepsilon = \frac{\alpha}{2}$ and determine $\delta(\frac{\alpha}{2})$ from the positive Liapunov stability. There will be points $p_n = f^{-t_n}(p)$ ($t_n > 0$), such that $\rho[p_n, q] < \frac{\delta}{2}$, and hence by what was already proved,

$$\rho[f^{t_n}(q), f^{t_n}(p_n)] = \rho[f^{t_n}(q), p] \leq \frac{\alpha}{2}.$$

But $f^{t_n}(q) \in M$, and by assumption $\rho[p, M] = \alpha$. The contradiction shows that $z = f^t(p)$ is B-recurrent. The theorem is thus proved. □

The above theorems characterize the property of almost periodic motion $z = f^t(p)$. For example, using Theorem 5.9, we can prove that the motion in the above Example of Cantorus $\mathcal{C}$ is not almost periodic.

Indeed, let the length of an interval (α_m, β_m) of the complement of F in the longitude circle $\mathbb{S}^1 \times \{0\}$ is $d_m = |\beta_m - \alpha_m| > 0$. However, we may take

---

[4] That is, Γ_p^- (or Γ_p^+) lies in a compact set.

two points $p_1 = (\theta_1, 0)$ and $p_2 = (\theta_2, 0)$ with $|\theta_2 - \theta_1| \ll 1$ of the second kind of the set F, between them there can be found an interval (α_j, β_j) of the complement of F, such that $T_2^k(\alpha_j, \beta_j) = (\alpha_m, \beta_m)$. Correspondingly, we have $\rho[f^k(p_1), f^k(p_2)] \geq d_m$ with arbitrarily small $\rho[p_1, p_2] > 0$. It follows that f^t is not Liapunov stable with respect to $\mathcal{C}$. Hence, the motion in the irregular minimal set $\mathcal{C}$ is not almost periodic.

Does it mean that the motion in an irregular minimal set is always not almost periodic?

The answer is no. The interested reader can find an irregular minimal set of almost periodic motion in the book [100] (i.e., the solenoid of Vietoris and van Dantzig).

Chapter 6

Fixed-Point Theorems

As shown above, there is a close relationship between the periodic solutions of periodic system and the fixed points of the associated Poincaré map. It is thus understandable why the existence theorems of fixed points play important roles in the theory of differential equations. We will introduce first the fixed-point theorems of Brouwer's type in this chapter, and then the fixed-point theorems of Poincaré's type in next chapter.

6.1 Poincaré Index

6.1.1 *Fundamental Theorem of Algebra*

The Poincaré index is the central part of the classical fixed-point theorems in topology. We begin with a simple proof of the fundamental theorem of algebra, which can be considered as the introduction of Poincaré index.

Theorem 6.1 *A polynomial $P(z)$ of degree $n \geq 1$ has at least a root in the complex field $\mathbb{C}$.*

Proof. Without loss of generality, assume

$$P(z) = z^n + a_1 z^{n-1} + \cdots + a_{n-1} z + a_n \qquad (n \geq 1)$$

is a general polynomial of degree n, where $a_1, \cdots, a_n$ are the coefficient constants in $\mathbb{C}$.

Given a point $z \in \mathbb{C}$, consider $P(z)$ as a vector on the complex plane $\mathbb{C}$ starting from the point z. It follows that the polynomial $P(z)$ defines a vector field in $\mathbb{C}$, which will be again denoted by $P(z)$ for simplicity.

Assume Theorem 6.1 is false. Then there is at least a polynomial

$$\tilde{P}(z) = z^n + \tilde{a}_1 z^{n-1} + \cdots + \tilde{a}_{n-1} z + \tilde{a}_n \neq 0, \qquad \forall\, z \in \mathbb{C}.$$

Hence, $\tilde{a}_n = \tilde{P}(0) \neq 0$.

Since $\tilde{P}(z)$ is non-singular in $\mathbb{C}$, we have

$$\tilde{P}(z) = |\tilde{P}(z)| e^{i\phi(z)},$$

where the module $|\tilde{P}(z)| > 0$ is one-valued and continuous in $z \in \mathbb{C}$, and the argument $\phi(z)$ is thus uniquely determined up to mod 2π, and is a locally continuous multi-valued function in $z \in \mathbb{C}$.

Let Γ be a Jordan closed curve in $\mathbb{C}$ (i.e., it is a simple continuous closed curve). Suppose we have a moving point z on Γ, such that it makes a counter-clockwise turn from an initial point z_0 to a terminal point $z_1 = z_0$. Therefore, the vector $\tilde{P}(z)$ varies continuously from the initial value $\tilde{P}(z_0)$ to the terminal value $\tilde{P}(z_1) = \tilde{P}(z_0)$. It is obvious that the module $|\tilde{P}(z)|$ varies continuously from $|\tilde{P}(z_0)|$ to $|\tilde{P}(z_1)| = |\tilde{P}(z_0)|$, while the argument $\phi(z)$ varies continuously from $\phi(z_0) = \phi_0$ to $\phi(z_1)$. Since $\phi(z)$ is determined up to a multiple of 2π, we have $\phi(z_1) = \phi(z_0) + 2k\pi$ with some integer $k = k(P, \Gamma)$. Note that the integer k may be equal to zero or not.

The above integer $k = k(P, \Gamma)$ is called the *Poincaré index* of the vector field $\tilde{P}(z)$ along the closed curve Γ. The above Poincaré index will be denoted by $\mathrm{Ind}\,[\tilde{P}, \Gamma]$ (i.e., $\mathrm{Ind}\,[\tilde{P}, \Gamma] = k(P, \Gamma)$).

In other words, we have

$$\mathrm{Ind}\,[\tilde{P}, \Gamma] = \frac{1}{2\pi} \Delta\phi \bigg|_{\Gamma},$$

where

$$\frac{1}{2\pi} \Delta\phi \bigg|_{\Gamma} = \frac{1}{2\pi} [\phi(z_1) - \phi(z_0)].$$

Now, given a constant $h > 0$, consider the circle

$$C_h : \qquad |z| = h > 0,$$

with a counter-clockwise direction. Since the vector field $\tilde{P}(z)$ is non-singular on $\mathbb{C}$, the Poincaré index of $\tilde{P}(z)$ along C_h varies continuously as h does. Let $I(h) = \mathrm{Ind}\,[\tilde{P}, C_h]$. It follows that $I(h)$ is continuous for $h > 0$. Since $I(h)$ is valued in integers, it is a constant integer for $h > 0$.

On the other hand, when h is sufficiently large, we have

$$I(h) = \mathrm{Ind}\,[\tilde{P}(z), C_h] = \mathrm{Ind}\,[z^n(1 + O(1/|z|)), C_h]$$

$$= \mathrm{Ind}\,[z^n, C_h] = n \geq 1,$$

and, for sufficiently small $h > 0$, we have

$$I(h) = \mathrm{Ind}\,[\tilde{P}(z), C_h] = \mathrm{Ind}\,[\tilde{P}(0), C_h] = \mathrm{Ind}\,[a_n, C_h].$$

It follows that

$$I(h) = \mathrm{Ind}\,[a_n, C_h] = n \geq 1, \qquad \forall\, h > 0. \tag{6.1}$$

However, since the argument $\phi(z)$ of a constant $a_n \neq 0$ is kept as a constant, we have $\mathrm{Ind}\,[a_n, C_h] = 0$, which is in conflict with (6.1).

Theorem 1.1 is thus proved by the contradiction. □

6.1.2 Planar Poincaré Index

Now, the Poincaré index for planar vector fields will be defined in a similar manner as that of the polynomial in $\mathbb{C}$.

Let V be a continuous vector field in the plane $\mathbb{R}^2$, and let Γ be a Jordan closed curve with a counter-clockwise direction. Assume V is non-singular on Γ.

Suppose a moving point $p \in \Gamma$ makes a counter-clockwise turn from an initial point p_0 to the terminal point $p_1 = p_0$. Then the vector $V(p)$ varies continuously from the vector $V(p_0)$ to the vector $V(p_1) = V(p_0)$, such that the module $|V(p)|$ varies continuously from $|V(p_0)|$ to $|V(p_1)| = |V(p_0)|$ and the argument $\phi(p)$ of V varies continuously from the initial value $\phi_0 = \phi(p_0)$ to the terminal value $\phi_1 = \phi(p_1)$. Since the argument $\phi(p)$ is determined up to a multiple of 2π, we get $\phi_1 = \phi_0 + 2m\pi$, where $m = m(V, \Gamma)$ is some fixed integer.

Then the integer m is defined as the *Poincaré index* of the vector field V along the Jordan closed curve Γ, denoted by $\mathrm{Ind}[V, \Gamma]$. That is,

$$\mathrm{Ind}\,[V, \Gamma] = \frac{1}{2\pi}\Delta\phi(p)\bigg|_\Gamma = \frac{1}{2\pi}(\phi_1 - \phi_0). \tag{6.2}$$

It is noticed that the Poincaré index $\mathrm{Ind}\,[V, \Gamma]$ is determined by the vector field V restricted on the closed curve Γ.

The general theory of Poincaré index is referred to the literature, for example, the book of [Krasnosel'skii and Zabreiko (1984)].

It can be easily seen that

$$\text{Ind}\,[-V, \Gamma] = \text{Ind}\,[V, \Gamma], \qquad \text{Ind}\,[V, \Gamma^-] = -\text{Ind}\,[V, \Gamma],$$

where Γ^- is the Jordan closed curve Γ with a clockwise direction.

Assume $V = (V_1, V_2)$ is a smooth vector field, and let Γ be a piece-wise smooth Jordan closed curve. Then using (6.2) yields an analytic formula of the planar Poincaré index

$$\text{Ind}\,[V, \Gamma] = \frac{1}{2\pi} \int_\Gamma d\phi = \frac{1}{2\pi} \int_\Gamma \frac{V_1 dV_2 - V_2 dV_1}{V_1^2 + V_2^2}. \tag{6.3}$$

However, we will be interested in the geometrical method for the most direct computation of the Poincaré index.

Let Γ_1 and Γ_2 be Jordan closed curves in D. If there is is a family of Jordan closed curves

$$H_\lambda = \{\, z = z(s, \lambda) \in D, \quad s \in \mathbb{S} \,\} \qquad (\mathbb{S} = \mathbb{R}/2\pi)$$

containing a parameter $\lambda \in [0, 1]$, such that $z = z(s, \lambda)$ is continuous in $(s, \lambda) \in \mathbb{S} \times [0, 1]$ and satisfies

$$H_0 = \Gamma_1 \qquad \text{and} \qquad H_1 = \Gamma_2,$$

then Γ_1 and Γ_2 are said to be *homotopic* in D, denoted by

$$\Gamma_1 \stackrel{D}{\sim} \Gamma_2\,;$$

and H_λ $(0 \leq \lambda \leq 1)$ is called a *homotopy* between Γ_1 and Γ_2. It is clear that "$\stackrel{D}{\sim}$" is an equivalence relation.

In a similar manner, let V_1 and V_2 be two non-singular vector fields on Γ. If there is a family of non-singular vector fields W_σ on Γ, containing a parameter $\sigma \in [0, 1]$, such that $W_\sigma(x)$ is continuous in $(x, \sigma) \in \Gamma \times [0, 1]$ and satisfies

$$W_0 = V_1 \qquad \text{and} \qquad W_1 = V_2,$$

then the vector fields V_1 and V_2 are said to be *homotopic* on Γ, denoted by

$$V_1 \stackrel{\Gamma}{\sim} V_2\,;$$

and W_σ is called a homotopy between V_1 and V_2. It is clear that "$\stackrel{\Gamma}{\sim}$" is an equivalence relation.

The following homotopic properties are useful for computing the Poincaré index.

Property 1. *If there is a homotopy H_λ between the Jordan curves Γ_1 and Γ_2 such that V is a continuous non-singular vector field on H_λ, then*

$$\mathrm{Ind}\,[V, \Gamma_2] = \mathrm{Ind}\,[V, \Gamma_1]. \tag{6.4}$$

Proof. Since V is a continuous non-singular vector field on H_λ,

$$I(\lambda) = \mathrm{Ind}\,[V, H_\lambda]$$

is a continuous function in $\lambda \in [0, 1]$.

Therefore, $I(\lambda)$ is a constant since it is valued in integers. It follows that $I(1) = I(0)$. Without loss of generality, assume $H_0 = \Gamma_1$ and $H_1 = \Gamma_2$. We have thus arrived at the conclusion (6.4). □

Property 2. *If there is a homotopy W_σ between the vector fields V_1 and V_2 such that W_σ is non-singular on the Jordan curve Γ, then*

$$\mathrm{Ind}\,[V_1, \Gamma] = \mathrm{Ind}\,[V_2, \Gamma]. \tag{6.5}$$

Proof. Let

$$I(\sigma) = \mathrm{Ind}\,[W_\sigma, \Gamma].$$

Then the proof of (6.5) is similar to that of (6.4). □

Property 3. *If V is a continuous non-singular vector field in a planar simple region D bounded by a closed Jordan curve Γ, then $\mathrm{Ind}\,[V, \Gamma] = 0$.*

Proof. Take a point $p \in D$, and let C_h be a circle centered at p with sufficiently small radius $h > 0$. Since $V(p) \neq 0$, we have

$$\mathrm{Ind}\,[V, C_h] = 0, \quad \text{when } h \text{ is sufficiently small.}$$

On the other hand, it is not hard to show that there is a homotopy H_λ between Γ and C_h in D. It follows that V is non-singular on H_λ. Hence, using Property 1 implies that $\mathrm{Ind}\,[V, \Gamma] = \mathrm{Ind}\,[V, C_h] = 0$. □

Property 4. *If Γ is a closed Jordan curve in the (x, y) plane with a continuously turning tangent (non-singular) vector field V on Γ, then $\mathrm{Ind}\,[V, \Gamma] = 1$.*

Proof. Let

$$U(x,y) = \frac{1}{|V(x,y)|} V(x,y), \qquad (x,y) \in \Gamma,$$

which is a unit tangent vector field on Γ. It is clear that

$$\text{Ind}\,[V,\Gamma] = \text{Ind}\,[U,\Gamma],$$

and so it suffices to prove the property for U.

Without loss of generality, assume that Γ lies entirely in the region $x > 0$, and that the points P of Γ are given by $P(t) = (x(t), y(t))$, ($0 \leq t \leq 1$), such that $U(t) = (x'(t), y'(t))$, and it can be further assumed that the positive x-axis is tangent to Γ at $P_0 = P(0)$, that is, $U(0)$ has the same direction as the positive x-axis (see Fig. 19).

We will prove Property 4 by constructing an auxiliary vector field W on the closed triangular region

$$\Delta: \qquad 0 \leq s \leq 1, \quad s \leq t \leq 1$$

in the (s,t) plane.

Define $W(s,s) = U(s)$ for $0 \leq s \leq 1$, $W(0,1) = -U(0)$, and for all other (s,t) in Δ define $W(s,t)$ to the unit vector in the direction from $P(s)$ to $P(t)$ on Γ. Let $\theta(s,t)$ be the angle that $W(s,t)$ makes with the positive x-axis. Clearly $\theta(0,0) = 0$, and since Γ remains in the region $x > 0$, $\theta(0,t)$ varies from 0 to π as t runs 0 to 1. Similarly, $\theta(s,1)$ varies from π to 2π as s varies from 0 to 1. From the definition of W it is clear that W is continuous on Δ, and $W \neq 0$ there. Hence, by the Property 3, applied to the boundary $\partial \Delta$ of T, $\text{Ind}\,[W, \partial \Delta] = 0$. This means that the variation of $\theta(s,s)$ as s goes from 0 to 1 is 2π. But this is precisely the variation in the angle that U makes with the positive x-axis as Γ is traversed once in the positive direction. Hence $\text{Ind}\,[U, \Gamma] = 1$, and the Property 4 is proved (see [21]). $\square$

The following result is an important consequence of the Property 4.

Corollary 6.1 *Assume Γ is a periodic orbit of a planar vector field V, then we have* $\text{Ind}\,[V, \Gamma] = 1$.

In what follows, the index of a singular point will be defined. Let q be an isolated singular point of the planar vector field V. Then there is a neighborhood $B(q)$ at q, such that q is a unique singular point of V in $B(q)$. Then $B = (B(q) \setminus \{q\})$ is a neighborhood at the deleted point q. Assume

J_1 and J_2 are Jordan closed curves in B. Then the indices $\text{Ind}\,[V, J_1]$ and $\text{Ind}\,[V, J_2]$ are well-defined. If J_1 and J_2 are non-contractible in B, then they are homotopic to each other in B; that is,

$$J_1 \stackrel{B}{\sim} J_2.$$

It follows from the Property 1 that

$$\text{Ind}\,[V, J_2] = \text{Ind}\,[V, J_1].$$

This implies that the value $\text{Ind}\,[V, J_1]$ does not depend on J_1, and is dependent on the singular point q only. Now, define the value $\text{Ind}\,[V, J_1]$ to be the *index of the singular point* q, denoted by $\text{Ind}\,(q, V)$. It is noted that the index of an isolated singular point q can be computed by

$$\text{Ind}\,(q, V) = \text{Ind}\,[V, J],$$

where J is a non-contractive Jordan closed curve in the neighborhood B at the deleted point q.

Property 5. *Let $\hat{D}$ be a closed domain bounded by a simple closed curve C. Assume V is a continuous vector field on $\hat{D}$, and has a finite number of singular points $\{q_j\}$ $(1 \leq j \leq m)$. If V is non-singular on C, then*

$$\text{Ind}\,[V, C] = \sum_{1 \leq j \leq m} \text{Ind}\,(q_j, V).$$

In particular, if V has no singular point in $\hat{D}$, then we have $\text{Ind}\,[V, C] = 0$.

(The proof is omitted here since it is similar to the proof of the residue theorem in the theory of complex variables.)

It is clear that the Corollary 6.1 together with the Property 5 yields the following result.

Corollary 6.2 *Assume C is a closed orbit of a planar vector field V. Then there is at least a singular point of V in the interior region of C.*

6.1.3 The Poincaré-Bendixson Formula

Assume V is a continuous vector field in the plane $\mathbb{R}^2$. Let Γ be a Jordan closed curve with a counter-clockwise direction. If V is non-singular on Γ, then the Poincaré index $\text{Ind}\,[V, \Gamma]$ is well-defined.

With a geometric interpretation, Poincaré gave the following formula

$$\text{Ind}\,[V, \Gamma] = \frac{I - E + 2}{2}, \tag{6.6}$$

where I is the number of the interior tangent points[1] of Γ, and E is the number of the exterior tangent points[2] of Γ.

The formula (6.6) provides a heuristic geometric method to compute the index, and its first analytic proof is due to Bendixson (see [6]). In literature, the formula (6.6) is sometimes used as the definition of the index $\text{Ind}\,[V, \Gamma]$ (see, for example, [3]).

Compared with the definition of $\text{Ind}[V, \Gamma]$, an extra assumption that the vector field V is defined in a neighborhood of Γ is added to the formula (6.6).

In what follows, we want to generalize the Poincaré-Bendixson Index Formula (6.6) to the case, where the vector field V is only defined on Γ. The key point is to define the interior and exterior tangent points in a new version, such that the new formula agrees with the old one aV defined in the nboooof *thit*Γ.

Since the index $\text{Ind}\,[V, \Gamma]$ is invariant under the small perturbations of the vector field V and the Jordan closed curve Γ, assume without destroying the generality that both V and Γ are analytic.

Let

$$z(s) = (x(s), y(s)), \quad (0 \leq s \leq 2\pi),$$

be a moving point on Γ, satisfying $z(0) = z(2\pi)$, where $z(s)$ is analytic in $s \in [0, 2\pi]$. Assume $z(s)$ moves along the direction of Γ (counter-clockwise) as s increases.

It follows that $\tau(s) = z'(s)/|z'(s)|$ is a unit tangent vector of Γ at the point $z(s)$. Denote by $\theta(s)$ the argument of the vector $\tau(s)$, satisfying the initial condition $0 \leq \theta(0) < 2\pi$.

Now, assume $\varphi(s)$ is the argument of $V(z(s))$, satisfying $0 \leq \varphi(0) < 2\pi$. Then $\psi(s) = \varphi(s) - \theta(s)$ measures the angle between the vectors $V(z(s))$ and $\tau(s)$ starting from the point $z(s)$.

Consider the vertical strip

$$S = \{(s, u) \in \mathbb{R}^2 \mid 0 \leq s \leq 2\pi, \ -\infty < u < \infty\},$$

---

[1] An interior tangent point p of Γ means that the integral curve of V passing through the point p is tangent to Γ and lies locally in the inside of Γ near the point p.

[2] An exterior tangent point p of Γ means that the integral curve of V passing through the point p is tangent to Γ and lies locally in the outside of Γ near the point p.

which contains the horizontal line-segments

$$C_j : \quad 0 \leq s \leq 2\pi, \quad u = j\pi \quad (j \in \mathbb{Z}),$$

and the analytic curve

$$\mathcal{J} : \quad u = \psi(s) \quad (0 \leq s \leq 2\pi).$$

Note that $\mathcal{J}$ has the left end-point $p_0 = (0, \psi(0))$ and the right end-point $p_{2\pi} = (2\pi, \psi(2\pi))$. If the curve $\mathcal{J}$ intersects the line-segment C_j at a point $(s^*, j\pi)$ (i.e., $\psi(s^*) = j\pi = \varphi(s^*) - \theta(s^*)$), then $V(z(s^*))$ is a tangent vector of Γ at the point $z(s^*)$.

If the curve $\mathcal{J}$ is strictly decreasing at the intersection point $(s^*, j\pi)$ (i.e., $\mathcal{J}$ passes through $(s^*, j\pi)$ from the above of C_j to the below of C_j as s increases), then $z(s^*)$ is called an "*exterior tangent point*" of Γ with respect to the vector field V.

If the curve $\mathcal{J}$ is strictly increasing at the intersection point $(s^*, j\pi)$ (i.e., $\mathcal{J}$ passes through $(s^*, j\pi)$ from the below of C_j to the above of C_j as s increases), then $z(s^*)$ is called an "*interior tangent point*" of Γ with respect to the vector field V.

Because the curve $\mathcal{J}$ is analytic, the number of "interior tangent points" and the number of "exterior tangent points" are finite.

Now, we are ready to prove the above Poincaré-Bendixson index formula in the new version.

Theorem 6.2 *Assume the vector field V on Γ is non-singular. Let I^* and E^* be the numbers of "interior tangent points" and "exterior tangent points" of Γ with respect to V, respectively. Then we have*

$$\mathrm{Ind}\,[V, \Gamma] = \frac{I^* - E^* + 2}{2}. \tag{6.6}*$$

Proof. It follows from Corollary 6.1 that $\mathrm{Ind}\,[\tau, \Gamma] = 1$. This means

$$\theta(2\pi) - \theta(0) = 2\pi. \tag{6.7}$$

On the other hand, the definition of $\mathrm{Ind}[V, \Gamma]$ implies

$$\varphi(2\pi) - \varphi(0) = 2\pi \mathrm{Ind}\,[V, \Gamma]. \tag{6.8}$$

It can be seen that the left boundary L of S is divided into the intervals

$$L_k : \quad s = 0, \quad k\pi \leq u < (k+1)\pi \quad (k \in \mathbb{Z}),$$

and the right boundary R of S is divided into the intervals

$$R_k : \quad s = 2\pi, \quad k\pi \leq u < (k+1)\pi \qquad (k \in \mathbb{Z}).$$

Let D_k be the rectangular region bounded by $L_k \cup C_k \cup R_k \cup C_{k+1}$.

Choose the parameter s, such that $0 \leq \psi(0) < \pi$ (i.e., the left end-point p_0 of $\mathcal{J}$ belongs to the interval L_0). On the other hand, it follows from $\psi(s) = \varphi(s) - \theta(s)$ together with (6.7) and (6.8) that

$$\psi(2\pi) - \psi(0) = [\varphi(2\pi) - \varphi(0)] - [\theta(2\pi) - \theta(0)]$$

$$= 2\pi(\operatorname{Ind}[V, \Gamma] - 1) = 2m\pi,$$

where

$$m = (\operatorname{Ind}[V, \Gamma] - 1). \tag{6.9}$$

Suppose a moving point on $\mathcal{J} \subset S$ changes continuously from its left end-point $p_0 \in L_0 \subset L$ to its right end-point $p_{2\pi} \in R_{2m}$ as s varies from $s = 0$ to $s = 2\pi$. It follows that the number j of the region D_j increases I^* times and decreases E^* times, where D_j are the regions visited by $\mathcal{J}$ during $0 \leq s \leq 2\pi$. Then, it follows from $p_0 \in L_0 \subset D_0$ and $p_{2\pi} \in R_{2\pi} \subset D_{2m}$ that $I^* - E^* = 2m$, which together with (6.9) yields (6.6)*.

Theorem 6.2 is thus proved. □

It is not hard to verify that the "exterior tangent points" and "interior tangent points" in the new version agree with the old ones considered by Poincaré-Bendixson in the case that the vector field V is defined in a neighborhood of Γ.

6.1.4 *Application of the Poincaré-Bendixson Formula*

We will apply the Poincaré-Bendixson formula to compute the indexes of the following singular points.

Definition 6.1 Let q be an isolated singular point of a vector field V in the plane. Then q is called, respectively,
  1) an *attractor* if the nearby orbits tends to q as $t \to +\infty$;
  2) a *repeller* if the nearby orbits tends to q as $t \to -\infty$;
  3) a *center* if the nearby orbits do not tend to q as $t \to \pm\infty$.

Let J be a Jordan closed curve surrounding the singular point q in a sufficient small neighborhood. Then, using the Poincaré-Bendixson index formula with a simple geometric intuition, we get

$$\text{Ind}\,(q,V) = \text{Ind}\,[V,J] = \begin{cases} 1, & \text{if } q \text{ is an attractor or a repellor;} \\ 1, & \text{if } q \text{ is a center;} \\ 1, & \text{if } q = \mathfrak{o} \text{ is the singular point in Fig. 2.} \end{cases}$$

Therefore, the index $\text{Ind}\,(q,V)$ is not sufficient to determine the type of the singular point q. For more details, the reader is referred to the literature (for example, [123]).

6.2 Vector Fields on Closed Surfaces

Assume $\mathfrak{F}$ is a *closed oriented C^1 - differentiabl surface*.[3] Therefore, at each point p of $\mathfrak{F}$, there is a neighborhood $\mathcal{N}_p$ in $\mathfrak{F}$, such that $\mathcal{N}_p$ is topologically equivalent to a planar open disk. For simplicity, assume $\mathcal{N}_p$ is itself a planar open disk. Since the planar vector field and its Poincaré indexes are, in fact, defined in the local sense, the definitions of vector fields and Poincaré indexes are generalized to the surfaces in a natural way. Then we can establish the theory of dynamical systems in the closed surfaces.

We have to introduce some basic knowledge.

6.2.1 The Euler Characteristic Formula

Let us begin with the familiar two-dimensional sphere $\mathbb{S}^2$. Assume there are two disjoint closed disks D_1 and D_2 on $\mathbb{S}^2$. Then

$$F_1 = \mathbb{S}^2 \setminus (D_1 \cup D_2)$$

is a surface having the boundary $\partial F_1 = C_1 \cup C_2$, where $C_1 = \partial D_1$ and $C_2 = \partial D_2$.

Consider in addition a cylindrical surface $H = \mathbb{S}^1 \times [0,1]$, which has the boundary circles $\Gamma_1 = \mathbb{S}^1 \times \{0\}$ and $\Gamma_2 = \mathbb{S}^1 \times \{1\}$. Let us slightly bend H such that it is pasted to F_1 with $\Gamma_1 = C_1$ and $\Gamma_2 = C_2$. Consequently, F_1 and H make up an oriented closed surface, denoted by $\mathcal{S}_1$, where H is called a *handle of the closed surface* $\mathcal{S}_1$. In this sense, $\mathcal{S}_1$ is a closed surface carrying a *handle*. It is clear that $\mathcal{S}_1$ is topologically equivalent to a torus

---
[3] That is, $\tilde{\mathcal{F}}$ is a two-dimensional oriented compact C^1- differentiable manifold without boundary.

$\mathbb{T}^2 = \mathbb{S}^1 \times \mathbb{S}^1$, which has a so-called *hole*. Therefore, $\mathcal{S}_1$ is also called a surface of a hole.

Similarly, suppose there are k pairs of disjoint closed disks

$$D_1, D_2; \quad \cdots \cdots \quad ; D_{2k-1}, D_{2k}$$

on $\mathbb{S}^2$. Denote by $C_i = \partial D_i$ the boundary circles, $i = 1, \cdots, 2k$. Then

$$F_k = \mathbb{S}^2 \setminus \sum_{j=1}^{k} (D_{2j-1} \cup D_{2j})$$

is a surface having the boundary

$$\partial F_k = \sum_{j=1}^{k} (C_{2j-1} \cup C_{2j}).$$

Consider in addition k cylindrical surfaces

$$H_1, \cdots, H_k,$$

which have respectively the boundaries

$$\partial H_j = \Gamma_{2j-1} \cup \Gamma_{2j} \qquad (j = 1, \cdots, k).$$

Paste $H_1, \cdots, H_k$ to F_k, such that $\Gamma_{2j-1} = C_{2j-1}$ and $\Gamma_{2j} = C_{2j}$. Then F_k and H_j ($j = 1, \cdots, k$) make up a closed surface $\mathcal{S}_k$, where $H_1, \cdots, H_k$ are called the *handles* of $\mathcal{S}_k$. In this sense, $\mathcal{S}_k$ is called a surface carrying k handles.

In geometry, associated to each handle H_j ($1 \leq j \leq k$) on the surface $\mathcal{S}_k$, there is a hole. Therefore, $\mathcal{S}_k$ is also called a surface of k holes.

As proved in the classic geometry, it seems clear that an oriented closed surface $\mathfrak{F}$ is topologically equivalent to a surface $\mathcal{S}_m$, where $m \geq 0$ is some integer. In literature, the number m is called the *genus* of the surface $\mathfrak{F}$, denoted by $g\,[\mathfrak{F}]$. It can be seen that $g\,[\mathbb{S}^2] = 0$, $g\,[\mathbb{T}^2] = 1$. In general, we have $g\,[\mathcal{S}_k] = k$ for $k = 0, 1, \cdots$.

Now, let $\mathfrak{F}$ be an oriented closed surface with genus $g\,[\mathfrak{F}]$. It follows from the local structure of $\mathfrak{F}$ that the closed surface $\mathfrak{F}$ can be divided into triangles[4], such that the intersection of two neighboring triangles is a common side or a common vertex. This kind of division is called a *triangle division* of $\mathfrak{F}$.

---

[4] A small triangle-like piece on $\mathfrak{F}$ can be understood as a planar triangle.

Let Δ be a triangle division of $\mathfrak{F}$, having a triangles, b sides and c vertexes. The formula

$$a - b + c = 2 - 2g\,[\mathfrak{F}] \tag{6.10}$$

together with its mathematical merit was first discovered by Euler. In literature, the number $(a - b + c)$ is called the *Euler characteristic* of the closed surface $\mathfrak{F}$. It can be seen from (6.10) that $\chi[\mathfrak{F}]$ is independent on the division Δ and is dependent only on the genus $g\,[\mathfrak{F}]$ (i.e., $\chi[\tilde{\mathcal{F}}] = 2 - 2g\,[\mathfrak{F}]$). But, a particular division Δ_0 is sufficient to compute the Euler characteristic $\chi[\mathfrak{F}]$.

6.2.2 Poincaré-Hopf Index Formula

Now, let V be a continuous vector field on a closed oriented surface $\mathfrak{F}$. Assume V has a finite number of singular points $\{q_k\}$ ($k \in \mathcal{I}$) with the indexes $\mathrm{Ind}\,(q_k, V)$ ($k \in \mathcal{I}$), respectively, where $\mathcal{I}$ is a finite set of integers (including the case for $\mathcal{I} = \emptyset$). Then the *Poincaré-Hopf index formula*

$$\sum_{k \in I} \mathrm{Ind}\,(q_k, V) = \chi[\mathfrak{F}] \tag{6.11}$$

holds on $\mathfrak{F}$. In literature, the Poincaré-Hopf index formula (6.11) is usually proved by the Euler characteristic formula (6.10). We will converse the proof. In other words, we will first prove the Poincaré-Hopf index formula (6.11) by the qualitative method of ordinary differential equations, and then apply it to prove the Euler characteristic formula (6.10) (see [51]).

Spacial Case for the Sphere.

Assume the sphere $\mathbb{S}^2$ is oriented by the outward normal vector, and then the direction of a closed curve on $\mathbb{S}^2$ is determined by the right-hand principle.

Let X be a continuous vector field on $\mathbb{S}^2$ having a finite number of singular points $\{q_m\}$, ($m \in \mathcal{I}$).

Take an ordinary point p_0 of X (i.e., $X(p_0) \neq 0$). Let D be a small closed disk on $\mathbb{S}^2$ centered at p_0, such that the vectors $\{X\}$ in D are non-degenerate and nearly parallel. Let C_0 be the boundary of D and let D_0 be the interior of D. It follows that there are two exterior tangent points y_1 and y_2 (with respect to D_0) for the vector field X on C_0.

Let $C_0(y_1, y_2)$ be the open arc of C_0 from y_1 to y_2 along the positive direction of C_0. Similarly, let $C_0(y_2, y_1)$ be the open arc from y_2 to y_1.

Therefore, the vectors $\{X\}$ on $C_0(y_1, y_2)$ point toward the inside or the outside of C_0 (with respective to D_0). For definiteness, assume the vectors X on $C_0(y_1, y_2)$ point toward the outside of C_0. Then the vectors X on $C_0(y_2, y_1)$ point toward the inside of C_0.

Without loss of generality, we assume the north pole N of $\mathbb{S}^2$ is a point in D_0. Let $G = (\mathbb{S}^2 \setminus D_0)$. It follows that the projection

$$\psi : \quad G \quad \to \quad G^* \subset \mathbb{R}^2,$$

initiating from the north pole N satisfies the property that $G^* = \psi(G)$ is a simple connected region and its boundary $C_0^* = \psi(C_0)$ is a Jordan closed curve. The projection ψ induces a vector field $X^* = \psi'(X)$ on G^*, such that X^* has two interior tangent points $y_1^* = \psi(y_1)$ and $y_2^* = \psi(y_2)$ on C_0^* (with respect to G^*). The vector field X^* on the arc $C_0^*(y_1^*, y_2^*)$ points toward the inside of C_0^*, while X^* on the arc $C_0^*(y_2^*, y_1^*)$ points toward the outside of C_0^*.

It follows that the number I of interior tangent points of X^* on C_0^* is equal to 2, and the number E of exterior tangent points is equal to 0. Using the Poincaré-Bendixson index formula, we have

$$\mathrm{Ind}\,[X^*, C_0^*] = \frac{2 + I - E}{2} = 2. \tag{6.12}$$

Since the singular points $q_1, \cdots, q_m$ of X belong to the interior of G, the planar vector field X^* has m singular points $q_1^*, \cdots, q_m^*$ in the interior of G^*. It follows from the property of planar vector fields that

$$\mathrm{Ind}\,[X^*, C_0^*] = \sum_{s=1}^{m} \mathrm{Ind}\,(q_s^*, X^*).$$

Consequently, using (6.12), we get

$$\sum_{s=1}^{m} \mathrm{Ind}\,(q_s^*, X^*) = 2. \tag{6.13}$$

On the other hand, the topological property of the projection ψ implies

$$\mathrm{Ind}\,(q_s^*, X^*) = \mathrm{Ind}\,(q_s, X) \quad (s \in I),$$

which together with (6.13) implies

$$\sum_{s \in I} \mathrm{Ind}\,(q_s, X) = 2. \tag{6.14}$$

Chapter 6. Fixed-Point Theorems

We have thus proved that *the algebraic sum of indexes of the singular points for a vector field on the sphere $\mathbb{S}^2$ is equal to 2*.

It follows that a vector field on $\mathbb{S}^2$ has at least a singular point.

General Case for the Closed Surfaces.

Let V be a vector field on a closed oriented surface $\mathfrak{F}$ having the genus $g[\mathfrak{F}]$. Assume V has a finite number of singular points $\{q_m\}$ ($m \in \mathcal{I}$). Let $\mathrm{Ind}\,(q_m, V)$ be the Poincaré index of the singular point q_m.

Consider a handle $H = \mathbb{S}^1 \times [0,1]$ of the surface $\mathfrak{F}$. Since V has only a finite number of singular points, there is a meridian circle $\Gamma_0 = \mathbb{S}^1 \times \{b\}$ ($0 < b < 1$) of H, such that V is non-singular on the 2ε-neighborhood $B_{2\varepsilon}$ at Γ_0, where $\varepsilon > 0$ is a small constant.

Since $\mathfrak{F}$ is oriented, the left-side and the right-side of Γ_0 are well-defined. For convenience, assume the meridian circle $\Gamma_{-\varepsilon} = \mathbb{S} \times \{b - \varepsilon\}$ is on the left-side of Γ_0, then the meridian circle $\Gamma_{+\varepsilon} = \mathbb{S} \times \{b+\varepsilon\}$ is on the right-side of Γ_0. It is noted that $\Gamma_{-\varepsilon} \cup \Gamma_{+\varepsilon} \subset B_{2\varepsilon}$.

Let $p \in \Gamma_0$. If the orbit Γ_p of V nearby the initial point p is locally seated at the left-side of Γ_0, then p is called a *left tangent point* of Γ_0 about V. Similarly, if the orbit Γ_p of V nearby the initial point p is locally seated at the right-side of Γ_0, then p is called a *right tangent point* of Γ_0 about V. Since V and Γ_0 are analytic, both the number N_l of left tangent points and the number N_r of right tangent points of Γ_0 about V are finite.

Let us cut off a handle H from $\mathfrak{F}$. Then we have the surface $\mathfrak{F}' = (\mathfrak{F} \setminus H)$, which has the boundary circles $C_1 = \mathbb{S}^1 \times \{0\}$ and $C_2 = \mathbb{S}^1 \times \{1\}$.

Consider again the sections $H_{+\varepsilon} = \mathbb{S}^1 \times [0, b+\varepsilon]$ and $H_{-\varepsilon} = \mathbb{S}^1 \times [b-\varepsilon, 1]$ on H, both of which contain the meridian circle Γ_0. To memorize the differences, denote the meridian circle $\Gamma_0 \subset H_{+\varepsilon}$ by Γ_0^1, and denote the meridian circle $\Gamma_0 \subset H_{-\varepsilon}$ by Γ_0^2, respectively. Note that $\partial H_{+\varepsilon} = C_1 \cup \Gamma_{+\varepsilon}$ and $\partial H_{-\varepsilon} = C_2 \cup \Gamma_{-\varepsilon}$.

Then, paste $H_{+\varepsilon}$ to $\mathfrak{F}'$ at C_1, and paste $H_{-\varepsilon}$ to $\mathfrak{F}'$ at C_2, such that $H_{-\varepsilon} \cap H_{+\varepsilon} = \emptyset$. The resulting surface will be denoted by $\mathfrak{F}^\alpha$, which has the boundary $\Gamma_{+\varepsilon} \cup \Gamma_{-\varepsilon}$. The vector field V on $\mathfrak{F}$ is correspondingly induced to a vector field V^α on $\mathfrak{F}^\alpha$.

Now, define a smooth function $\eta(x)$ on $\tilde{\mathcal{F}}^\alpha$, such that

$$\eta(x) = 0, \qquad x \in \Gamma_{+\varepsilon} \cup \Gamma_{-\varepsilon},$$

$$0 < \eta(x) \leq 1, \qquad x \notin \Gamma_{+\varepsilon} \cup \Gamma_{-\varepsilon},$$

and $\eta(x) = 1$ in the neighborhoods of the singular points of V^α.

Therefore, we have a vector field $\eta(x)V^\alpha(x)$ on $\mathfrak{F}^\alpha$, which equals the zero vector on $\Gamma_{+\varepsilon} \cup \Gamma_{-\varepsilon}$. Then let $\Gamma_{+\varepsilon}$ be continuously contracted to a point P^+, and $\Gamma_{-\varepsilon}$ be continuously contracted to a point P^-. It follows that $\tilde{\mathcal{F}}^\alpha$ is deformed into a closed surface $\tilde{\mathcal{F}}^\beta$, with the genus $g[\mathfrak{F}^\beta] = g[\mathfrak{F}] - 1$. The vector field $\eta(x)V^\alpha(x)$ is also induced to a vector field V^β on $\mathfrak{F}^\beta$. It can be seen that the singular points of V together with the points P^+ and P^- become the singular points of V^β.

It is noted that Γ_0^1 surrounds the singular point P^+ in a small neighborhood. Similarly, Γ_0^2 surrounds the singular point P^- in a small neighborhood. Hence, we have

$$\begin{cases} \text{Ind}\,(P^+, V^\beta) = \text{Ind}\,[V^\beta, \Gamma_0^1], \\ \text{Ind}\,(P^-, V^\beta) = \text{Ind}\,[V^\beta, \Gamma_0^2]. \end{cases} \quad (6.15)$$

It follows from the property of Γ_0 that a right tangent point of Γ_0 is an interior tangent point of Γ_0^1 and a left tangent point of Γ_0 is an exterior tangent point of Γ_0^1. Similarly, a right tangent point of Γ_0 is an exterior tangent point of Γ_0^2, and a left tangent point of Γ_0 is an interior tangent point of Γ_0^1. Therefore, Γ_0^1 has N_r interior tangent points and N_l exterior tangent points, and Γ_0^2 has N_l interior tangent points and N_r exterior tangent points. The index formula of Poincaré-Bendixson implies

$$\text{Ind}\,[V^\beta, \Gamma_0^1] = \frac{N_r - N_l + 2}{2}, \quad \text{Ind}\,[V^\beta, \Gamma_0^2] = \frac{N_l - N_r + 2}{2}.$$

It follows that

$$\text{Ind}\,[V^\beta, \Gamma_0^1] + \text{Ind}\,[V^\beta, \Gamma_0^2] = 2,$$

which together with (6.15) implies

$$\text{Ind}\,(P^+, V^\beta) + \text{Ind}\,(P^-, V^\beta) = 2. \quad (6.16)$$

Now, if we repeat the above process g times, all the g handles of the closed surface $\mathfrak{F}$ are eliminated, and the surface $\mathfrak{F}$ is changed into a closed surface $\mathcal{S}_0$ without a hole (i.e., it is a sphere), and the original vector field V on $\mathfrak{F}$ is induced to a vector field U on $\mathcal{S}_0$. It can be seen that the singular points of U consist of the singular points $\{q_m\}$ $(m \in J)$ of V and, in addition, $2g$ singular points

$$P_1^+, P_1^-, \cdots, P_g^+, P_g^-,$$

which are introduced in the above processes for eliminating the handles $H_1, \cdots, H_g$, respectively. Then, using the formula (6.14), we get

$$\sum_{i \in I} \text{Ind}\,(q_i, U) + \sum_{j=1}^{g} [\text{Ind}\,(P_j^+, U) + \text{Ind}\,(P_j^-, U)] = 2. \qquad (6.17)$$

Notice that

$$\text{Ind}\,(q_i, U) = \text{Ind}\,(q_i, V) \qquad (i \in I). \qquad (6.18)$$

It follows from (6.16) that

$$\text{Ind}\,(P_j^+, U) + \text{Ind}\,(P_j^-, U) = 2 \qquad (j = 1, \cdots, g). \qquad (6.19)$$

Finally, using (6.17), (6.18) and (6.19) yields

$$\sum_{i \in I} \text{Ind}\,(q_i, V) = 2 - 2g. \qquad (6.20)$$

We have thus proved the following

Theorem 6.3 *The Poincaré-Hopf index formula (6.20) holds on a closed oriented surface $\mathfrak{F}$.*

Finally, it is noted from the Poincaré-Hopf index formula (6.20) that a continuous vector field on a closed surface $\mathfrak{F}$ has at least a singular point if $g[\mathfrak{F}] \neq 1$ (i.e., $\mathfrak{F}$ is not a torus). In fact, a torus $\mathbb{T}^2$ admits of non-singular vector fields. For example, the vector fields

$$V_1(x, y) = (1, \sqrt{2}), \quad \text{and} \quad V_2(x, y) = (\sin(x+y), \cos(x+y))$$

are non-singular on $\mathbb{T}^2$.

6.2.3 *Proof of the Euler Characteristic Formula*

Now, we are prepared to prove the Euler characteristic formula (6.10) via the Poincaré-Hopf index formula (6.20).

In fact, assume $\mathfrak{F}$ is an oriented closed surface with genus g, and assume Δ is a triangle division of $\mathfrak{F}$ having a triangles, b sides and c vertexes. Using an elementary geometrical method, we can construct a continuous vector field V on $\mathfrak{F}$, such that all the vertexes P_i $(i = 1, \cdots, c)$ of the division Δ are attractors of V, all the middle point Q_j $(j = 1, \cdots, b)$ of the sides are saddles of V, and all the centers R_k $(k = 1, \cdots, a)$ of the triangles are repellers of V. Assume V has no other singular points. Then, using the Poincaré-Hopf index formula (6.20), we have

$$\sum_{i=1}^{c} \text{Ind}\,(P_i, V) + \sum_{j=1}^{b} \text{Ind}\,(Q_j, V) + \sum_{k=1}^{a} \text{Ind}\,(R_k, V) = 2 - 2g,$$

which together with

$$\text{Ind}\,(P_i, V) = +1 \qquad (i = 1, \cdots, c),$$

$$\text{Ind}\,(Q_j, V) = -1 \qquad (j = 1, \cdots, b),$$

$$\text{Ind}\,(R_k, V) = +1 \qquad (k = 1, \cdots, a),$$

yields

$$c - b + a = 2 - 2g.$$

We have thus proved the Euler characteristic formula (6.10). □

Remark 6.1 *Using the property of oriented or non-oriented surfaces, we can in fact prove a general Poincaré-Hopf index formula*

$$\sum_{k \in I} \text{Ind}\,(q_k, V) = \chi[\mathfrak{F}],$$

where the Euler characteristic

$$\chi[\mathfrak{F}] = \begin{cases} 2 - 2g, & \text{if } \mathfrak{F} \text{ is orientable;} \\ 2 - g, & \text{if } \mathfrak{F} \text{ is non-orientable.} \end{cases}$$

Then the general Euler characteristic formula

$$a - b + c = \chi[\mathfrak{F}]$$

(see [Aranson et al (1996)]) *can be proved in a similar manner.*

6.3 Spatial Vector Fields

6.3.1 *Indexes of Spatial Vector Fields*

Now, we begin to generalize the Poincaré indexes to the vector fields in higher dimensional spaces.

For this aim, let us review the planar Poincaré index $\text{Ind}\,[V, \Gamma]$ with a different point-view. Assume $\text{Ind}\,[V, \Gamma] = m$, where the vector field V is non-singular on a Jordan closed curve Γ and m is an integer. Without loss of generalization, assume V and Γ are sufficiently smooth.

Clearly,
$$E(x) = \frac{1}{|V(x)|}V(x), \qquad (x \in \Gamma)$$
is a unit-vector field on Γ. If $E(x)$ is considered as a vector starting from the origin $\mathbf{0} \in \mathbb{R}^2$, then the terminal point of $E(x)$, denoted again by $E(x)$, lies on the unit-circle $\mathbb{S}^1$ centered at $\mathbf{o}$ (i.e., $E(x) \in \mathbb{S}^1$). Hence, we have a mapping
$$\Theta : \quad \Gamma \to \mathbb{S}^1; \qquad x \mapsto E(x)$$
(from the closed curve Γ to the unit-circle $\mathbb{S}^1$). It follows from the definition of $\text{Ind}\,[V,\Gamma] = m$ that when a moving point x on Γ makes up a counter clockwise turn, the point $E(x)$ will make up m (algebraic) turns on $\mathbb{S}^1$. In other words, the image-set $\Theta(\Gamma)$ continuously sweeps the circle $\mathbb{S}^1$ m times. More exactly, if an (oriented) increment ds is given at the moving point x on Γ, there is an induced (oriented) increment $dy = \det[E'(x)]ds$ at the image point $y = E(x) \in \mathbb{S}^1$, such that the algebraic number of the layers of $\Theta(\Gamma)$ on $\mathbb{S}^1$ is equal to
$$m = \frac{1}{2\pi}\int_{\mathbb{S}^1} dy = \frac{1}{2\pi}\int_\Gamma \det[E'(x)]\,ds.$$
It follows that
$$\text{Ind}\,[V,\Gamma] = \frac{1}{2\pi}\int_\Gamma \det[E'(x)]\,ds. \tag{6.21}$$

This formula is suitable to generalize the Poincaré index from the plane to the higher-dimensional spaces.

Let V be a smooth vector field on an $(n-1)$-dimensional compact smooth manifold Ω without boundary (for simplicity, such a manifold Ω is sometimes called an $(n-1)$-dimensional closed smooth surface).

Assume $V(x)$ is a non-degenerate vector field on Ω. Then
$$E(x) = \frac{1}{|V(x)|}V(x) \qquad (x \in \Omega)$$
is a smooth unit-vector field on Ω, which defines a smooth mapping
$$\Phi : \quad \Omega \to \mathbb{S}^{n-1}; \qquad x \mapsto E(x),$$

where $\mathbb{S}^{n-1}$ is the $(n-1)$-dimensional unit-sphere centered at the origin $\mathfrak{o} \in \mathbb{R}^n$. When the point x sweeps on Ω, the image point $y = E(x)$ varies on the sphere $\mathbb{S}^{n-1}$. If an (oriented) increment dx is given to the moving point x on Ω, then an increment $dy = \det[E'(x)]dx$ is induced at the image-point $y = U(x)$ on $\mathbb{S}^{n-1}$, where $E'(x)$ denotes the derivative of $y = E(x)$ with respect to $x \in \Omega$.

Now, define the *Poincaré index of V on Ω* by the formula

$$\text{Ind}\,[V, \Omega] = \frac{1}{\|\mathbb{S}^{n-1}\|} \int_{\Phi(\Omega)} dy,$$

where $\|\mathbb{S}^{n-1}\|$ denotes the "area" of the $(n-1)$-dimensional unit-sphere $\mathbb{S}^{n-1}$. In other words, Poincaré index of V on Ω is defined by

$$\text{Ind}\,[V, \Omega] = \frac{1}{\|\mathbb{S}^{n-1}\|} \int_{\Omega} \det[E'(x)]dx, \qquad (6.22)$$

It can be seen that the geometrical meaning of the Poincaré index Ind $[V, \Omega]$ is the algebraic number of the (oriented) layers of $\Phi(\Omega)$ covering the sphere $\mathbb{S}^{n-1}$.

Similar to the planar Poincaré indices, the following properties hold for the spatial Poincaré indices:

(1) Let F_1 and F_2 be smooth oriented $(n-1)$-dimensional closed surfaces in $\mathbb{R}^n$. Assume they are homotopic with a smooth homotopy H_λ ($0 \le \lambda \le 1$), such that $F_1 = H_0$ and $F_2 = H_1$. If a smooth vector field V is non-singular on H_λ ($\lambda \in [0, 1]$), then we have

$$\text{Ind}\,[V, F_2] = \text{Ind}\,[V, F_1].$$

(2) Let V_1 and V_2 be smooth vector fields on an oriented $(n-1)$-dimensional closed surface $F \subset \mathbb{R}^n$. Assume there is a homotopy of vector-fields X_λ ($\lambda \in [0, 1]$) between V_1 and V_2. If the vector field X_λ is non-singular on F for $\lambda \in [0, 1]$, then we have

$$\text{Ind}\,[V_1, F] = \text{Ind}\,[V_2, F].$$

(3) Assume $G \subset \mathbb{R}^n$ is a compact region, and its boundary $F = \partial G$ is a smooth $(n-1)$-dimensional oriented closed surface. If V is a continuous vector field in G with $\text{Ind}\,[V, F] \ne 0$, there is at least a singular point of V in the region G.

Similar to the index of a planar singular point, we can define the index of a spatial singular point as follows:

$$\text{Ind}(q, V) = \text{Ind}[V, \Omega],$$

where q is an isolated singular point of a continuous vector field V in the region G, and $\Omega (\subset G)$ is a closed surface surrounding the unique singular point q.

Example 6.1   Consider the vector field

$$V(x) = Ax, \qquad x \in \mathbb{R}^n,$$

where A is a constant $n \times n$-matrix with $\det[A] \neq 0$. It is noted that $y = Ax$ is a non-singular linear transformation and $\mathbf{o}$ is an isolated singular point of V. Let $\mathcal{F}$ be an $(n-1)$-dimensional closed surface surrounding the origin $\mathbf{o}$. Then we have the mapping

$$x \quad \mapsto \quad y = E(x) = \frac{1}{|Ax|} Ax$$

from Ω to $\mathbb{S}^{n-1}$. It follows that

$$\text{Ind}(\mathbf{0}, V) = \text{Ind}[V, \Omega] = \frac{1}{\|\mathbb{S}^{n-1}\|} \int_\Omega \det[E'(x)]dx.$$

It follows from the geometrical meaning of $y = E(x)$ on Ω (or the meaning of the non-degenerate linear transformation $y = Ax$) that

$$\text{Ind}(\mathbf{o}, V) = \begin{cases} +1, & \text{if } \det[A] > 0; \\ -1, & \text{if } \det[A] < 0. \end{cases} \tag{6.23}$$

6.3.2   The Topological Degree

In the application of differential equations, sometimes, it is convenient to use the topological degrees of mappings instead of the Poincaré indices of vector fields. For example, consider a continuous mapping

$$f : \quad G \to \mathbb{R}^n,$$

where G is a compact connected region. Then, associated to the mapping f, we have a continuous vector field

$$U(x) := f(x) - x, \qquad x \in G.$$

It is clear that a fixed point q of the mapping f is equivalent to a singular point q of the vector field U.

Assume the boundary of G is an $(n-1)$-dimensional piecewise smooth oriented surfaces, say $\Omega = \partial G$. If the mapping f has no fixed point on Ω, then the vector field is non-singular on Ω. Hence, the Poincaré index $\text{Ind}\,[U, \Omega]$ is well-defined. Then define

$$\deg_G(f) := \text{Ind}\,[U, \partial G]$$

as the *topological degree* of the mapping f on the region G.

So, it is clear that the method of topological degree is in fact equivalent to the method of Poincaré index.

Theorem 6.4 *If $\deg_G(f) \neq 0$, then f has at least a fixed point in G.*

Proof. In fact, it is a direct consequence of the property of the index $\text{Ind}\,[U, \partial G] \neq 0$ (i.e., the above-mentioned Property (3)). $\square$

6.4 Fixed-Point Theorems of Brouwer Type

The following fixed-point theorems are well-known in the classic topology for their important applications.

(1) Brouwer Fixed-Point Theorem

A bounded closed region D in $\mathbb{R}^n$ is called a *topological n-ball* if it is topologically equivalent to the unit n-ball

$$B_1: \quad x_1^2 + \cdots + x_n^2 \leq 1, \quad (x_1, \cdots, x_n) \in \mathbb{R}^n.$$

In other words, for a topological n-ball D, there is a topological mapping h from a unit n-ball B_1 onto D (i.e., $h(B_1) = D$).

The following result is well-known as the *Brouwer Fixed-Point Theorem*.

Theorem 6.5 *If f is a continuous mapping from a topological n-ball D into itself (i.e., $f(D) \subset D$), then it has at least a fixed-point in D.*

Proof. Since $D \subset \mathbb{R}^n$ is a topological n-ball, there is a topological map

$$h: \quad B_1 \to D,$$

such that $h(B_1) = D$. Then $g = (h^{-1} \circ f \circ h)$ is a continuous mapping from B_1 to itself (i.e., $g(B_1) \subset B_1$).

We first prove that g has at least a fixed point in B_1.

If g has a fixed point on the boundary $\mathcal{S}_1 = \partial B_1$, the desired conclusion holds true.

Now assume g has no fixed point on the boundary $\mathcal{S}_1$. Then the vector field
$$W(z) = g(z) - z$$
is non-singular on the boundary $\mathcal{S}_1$. It follows that
$$V_\alpha(z) = (1-\alpha)W(z) - \alpha z \qquad (0 \leq \alpha \leq 1)$$
is a homotopy between the vector fields $V_0(z) = W(z)$ and $V_1(z) = -z$. It is noted that $V_\alpha(z) \neq 0$ for $z \in B_1$ and $0 \leq \alpha \leq 1$, which together with (6.23) implies
$$\deg_{B_1}(g) = \text{Ind}\,[W, \mathcal{S}_1] = \text{Ind}\,[V_1, \mathcal{S}_1] = \text{Ind}\,(\mathfrak{o}, -z) = (-1)^n \neq 0.$$

Then, using Theorem 6.4, we conclude that the mapping g has at least a fixed point $\xi \in B_1$ (i.e., $g(\xi) = \xi$).

Finally, it follows from $\xi = g(\xi) = h^{-1} \circ f \circ h(\xi)$ that $f(h(\xi)) = h(\xi)$. That is, $\eta = h(\xi)$ is a fixed point of the mapping f in D.

Theorem 6.5 is thus proved. □

It is remarked here that the condition $f(D) \subset D$ in the Brouwer fixed-point theorem means that the region D is contracted under the mapping f. This is the precondition for the application of Brouwer theorem. Despite of this severe condition, there are many examples, where the Brauwer fixed-point theorem is successfully applied to the theory of differential equations (see, for examples, [Coddington and Levinson (1955)], [Ding (1980)] and [Zhang et al (1997)]).

(2) Poincaré-Bohl Fixed-Point Theorem

Assume $D \subset \mathbb{R}^n$ is a bounded closed region. Let q_0 be an interior point of D. If any ray L in $\mathbb{R}^n$ initiating from q_0 intersects the boundary Ω of D at one and only one point, then the region D (and its boundary Ω) is said to be *starlike* (with respect to the point q_0).

Now, assume D is a compact region D starlike with respect to the origin $\mathfrak{o}$ of $\mathbb{R}^n$.

Let L be a ray initiating from $\mathfrak{o}$, and let $p = L \cap \Omega$. Then the closed interval $L[\mathfrak{o}, p]$ of L between the points $\mathfrak{o}$ and p is seated in the region D, and the open interval $L(p, \infty)$ of L from p to the infinity is seated at the outside of D.

The following theorem is called the *Poincaré-Bohl fixed-point theorem*.

Theorem 6.6 *Consider a continuous mapping*

$$f : D \to \mathbb{R}^n,$$

where $D \subset \mathbb{R}^n$ is a closed starlike region, satisfying the condition

$$f(p) \notin L(p, \infty), \quad \forall\, p \in \partial D \tag{6.24}$$

$$(\text{or} \quad f(p) \notin L[\mathfrak{o}, p], \quad \forall\, p \in \partial D\,),$$

then f has at least a fixed-point ξ in D.

Proof. Consider the vector field

$$W(x) = f(x) - x, \quad x \in D.$$

If f has a fixed point in ∂D, then the desired theorem holds. So, it suffices to consider the case when f has no fixed point on the boundary ∂D. In this case, the vector field W is non-degenerate on ∂D.

For definiteness, assume (6.24) holds. Then the vector field W is non-singular on ∂D, and

$$V_\alpha(z) = \alpha W(z) + (1 - \alpha) V_0(z) \quad (z \in \partial D), \quad 0 \leq \alpha \leq 1,$$

is a homotopy between the vector fields $W(z)$ and $V_0(z) = -z$ on ∂D (with $V_\alpha(z) \neq 0$ for $z \in \partial D$ and $\alpha \in [0,1]$). It follows that

$$\deg_D(f) = \text{Ind}\,[W, \partial D] = \text{Ind}\,[V_0, \partial D] = (-1)^{n+1} \neq 0.$$

Then, using Theorem 6.4, we prove Theorem 6.6 under the condition (6.24).

Indeed, we can prove Theorem 6.6 in a similar way under the condition below (6.24). □

Corollary 6.3 *The above Poincaré-Bohl fixed-point theorem holds true under the condition*

$$f(p) \notin L[\mathfrak{o}, p], \quad \forall\, p \in \partial D. \tag{6.25}$$

Roughly speaking, the condition (6.25) means that the mapping f does not make a complete turn (i.e., a $2k\pi$-increment) for the angular of the point on the boundary ∂D. It is this property that makes the Poincaré-Bohl theorem very useful to proving the existence of periodic solutions of ordinary differential equations (see the Chapters in the sequel).

(3) Bernstein Fixed-Point Theorem

The following Bernstein fixed-point theorem needs the "hyperbolic" boundary condition (i.e., the boundary is compressed in some directions and extended in the other directions under the action of the mapping).

Let
$$Q: \quad |x_1| \leq a_1, \cdots, |x_n| \leq a_n,$$
be a rectangular region. Then the boundary of Q has the branches
$$\begin{cases} L_i = \{(x_1, \cdots, x_n) \in Q: \quad x_i = -a_i \}, \\ R_i = \{(x_1, \cdots, x_n) \in Q: \quad x_i = +a_i \}, \end{cases} \quad i = 1, \cdots, n.$$

Theorem 6.7 *If a continuous mapping*
$$f: \quad Q \to \mathbb{R}^n$$
satisfies the "hyperbolic" boundary condition:
$$(f_i(x) - x_i)|_{L_i} \cdot (f_i(x) - x_i)|_{R_i} < 0, \quad i = 1, \cdots, n, \qquad (6.26)$$
then f has at least a fixed-point in Q.

Proof. Let us consider the continuous vector field $V(x) = f(x) - x$ on Q. It follows from the condition (6.26) that the i-th component $V_i(x) = f_i(x) - x_i$ assumes opposite signs on the opposite branches L_i and R_i of the boundary ∂Q. Now, consider an auxiliary vector field $W(x) = Ax$, where A is a diagonal matrix $\{\sigma_1, \cdots, \sigma_n\}$ with
$$\sigma_i = \text{sign}((f_i(x) - x_i)|_{R_i}) \neq 0, \quad (i = 1, \cdots, n).$$
It follows from (6.26) that the vector fields V and W are non-singular on ∂Q and satisfy
$$U_\lambda(x) = \lambda V(x) + (1-\lambda)W(x) \neq 0 \quad (x \in \partial Q), \quad \text{for } \lambda \in [0,1].$$
Therefore, U_λ is a homotopy between the vector fields V and W on ∂Q, and thus we have
$$\deg_Q(f) = \text{Ind}[V, \partial Q] = \text{Ind}[W, \partial Q] = \text{sgn}(\sigma_1 \cdots \sigma_n) \neq 0,$$
which together with Theorem 6.4 completes the proof of the theorem. □

Chapter 7

Bend-Twist Theorem

7.1 Generalized Poincaré-Birkhoff Twist Theorem

In this chapter, we are concerned with the existence theorems of at least two fixed points for the map in annulus. This type of fixed-point theorem was originally formulated by Poincaré as a conjecture with proof in special cases [103], shortly before his death in 1912. In literature, this conjecture is called *the last geometric theorem of Poincaré*. It can be sketched in the following form:

"If f is an orientation and area preserving topological transformation from an annulus $\mathcal{A}$ onto itself such that the inner boundary and the outer boundary of $\mathcal{A}$ are rotated in opposite directions, then f has at least two fixed points in the annulus $\mathcal{A}$."

Birkhoff first gave a proof for the Poincaré last geometrical theorem in [9] and then generalized it [10] to the result called the *the Poincaré-Birkhoff twist theorem* in literature. However, the Birkhoff's proofs seem to be not widely accepted by mathematicians. Indeed, even his proof for the Poincaré last geometric theorem was reverified by the late mathematicians (see the papers [16] and [98]).

Nonetheless, the *generalized Poincaré-Birkhoff twist theorem* has been developed by other mathematicians along the brilliant idea of Poincaré-Birkhoff (see, for examples, [63], [64] and [65]). A short history of development is reported in the references [34] and [105].

Now, let us begin to introduce the generalized Poincaré-Birkhoff twist theorem.

Let z be the point in the plane $\mathbb{R}^2$. Denote by $z = (x, y)$ the Cartesian coordinates and by $\langle r, \theta \rangle$ the polar coordinates. So, we have

$$(x, y) = (r\cos\theta,\ r\sin\theta) = \langle r,\ \theta \rangle,$$

where

$$r = \sqrt{x^2 + y^2} \geq 0 \quad \text{and} \quad \theta = \arctan \frac{y}{x}.$$

A Jordan closed curve J in $\mathbb{R}^2$ is called *starlike* (with respect to the origin $\mathfrak{q}$) if J intersects each ray issuing from $\mathfrak{q}$ at one and only one point. If $\mathfrak{q}$ is the origin of the plane, the starlike closed curve J can be represented by a continuous function $r = R(\theta)$ in polar coordinates; that is,

$$J: \quad r = R(\theta) > 0, \quad \theta \in \mathbb{R}^1,$$

where $R(\theta)$ is a periodic function with period 2π.

Let

$$J_1: \quad r = R_1(\theta) > 0, \quad \theta \in \mathbb{R}^1$$

and

$$J_2: \quad r = R_2(\theta) > 0, \quad \theta \in \mathbb{R}^1$$

be two starlike closed curves, satisfying the condition

$$R_1(\theta) < R_2(\theta), \quad \theta \in \mathbb{R}^1.$$

Then the region

$$\mathcal{A} = \{\langle r, \theta \rangle \in \mathbb{R}^2 : \quad R_1(\theta) \leq r \leq R_2(\theta), \ \theta \in \mathbb{R}^1\}$$

is called a *starlike annulus* (with respect to $\mathfrak{o}$). Note that J_1 and J_2 are the *inner boundary* and the *outer boundary* of $\mathcal{A}$, respectively.

Assume $\mathcal{A}$ is a starlike annulus contained in another starlike annulus $\mathcal{A}^*$ (i.e., $\mathcal{A} \subset \mathcal{A}^*$). Consider a continuous map

$$f: \quad \mathcal{A} \to \mathcal{A}^*; \quad \langle r, \theta \rangle \mapsto \langle \rho, \phi \rangle,$$

defined by

$$\rho = \rho(r, \theta) > 0, \quad \phi = \theta + \Phi(r, \theta),$$

where $\rho(r, \theta)$ is a continuous function in $\mathcal{A}$, and $\Phi(r, \theta)$ is a multi-valued function in $\mathcal{A}$ determined up to the multiples of 2π, such that each branch $\Phi(r, \theta)$ is locally continuous.

Definition 7.1 If a continuous branch $\Phi(r,\theta)$ satisfies the condition

$$\begin{cases} \Phi(r,\theta) < 2j\pi, & \text{for } \langle r,\theta \rangle \in J_1; \\ \Phi(r,\theta) > 2j\pi, & \text{for } \langle r,\theta \rangle \in J_2, \end{cases} \quad (7.1)$$

for some integer j, then f is called a *twist map*. Correspondingly, (7.1) is called a *twist condition*.

Definition 7.2 A map f is called *area-preserving* in $\mathcal{A}$ if it satisfies the condition

$$\mu[f(\mathfrak{D})] = \mu[\mathfrak{D}],$$

for any open set $\mathfrak{D}$ in $\mathcal{A}$, where $\mu[\cdot]$ is the Lebesgue planar measure.

If f is an area-preserving map in $\mathcal{A}$ and satisfies the boundary condition

$$\int_{\langle \rho,\phi \rangle \in f(J_1)} \rho^2 \, d\phi = \int_{\langle r,\theta \rangle \in J_1} r^2 \, d\theta \quad (7.2)$$

on the inner boundary J_1, then f is called a *strictly area-preserving* map in $\mathcal{A}$.

The exact symplectic map defined in the paper [65] is indeed a strictly area-preserving diffeomorphism from $\mathcal{A}$ to $\mathcal{A}^*$.

On the other hand, it can be proved by the method of [64] that a strictly area-preserving C^1-map f on $\mathcal{A}$ can be extended to $\mathbb{R}^2$ as an area-preserving C^1-map having a fixed point $\mathfrak{p}$ surrounded by J_1. Conversely, let

$$f: \quad \mathbb{R}^2 \to \mathbb{R}^2,$$

be a continuous area-preserving map having a fixed point $\mathfrak{p}$. Then the map f is strictly area-preserving on each annulus $\mathcal{A}$, which is starlike with respect to the point $\mathfrak{p}$.

We are ready now to introduce the generalized Poincaré-Birkhoff twist theorem as follows (see [63], [64] and [65]).

Theorem 7.1 *If f is a strictly area-preserving twist diffeomorphism from $\mathcal{A}$ to $\mathcal{A}^*$, then f has at least two distinct fixed points in $\mathcal{A}$.*

Example 7.1 It is obvious that

$$\mathcal{A}_0 = \{\langle r,\theta \rangle \in \mathbb{R}^2 : \quad 0 < 1 \leq r \leq 4, \ \theta \in \mathbb{S}^1 \}$$

is a starlike annulus contained in the starlike annulus

$$\mathcal{A}_0^* = \{ \langle r,\theta \rangle : \quad 2^{-1} \leq r \leq 6, \ \theta \in \mathbb{S}^1 \}.$$

Consider the map
$$g: \quad \langle r, \theta \rangle \mapsto \langle \rho, \phi \rangle,$$
defined by
$$\rho = r + p(r), \quad \phi = \theta + 2\pi r, \quad \langle r, \theta \rangle \in \mathcal{A},$$
where
$$p(r) = \frac{\varepsilon}{r + \sqrt{r^2 + \varepsilon}}, \quad \varepsilon > 0 \text{ is a parameter in } (0, 1).$$
Then g is an analytic map from $\mathcal{A}_0$ to $\mathcal{A}_0^*$.

It follows from $\Phi(r, \theta) = 2\pi r$ that g satisfies the twist condition (7.1) with $j = 2$. Hence g is a twist map.

Moreover, since
$$\rho d\rho d\phi = \rho \begin{vmatrix} \dfrac{\partial \rho}{\partial r} & \dfrac{\partial \rho}{\partial \theta} \\ \dfrac{\partial \phi}{\partial r} & \dfrac{\partial \phi}{\partial \theta} \end{vmatrix} dr d\theta = r dr d\theta, \quad \text{for } \langle r, \theta \rangle \in \mathcal{A}_0,$$

the map g is area-preserving in $\mathcal{A}_0$. However, g has no fixed point in $\mathcal{A}_0$ because of $\rho > r$.

It is noted that the inner boundary circle J_1 ($r = 1$) of $\mathcal{A}_0$ bounds a disk of radius 1, and its image $g(J_1)$ bounds a disk of radius > 1. It follows that the map g does not satisfy the boundary condition (7.2) on J_1. In other words, the map g is not strictly area-preserving in $\mathcal{A}_0$.

The above example expresses the fact that an area-preserving analytic twist map from $\mathcal{A}$ to $\mathcal{A}^*$ may have no fixed point. It follows that the strictly area-preserving condition is an essential assumption in Theorem 7.1 (i.e., the generalized Poincaré-Birkhoff twist theorem). Obviously, the strictly area-preserving assumption is a severe restriction to the application of the twist theorem. We are interested to find some new flexile condition instead.

7.2 Analytic Bend-Twist Theorem

Assume f is a twist map satisfying the twist condition (7.1). Let
$$\Omega_f = \{\langle r, \theta \rangle \in \mathcal{A}: \quad \Phi(r, \theta) = 2j\pi \},$$
where j is a given integer given in the twist condition (7.1).

Lemma 7.1 *If f is a twist map satisfying the twist condition (7.1), then Ω_f is a closed nonempty subset contained in the interior of $\mathcal{A}$.*

Proof. It is obvious that Ω_f is a closed subset in $\mathcal{A}$ and the condition (7.1) implies that Ω_f belongs to the interior of $\mathcal{A}$.

Therefore, we only need to prove that Ω_f is non-empty. In fact, let

$$r_1 = R_1(0) \quad \text{and} \quad r_2 = R_2(0).$$

Then we have a closed interval $[r_1, r_2]$ on the ray $L_0 : \theta = 0$, such that $\Phi(r, 0)$ is continuous on the interval $[r_1, r_2]$. Using the twist condition (7.1) implies that $\Phi(r_1, 0) < 2j\pi$ and $\Phi(r_2, 0) > 2j\pi$. It follows from the continuity of $\Phi(r, 0)$ that there is at least an interior point r^* of $[r_1, r_2]$, satisfying $\Phi(r^*, 0) = 2j\pi$. It means that $\langle r^*, 0 \rangle$ is a point of Ω_f. Hence, Ω_f is non-empty.

The proof of Lemma 7.1 is thus complete. □

Generally speaking, the topological structure of the set Ω_f may be very pathological even if the twist map f is C^∞-differentiable. However, it is regular if f is analytic.

Lemma 7.2 *If the twist map f is analytic in $\mathcal{A}$, then Ω_f consists of finitely many analytic arcs and it contains at least a non-contractible Jordan closed curve Γ.*

Proof. Since f is analytic in $\mathcal{A}$, each branch $\Phi(r, \theta)$ is analytic in $\mathcal{A}$. It follows from the Weierstrass's preparation lemma (see [86]) that Ω_f consists of finitely many analytic arcs.

If Ω_f contains a Jordan closed curve Γ which is non-contractible in $\mathcal{A}$, then we have

$$I \cap \Gamma \neq \emptyset,$$

where I is any continuous curve joining S_1 and S_2 in $\mathcal{A}$. It follows that

$$I \cap \Omega_f \neq \emptyset. \tag{7.3}$$

On the other hand, if there is some continuous curve I joining S_1 and S_2, such that (7.3) does not hold (i.e., if

$$I \cap \Omega_f = \emptyset,$$

for some continuous curve I joining S_1 and S_2 in $\mathcal{A}$), then each component of Ω_f is contractible in $\mathcal{A}$.

Hence, to prove the existence of a non-contractible closed curve in Ω_f, it suffices to prove that the inequality (7.3) holds for any continuous curve I joining S_1 and S_2 in $\mathcal{A}$.

In fact, assume the contrary. Then there is at least a continuous curve K connecting S_1 and S_2 in $\mathcal{A}$, such that

$$K \cap \Omega_f = \emptyset. \tag{7.4}$$

Since the continuous curve K connects S_1 and S_2, the twist condition (7.1) implies that there are points

$$\langle r_1, \theta_1 \rangle \in K \cap S_1 \quad \text{and} \quad \langle r_2, \theta_2 \rangle \in K \cap S_2$$

satisfying $\Phi(r_1, \theta_1) < 2j\pi$ and $\Phi(r_2, \theta_2) > 2j\pi$. It follows that

$$K^- = \{\langle r, \theta \rangle \in K : \quad \Phi(r_1, \theta_1) < 2j\pi\}$$

and

$$K^+ = \{\langle r, \theta \rangle \in K : \quad \Phi(r_1, \theta_1) > 2j\pi\}$$

are two disjoint non-empty open sub-arcs of K. Using (7.4) yields that

$$K = K^- \cup K^+.$$

It means that K is a union of two disjoint nonempty open sub-arcs. This conclusion contradicts, however, the continuity of K. We have thus proved that Ω_f contains at least a non-contractible closed curve Δ in $\mathcal{A}$. Since Δ is a non-contractible closed curve consisting of finitely many analytic arcs in $\mathcal{A}$, it is easy to select a non-contractible Jordan closed curve Γ as a branch of Δ.

The proof of Lemma 7.2 is thus completed. $\square$

Corresponding to the above-mentioned twist map f, define the function

$$\Upsilon(r, \theta) = \rho(r, \theta) - r, \quad \text{for } \langle r, \theta \rangle \in \mathcal{A}.$$

Lemma 7.3 *If f is an analytic twist map and $\Upsilon(r, \theta)$ has a root*[1] *$\langle r_0, \theta_0 \rangle$ in Ω_f, then $\langle r_0, \theta_0 \rangle$ is a fixed point of f.*

Proof. Since $z_0 = \langle r_0, \theta_0 \rangle$ is a root of $\Upsilon(r, \theta)$ in Ω_f, we have

$$\Upsilon(r_0, \theta_0) = 0 \quad \text{and} \quad \Phi(r_0, \theta_0) = 2j\pi.$$

---
[1] A point $\langle r_0, \theta_0 \rangle$ is called a root of $\Upsilon(r, \theta)$ if $\Upsilon(r_0, \theta_0) = 0$.

It follows from
$$f(z_0) = \langle r_0 + \Upsilon(r_0, \theta_0), \theta_0 + \Phi(r_0, \theta_0) \rangle$$
that $f(z_0) = z_0$. Hence, $z_0 = \langle r_0, \theta_0 \rangle$ is a fixed point of f. □

Assume f is an analytic twist map in $\mathcal{A}$. Let Γ be a non-contractible Jordan closed curve in Ω_f defined above. Then the following Corollaries 7.1 and 7.2 can be easily derived from Lemma 7.3.

Corollary 7.1 *If there are n disjoint continuous curves Λ_i ($i = 1, \cdots, n$) connecting the inner boundary J_1 and the outer boundary J_2, respectively, such that*
$$\Upsilon(r, \theta) = 0, \qquad \text{for } \langle r, \theta \rangle \in \Lambda_i \ (i = 1, \cdots, n),$$
then f has at least n distinct fixed points in Γ. □

Corollary 7.2 *If $\Upsilon(r, \theta)$ is identically equal to zero on Γ, then Γ is an invariant closed curve of f.* □

Example 7.2 Let m be a positive integer (≥ 1). It is obvious that
$$h: \quad \rho = r + \sin m\theta, \quad \phi = \theta + 2\pi r,$$
is an analytic map defined on the annulus
$$\mathcal{A}_0: \quad 2 \leq r \leq 4, \quad \theta \in \mathbb{S}^1.$$
It follows from
$$\Phi(r, \theta) = (\theta + 2\pi r) - 6\pi \quad \text{and} \quad \Phi(2, \theta) \cdot \Phi(4, \theta) < 0$$
that the map h is twist on $\mathcal{A}_0$. On the other hand, since
$$\rho d\rho d\phi = (r + \sin m\theta)(1 - 2m\pi \cos \theta) dr d\theta \neq r dr d\theta,$$
the map h is not area-preserving.

On the other hand, consider the line-segments
$$\Lambda_j := \left\{ \langle r, \theta \rangle \in \mathcal{A}_2 \ \Big| \ \theta = \frac{j}{m}\pi, \ 2 \leq r \leq 4 \right\} \qquad (-m \leq j < m),$$
which are disjoint from each other and connect the inner boundary circle and the outer boundary circle of the annulus $\mathcal{A}_0$ respectively. It can be seen that
$$\Upsilon(r, \theta) = \sin m\theta = 0, \qquad \forall \langle r, \theta \rangle \in \Lambda_j \qquad (-m \leq j < m).$$

Therefore, it follows from Corollary 7.1 that h has at least $2m$ fixed points in the annulus $\mathcal{A}_0$. The remarkable feature of this example is that the map h is not area-preserving and the number $2m$ of fixed points can be chosen as large as you please.

Definition 7.3 If f is an analytic twist map, such that $\Upsilon(r,\theta)$ changes signs on the non-contractible Jordan closed curve Γ in Ω_f, then f is called a *bend-twist map*.

It can be seen later that the bend-twist condition is indeed at the heart of fixed-point theorem for the twist map.

Now, we are in position to prove the *bend-twist theorem* as follows.

Theorem 7.2 *If f is a bend-twist analytic map in $\mathcal{A}$, then it has at least two distinct fixed points in $\mathcal{A}$.*

Proof. Since f is a bend-twist analytic map in $\mathcal{A}$, $\Upsilon(r,\theta)$ changes signs on some non-contractible Jordan closed curve Γ in Ω_f. Hence, there are at least two points $\langle r_1, \theta_1 \rangle$ and $\langle r_2, \theta_2 \rangle$ in Γ, such that

$$\Upsilon(r_1,\theta_1) < 0 \quad \text{and} \quad \Upsilon(r_2,\theta_2) > 0. \tag{7.5}$$

Now, suppose that we are moving clockwise on Γ starting from the point $\langle r_1, \theta_1 \rangle \in \Gamma$. Then we have to meet a point $\langle r_1^*, \theta_1^* \rangle \in \Gamma$ first, such that

$$\Upsilon(r_1^*, \theta_1^*) = 0,$$

before arriving at the point $\langle r_2, \theta_2 \rangle \in \Gamma$. Similarly, if we continue to move clockwise on Γ starting from the point $\langle r_2, \theta_2 \rangle \in \Gamma$, we have to meet another point $\langle r_2^*, \theta_2^* \rangle \in \Gamma$, such that

$$\Upsilon(r_2^*, \theta_2^*) = 0,$$

before coming back to the initial point $\langle r_1, \theta_1 \rangle \in \Gamma$. Hence, we obtain two distinct fixed points $\langle r_1^*, \theta_1^* \rangle$ and $\langle r_2^*, \theta_2^* \rangle$ of f.

Theorem 7.2 is thus proved. $\square$

The following corollary is suggestive to apply Theorem 7.2 (see the final Example of this chapter).

Corollary 7.3 *Let f be an analytic twist map in $\mathcal{A}$. If there are two disjoint continuous curves Π_1 and Π_2 in $\mathcal{A}$ connecting respectively the inner* boundary J_1 and the outer boundary J_2, such that

$$\begin{cases} \Upsilon(r,\theta) < 0, & \langle r,\theta \rangle \in \Pi_1; \\ \Upsilon(r,\theta) > 0, & \langle r,\theta \rangle \in \Pi_2, \end{cases}$$

then f is a bend-twist map in $\mathcal{A}$.

7.3 Analytic Poincaré-Birkhoff Twist Theorem

It follows from the following theorem that when the map is analytic, the generalized Poincaré-Birhkoff twist theorem can be derived from the bend-twist theorem (i.e., Theorem 7.2).

Theorem 7.3 *If f is an analytic twist map satisfying the strictly area-preserving condition in $\mathcal{A}$, then f has at least two fixed points in $\mathcal{A}$.*

Proof. If $\Upsilon(r,\theta) \equiv 0$ for $\langle r,\theta \rangle \in \Gamma$, then each point of Γ is a fixed point of f. Theorem 7.3 is naturally true in this case.

Then, consider the case

$$\Upsilon(r,\theta) \not\equiv 0, \qquad \langle r,\theta \rangle \in \Gamma.$$

We claim that $\Upsilon(r,\theta)$ changes signs on $\langle r,\theta \rangle \in \Gamma$.

Assume the contrary. It means that $\Upsilon(r,\theta)$ does not change signs on Γ. Without loss of generalization, let

$$\Upsilon(r,\theta) \geq 0, \quad \text{with} \quad \Upsilon(r,\theta) \not\equiv 0, \quad \langle r,\theta \rangle \in \Gamma. \tag{7.6}$$

On the other hand, since Γ is a piecewise analytic Jordan closed curve in $\mathcal{A}$, it is rectifiable. Moreover, it can be represented by the formula

$$r = \tilde{r}(\theta), \qquad \theta \in \mathbb{R}^1, \tag{7.7}$$

where $\tilde{r}(\theta) > 0$ is a multi-valued and locally continuous periodic function of minimal period 2π.

Note that Γ is a non-contractible Jordan closed curve in $\mathcal{A}$ surrounding the inner boundary J_1. Since f is strictly area-preserving in $\mathcal{A}$, then the boundary condition (7.2) implies the condition

$$\frac{1}{2} \int_{\langle \rho,\phi \rangle \in f(\Gamma)} \rho^2 \, d\phi = \frac{1}{2} \int_{\langle r,\theta \rangle \in \Gamma} r^2 \, d\theta, \tag{7.8}$$

where
$$\phi = \theta + \Phi(r,\theta), \qquad \Phi(r,\theta) \equiv 2j\pi, \qquad \text{for } \langle r,\theta\rangle \in \Gamma.$$

Take points $\langle r_i, \theta_i\rangle$ and $\langle r_{i+1}, \theta_{i+1}\rangle$ in Γ and let
$$\langle \rho_i, \phi_i\rangle = f(\langle r_i, \theta_i\rangle) \qquad \text{and} \qquad \langle \rho_{i+1}, \phi_{i+1}\rangle = f(\langle r_{i+1}, \theta_{i+1}\rangle).$$

Then define
$$\Delta\theta_i = \theta_{i+1} - \theta_i \qquad \text{and} \qquad \Delta\phi_i = \phi_{i+1} - \phi_i.$$

It follows from $\phi_{i+1} = \theta_{i+1} + 2j\pi$ and $\phi_i = \theta_i + 2j\pi$ that $\Delta\phi_i = \Delta\theta_i$ on Γ. Hence, the differential operation $d\phi = d\theta$ holds on Γ. Hence we have
$$\frac{1}{2}\int_{\langle\rho,\phi\rangle \in f(\Gamma)} \rho^2 \, d\phi = \frac{1}{2}\int_{\langle r,\theta\rangle \in \Gamma} \rho^2(r,\theta)\, d\theta,$$

which together with (7.8) yields
$$\frac{1}{2}\int_{\langle r,\theta\rangle \in \Gamma} \rho^2(r,\theta)\, d\theta = \frac{1}{2}\int_{\langle r,\theta\rangle \in \Gamma} r^2 \, d\theta;$$

that is,
$$\frac{1}{2}\int_0^{2\pi} \rho^2(r,\theta)\, d\theta = \frac{1}{2}\int_0^{2\pi} r^2 \, d\theta,$$

where the variable r in the integrands is given by (7.7) (i.e., $r = \tilde{r}(\theta)$). It follows that
$$\int_0^{2\pi} h(r,\theta)\, d\theta = 0, \tag{7.9}$$

where the integrand $h(r,\theta)$ is continuous and defined by
$$h(r,\theta) = \frac{1}{2}(\rho^2(r,\theta) - r^2) = \frac{\rho(r,\theta) + r}{2}\Upsilon(r,\theta) \geq 0, \qquad \text{with } r = \tilde{r}(\theta).$$

However, the equality (7.9) is evidently in conflict with the property (7.6). This contradiction proves the desired claim. It follows that the analytic map f is bend-twist in $\mathcal{A}$.

Therefore, Theorem 7.3 is proved by using Theorem 7.2. $\square$

7.4 Application of the Bend-Twist Theorem

It can be seen from Theorem 7.3 that for the analytic maps the bend-twist theorem can replace the generalized Poincaré-Birkhoff twist theorem. On the other hand, there are examples showing that the bend-twist theorem is still applicable to certain cases where the generalized Poincaré-Birkhoff twist theorem does not work. Therefore, the bend-twist theorem can not be replaced by the generalized Poincaré-Birkhoff twist theorem.

Example 7.3 Consider the nonlinear Duffing differential equation

$$\frac{d^2x}{dt^2} + \varepsilon c\frac{dx}{dt} + (ax + bx^3) = Ep(t), \qquad (7.10)$$

which is equivalent to the system

$$\begin{cases} \dot{x} = y, \\ \dot{y} = -(ax + bx^3) - \varepsilon cy + Ep(t), \end{cases} \qquad (7.11)$$

where $a > 0$, $b > 0$, $c >$ are constants, $p(t)$ is a periodic function with minimal period $T > 0$, and E and ε are small parameters. It is noted that (7.11) is a dissipative Duffing equation when $\varepsilon c > 0$.

When $\varepsilon = 0$ and $E = 0$, (7.11) becomes an autonomous system

$$\dot{x} = y, \qquad \dot{y} = -(ax + bx^3), \qquad (7.12)$$

which has the first integral

$$J_\lambda: \qquad y^2 + \left(ax^2 + \frac{b}{4}x^4\right) = \lambda,$$

where λ is an integral constant. It is obvious that J_λ is a closed orbit if $\lambda > 0$, and J_0 is degenerated to the singular point $\mathbf{o}$. Let $\tau(\lambda)$ be the minimal period of the closed orbit J_λ ($\lambda > 0$). It is well-known in the theory of differential equations that $\tau(\lambda)$ is analytic and monotonously decreasing in $\lambda \in (0, \infty)$, satisfying

$$\lim_{\lambda \to 0^+} \tau(\lambda) = \frac{2\pi}{\sqrt{a}} \qquad \text{and} \qquad \lim_{\lambda \to \infty} \tau(\lambda) = 0.$$

It follows that for a given integer $k > 0$, there exist a large prime integer $p > 0$ and a constant $\lambda_0 > 0$, such that

$$p\tau(\lambda_0) = kT, \qquad (7.13)$$

with
$$\tau(\lambda_0) < \frac{T}{2}. \tag{7.14}$$

In what follows, assume $\mathcal{A}_0$ is the annulus bounded by the starlike closed curves $S_1 = J_{\lambda_0-\sigma}$ and $S_2 = J_{\lambda_0+\sigma}$, where $\sigma > 0$ is a sufficiently small constant to be determined later.

Assume $\mathcal{P} = \mathcal{P}(\varepsilon, E)$ is the Poincaré map of (7.11) on $\mathcal{A}_0$ for small parameters ε and E.

Denote by
$$z = \langle \rho(t, r_0, \theta_0, \varepsilon, E), \phi(t, r_0, \theta_0, \varepsilon, E) \rangle \tag{7.15}$$

the solution of (7.11) in polar coordinates passing through the initial point $\langle r_0, \theta_0 \rangle$. It follows that $f = \mathcal{P}^k$ is an analytic map defined by
$$f: \quad \langle r, \theta \rangle \mapsto \langle \rho, \phi \rangle,$$
with
$$\begin{cases} \rho = \rho(kT, r, \theta, \varepsilon, E) = r + \Upsilon(r, \theta, \varepsilon, E), \\ \phi = \phi(kT, r, \theta, \varepsilon, E) = \theta + \Phi(r, \theta, \varepsilon, E), \end{cases}$$

where Υ and Φ are analytic functions in $(r, \theta, \varepsilon, E)$ for the initial point $\langle r, \theta \rangle \in \mathcal{A}_0$ and the small parameters ε and E. It can be seen from the autonomous system (7.12) (unperturbed) that f is a twist map on $\mathcal{A}_0$ when $\varepsilon = 0$ and $E = 0$. Hence, f is still a twist map on $\mathcal{A}_0$ for $(\varepsilon, E) \neq (0, 0)$ whenever ε and E are small enough.

Now, consider the case where $\varepsilon = 0$ and E is small enough. Then we have the corresponding Duffing system
$$\dot{x} = y, \quad \dot{y} = -(ax + bx^3) + Ep(t). \tag{7.16}$$

Let
$$\begin{cases} x = x(t, \xi, \eta, E) = x_0(t) + x_1(t)E + x_2(t)E^2 + \cdots, \\ y = y(t, \xi, \eta, E) = y_0(t) + y_1(t)E + y_2(t)E^2 + \cdots, \end{cases} \tag{7.17}$$

be the solution of (7.16) satisfying the initial condition
$$(x(0, \xi, \eta, E), y(0, \xi, \eta, E)) = (\xi, \eta) \in J_{\lambda_0},$$

Chapter 7. Bend-Twist Theorem

such that the coefficients $(x_j(t), y_j(t))$ $(j = 0, 1, 2, \cdots)$ are independent of E, satisfying

$$(x_0(0), y_0(0)) = (\xi, \eta), \qquad (x_j(0), y_j(0)) = (0, 0) \qquad (j \geq 1).$$

Since E is a small parameter, the solution (7.17) in the finite interval $[0, kT]$ can be determined by the perturbation method.

It can be shown that $(x_0(t), y_0(t))$ is kT-periodic and satisfies

$$\dot{x}_0(t) = y_0(t), \qquad \dot{y}_0(t) = -(ax_0(t) + bx_0^3(t)), \tag{7.18}$$

and $(x_1(t), y_1(t))$ satisfies

$$\dot{x}_1(t) = y_1(t), \qquad \dot{y}_1(t) = -[a + 3bx_0^2(t)]x_1(t) + p(t). \tag{7.19}$$

Let $U(t)$ be the matrix solution of the linear homogeneous system

$$\dot{x} = y, \qquad \dot{y} = -[a + 3bx_0^2(t)]x,$$

satisfying the initial condition $U(0) = I_0$, where I_0 is the 2×2-unit-matrix. If b is sufficiently small, we have the approximation

$$U(t) \simeq \begin{pmatrix} \cos\sqrt{a}t & \dfrac{1}{\sqrt{a}}\sin\sqrt{a}t \\ -\sqrt{a}\sin\sqrt{a}t & \cos\sqrt{a}t \end{pmatrix},$$

which yields

$$U^{-1}(s) \simeq \begin{pmatrix} \cos\sqrt{a}s & -\sqrt{a}\sin\sqrt{a}s \\ \dfrac{1}{\sqrt{a}}\sin\sqrt{a}s & \cos\sqrt{a}s \end{pmatrix}.$$

It follows from

$$\begin{pmatrix} x_1(t) \\ y_1(t) \end{pmatrix} = \int_0^t U(t)U^{-1}(s) \begin{pmatrix} 0 \\ p(s) \end{pmatrix} ds$$

that

$$\begin{cases} x_1(t) \simeq \dfrac{1}{\sqrt{a}} \int_0^t \sin\sqrt{a}(t-s)p(s)\, ds, \\ y_1(t) \simeq \int_0^t \cos\sqrt{a}(t-s)p(s)\, ds, \end{cases}$$

which implies

$$\begin{cases} x_1(kT) \simeq \dfrac{1}{\sqrt{a}} \displaystyle\int_0^{kT} \sin\sqrt{a}(kT-s)p(s)\,ds, \\ y_1(kT) \simeq \displaystyle\int_0^{kT} \cos\sqrt{a}(kT-s)p(s)\,ds. \end{cases}$$

Assume the condition

$$\left(\frac{1}{\sqrt{a}} \int_0^{kT} \sin\sqrt{a}(kT-s)p(s)\,ds,\ \int_0^{kT} \cos\sqrt{a}(kT-s)p(s)\,ds \right) \neq (0,0).^2$$

It follows that

$$(x_1(kT), y_1(kT)) \neq (0, 0), \tag{7.20}$$

whenever b is sufficiently small.

Consider the function

$$\Upsilon(\xi, \eta, 0, E) = \sqrt{x^2(kT, \xi, \eta, 0, E) + y^2(kT, \xi, \eta, 0, E)} - \sqrt{\xi^2 + \eta^2},$$

which satisfies the condition

$$\Upsilon(\xi, \eta, 0, 0) = 0, \quad \text{for } (\xi, \eta) \in J_{\lambda_0}. \tag{7.21}$$

On the other hand, we have

$$\frac{\partial \Upsilon}{\partial E}(\xi, \eta, 0, E) = \frac{x(kT)\dfrac{\partial x}{\partial E}(kT) + y(kT)\dfrac{\partial y}{\partial E}(kT)}{\sqrt{x^2(kT, \xi, \eta, E) + y^2(kT, \xi, \eta, E)}}.$$

$$= \frac{x_0(kT)x_1(kT) + y_0(kT)y_1(kT) + \mathcal{O}(E)}{\sqrt{\xi^2 + \eta^2} + \mathcal{O}(E)}.$$

$$= \frac{\xi x_1(kT) + \eta y_1(kT)}{\sqrt{\xi^2 + \eta^2}} + \mathcal{O}(E).$$

It follows from (7.20) that there are at least two points (ξ_1, η_1) and (ξ_2, η_2) in J_{λ_0}, such that

$$\frac{\partial \Upsilon}{\partial E}(\xi_1, \eta_1, 0, 0) < 0 \quad \text{and} \quad \frac{\partial \Upsilon}{\partial E}(\xi_2, \eta_2, 0, 0) > 0. \tag{7.22}$$

---
[2] It can be easily shown that such a function $p(t)$ is really existent.

It can be seen from (7.21) that

$$\Upsilon(\xi_1, \eta_1, 0, 0) = 0 \quad \text{and} \quad \Upsilon(\xi_2, \eta_2, 0, 0) = 0. \tag{7.23}$$

Then, using (7.23) and (7.22), we obtain a small constant $E_0 > 0$, such that

$$\Upsilon(\xi_1, \eta_1, 0, E_0) < 0 \quad \text{and} \quad \Upsilon(\xi_2, \eta_2, 0, E_0) > 0. \tag{7.24}$$

Since $\Upsilon(x, y, , \varepsilon, E)$ is continuous in $(x, y, , \varepsilon, E)$, then (7.24) implies that there exist a small constant $\varepsilon_0 > 0$ and two small open disks $\mathcal{N}_i$ centered at the points (ξ_i, η_i) $(i = 1, 2)$, respectively, such that

$$\begin{cases} \Upsilon(x_1, y_1, \varepsilon, E_0) < 0, & \text{for } (x_1, y_1) \in \mathcal{N}_1, \ 0 < \varepsilon < \varepsilon_0, \\ \Upsilon(x_2, y_2, \varepsilon, E_0) > 0, & \text{for } (x_2, y_2) \in \mathcal{N}_1, \ 0 < \varepsilon < \varepsilon_0. \end{cases} \tag{7.25}$$

It is clear that for each $i = 1, 2$, the open disk $\mathcal{N}_i$ intersects the inner boundary S_1 and the outer boundary S_2 of $\mathcal{A}_0$ whenever $\sigma > 0$ is small enough.

It follows from (7.25) and Corollary 7.3 that the analytic map f is bend-twist in $\mathcal{A}_0$ if $0 < \varepsilon < \varepsilon_0$ and $E = E_0$. Then, Theorem 7.2 implies that if $0 < \varepsilon < \varepsilon_0$ and $E = E_0$, f has at least two distinct fixed points $\zeta_i = \zeta_i(\varepsilon, E_0)$ $(i = 1, 2)$ in $\mathcal{A}_0$.

Then, if $0 < \varepsilon < \varepsilon_0$ and $E = E_0$, the Duffing equation (7.11) has two kT-periodic solutions

$$z = \langle \rho(t, \zeta_i, \varepsilon, E), \phi(t, \zeta_i, \varepsilon, E) \rangle, \quad i = 1, 2. \tag{7.26}$$

Now, we claim that the kT-periodic solutions (7.26) are subharmonic of order k. In other words, for each i, the minimal period of the periodic solution in (7.26) is kT.

Assume the contrary. Then, the minimal period of the solution (7.26) (for $i = 1$ or $i = 2$) is nT where n is an integer such that $0 < n < k$. It follows that ζ_i is a minimal n-periodic point of f for $0 < \varepsilon < \varepsilon_0$ and $0 < E = E_0 \ll 1$. Let $k = sn + r$ where $s \geq 1$ and r are integers, such that $0 \leq r < n$. Then we have

$$\zeta_i = f^k(\zeta_i) = f^{r+sn}(\zeta_i) = f^r(f^n \circ \cdots \circ f^n(\zeta_i)) = f^r(\zeta_i),$$

which means that ζ_i is a r-periodic point of f. Since n is the minimal period with $0 \leq r < n$, then $r = 0$. Hence we have $k = sn$. Moreover, it follows from $0 < n < k$ that $s \geq 2$.

Since $\zeta_i(\varepsilon, E_0)$ is an n-periodic point of f, we have

$$\phi(nT, \zeta_i(\varepsilon, E_0), \varepsilon, E_0) - \phi(0, \zeta_i(\varepsilon, E_0), \varepsilon, E_0) = -2m\pi, \qquad (7.27)$$

for some integer $m > 0$, which depends in principle on (ε, E_0). Now, let $(\varepsilon, E_0) = (\varepsilon_j, E_{0j})$ in (7.27), with a sequence $(\varepsilon_j, E_{0j}) \to (0, 0)$ as $j \to \infty$. Therefore, we have

$$\Delta_i(\varepsilon_j, E_{0j}) = -2m(\varepsilon_j, E_{0j})\pi, \qquad (7.28)$$

where

$$\Delta_i(\varepsilon_j, E_{0j}) = \phi(nT, \zeta_i(\varepsilon_j, E_{0j}), \varepsilon_j, E_{0j}) - \phi(0, \zeta_i(\varepsilon_j, E_{0j}), \varepsilon_j, E_{0j}).$$

Since $\zeta_i(\varepsilon_j, E_{0j}) \in \mathcal{A}_0$ is a bounded sequence, there is a subsequence of $\zeta_i(\varepsilon_j, E_{0j}) \in \mathcal{A}_0$ converging to a point $\zeta_i^* \in \mathcal{A}_0$. Without destroying the generality, assume

$$\lim_{j \to \infty} \zeta_i(\varepsilon_j, E_{0j}) = \zeta_i^*.$$

Note that $\Delta_i(\varepsilon_j, E_{0j})$ in (7.28) is continuous with respect to the sequence $(\varepsilon_j, E_{0j}) \to (0, 0)$ and $m(\varepsilon_j, E_{0j})$ is an integer. It follows that $m(\varepsilon_j, E_{0j})$ is a constant integer $m > 0$, and (7.28) implies

$$\Delta_i(0, 0) = \phi(nT, \zeta_i^*, 0, 0) - \phi(0, \zeta_i^*, 0, 0) = -2m\pi. \qquad (7.29)$$

Since $\zeta_i^* \in \mathcal{A}_0$, then ζ_i^* is sufficiently near J_{λ_0} if $\sigma > 0$ is small enough. It follows from (7.29) that $\zeta_i^* \in J_{\lambda_0}$. Then, using (7.29), we obtain

$$nT = m\tau(\lambda_0),$$

which together with (7.14) implies

$$T \leq nT = m\tau(\lambda_0) < m\frac{T}{2}.$$

Hence, $m > 2$. On the other hand, using

$$p\tau(\lambda_0) = kT = snT = sm\tau(\lambda_0)$$

yields that $p = sm$, where $s \geq 2$ and $m \geq 2$ are integers. However, this is a contradiction since p is a prime integer.

We have thus proved that when $0 < \varepsilon < \varepsilon_0 \ll 1$ and $E = E_0$ ($0 < E_0 \ll 1$), the Duffing equation (7.11) has two subharmonic solutions (7.26) of order k.

Remark 1. In history, the earliest strange attractor in the theory of nonlinear oscillations was found in the numerical analysis of the dissipative Duffing equation (10.2) performed by the experiment during the 1950's in Tokyo University. According to Hayashi's viewpoint, the complicated strange attractor is caused by the "subharmonic motions", however, his student Ueda considered the cause of "chaotic motions". Either way, the existence of subharmonic motions for the dissipative Duffing equation (10.2) is now proved as above.

Remark 2. In Chapter 9 of this book, the bend-twist theorem will be furthermore used to prove the existence of periodic solutions of a forced polynomial system of non-conservative differential equation.

Chapter 8

Chaotic Motions

8.1 Definition of Chaotic Motion

Roughly speaking, if a motion is Liapunov unstable, a small disturbance of initial condition may produce a large deviation to the motion, such that it is difficult to predict the future position of motion in a satisfactory way. In such a case, we are lead to the theory of the so-called "random" or "chaos". Therefore, "Liapunov unstability" or "un-predicability" is the characteristic property of "chaos". In other words, "Liapunov stability" or "predicability" is the characteristic property of "non-chaos".

It follows that the almost periodic motion is "non-chaotic". In fact, assume $z = f^t(p)$ is an almost periodic motion in a complete metric space $\mathcal{R}$. Denote by Γ_p the orbit of the motion $z = f^t(p)$. Then the orbit-closure $A = \widehat{\Gamma}_p$ is a compact invariant set in $\mathcal{R}$. Each motion $z = f^t(x)$ starting from any given point $x \in A$ is dense in A and Liapunov stable (with respect to A).

More generally, assume Λ is a compact invariant set of a flow f^t, such that it is topologically transitive (i.e., f^t has a dense orbit in Λ). Assume all motions in Λ are Liapunov stable (i.e., predicable). It follows that all motions in Λ are Liapunov uniformly stable. Hence, we have the following condition

($\mathfrak{P}$): $\forall \, x \in \Lambda$ and $\forall \, \varepsilon > 0$, $\exists \delta = \delta(\varepsilon) > 0$, such that $\forall \, y \in B_\delta(x)$ and $\forall \, t > 0$, satisfying

$$\rho[f^t(y), f^t(x)] < \varepsilon.$$

In this sense, Λ is "predicable" if and only if ($\mathfrak{P}$) is satisfied.

Therefore, Λ is "un-predicable" if and only if we have the condition ($\mathfrak{P}'_i$), which is negative to ($\mathfrak{P}$), for certain index $i \in \mathcal{I}$. For example, the condition

($\mathfrak{P}'_1$): $\forall\, x \in \Lambda$, $\exists\, \varepsilon_0 > 0$, $\forall \delta > 0$, such that $\exists\, y \in B_\delta(x)$ and $\exists\, s > 0$, satisfying
$$\rho[f^s(y), f^s(x)] \geq \varepsilon_0,$$
means that all motions in Λ are uniformly Liapunov unstable.

Furthermore, we have the conditions, say

($\mathfrak{P}'_2$): All motions in Λ are Liapunov unstable;

($\mathfrak{P}'_3$): (Liapunov) unstable motions coexist with stable motions in Λ;

and so on.

It follows that each condition ($\mathfrak{P}'_i$) for $i \in \mathcal{I}$ describes the motion, which is more complicated than the predicable motion. Accordingly, it can be used to define the "chaotic motion".

On the other hand, it is obvious that
$$(\mathfrak{P}'_1) \quad \Rightarrow \quad (\mathfrak{P}'_2),$$
but we do not know whether the converse is true or not. In fact, we know very little about the above condition ($\mathfrak{P}'_i$) for $i \neq 1$.

With this consideration in mind, we would like to choose the following definition of chaos, which is widely accepted in literature (for examples, see [23], [79] and [118]).

Definition 8.1 Let Λ be a topologically transitive and compact invariant set of f^t. The dynamical system f^t is said to have *sensitive dependence on initial conditions* on Λ if the condition ($\mathfrak{P}'_1$) is satisfied.

Definition 8.2 Let Λ be a non-trivial quasi-minimal set of f^t. The dynamical system f^t is said to be *chaotic* on Λ if it has sensitive dependence on initial conditions on Λ. Correspondingly, Λ is called a *chaotic set* of f^t, and the motion having a dense orbit in Λ is called a *chaotic motion* of f^t.

By definition, a chaotic set must be a quasi-minimal set. Therefore, Theorem 5.2 implies that the set of chaotic points[1] is a residual set in the chaotic set.

It can be seen that the condition ($\mathfrak{P}'_i$) for $i \neq 1$ involves "more chaotic" property of motion.

---
[1] A point ξ is said to be chaotic if the motion $f^t(\xi)$ is chaotic.

8.2 Chaotic Quasi-Minimal Set

It is not interesting to analyze the relationship between the quasi-minimal sets and the chaotic sets, for examples:
1) When is a quasi-minimal set chaotic?
2) Can a minimal set be chaotic?

For these problems, let us refer back to the relation

$$P \subset A \subset B \subset Q \subset G,$$

stated in (5.21).

Since a chaotic motion must be P- recurrent, there is no chaotic motion in $(G \setminus Q)$.

Lemma 8.1 *If $z = f^t(p)$ is an almost periodic motion (i.e., $f^t(p) \in A$), then all motions starting from $\widehat{\Gamma}_p$ are uniformly Liapunov stable and almost periodic.*

Proof. This lemma is in fact the general property of almost periodic motions. □

Using Lemma 8.1, we assert that a chaotic motion $z = f^t(p)$ must be P-recurrent but not almost periodic (i.e., $f^t(p) \in (Q \setminus A)$).

Lemma 8.2 *If $z = f^t(p)$ is B-recurrent but not almost periodic (i.e., $f^t(p) \in (B \setminus A)$), then all motions starting from $\widehat{\Gamma}_p$ is Liapunov unstable.*

Proof. Assume the contrary. Then there is a point $\xi \in \widehat{\Gamma}_p$, such that the motion $z = f^t(\xi)$ is Liapunov stable.

On the other hand, since $\widehat{\Gamma}_p$ is a minimal set and $\xi \in \widehat{\Gamma}_p$, then $z = f^t(\xi)$ is a B-recurrent motion.

Hence, the motion $z = f^t(\xi)$ is B-recurrent and Liapunov stable. Using Theorem 5.10, we assert that $z = f^t(\xi)$ is an almost periodic motion. Hence, every motion starting from $\widehat{\Gamma}_\xi$ is almost periodic. It follows from $\widehat{\Gamma}_\xi = \widehat{\Gamma}_p$ that $z = f^t(p)$ is almost periodic. However, this contradicts the assumption: $f^t(p) \in (B \setminus A)$. We have thus completed the proof of Lemma 8.2. □

In history, Birkhoff first found that there are recurrent motions, which are not almost periodic. It follows from Lemma 8.2 that each motion $f^t(p) \in (B \setminus A)$ is Liapunov unstable, but we do not know if it is chaotic. More generally, it is a question whether or not the motion $f^t(p) \in (Q \setminus A)$ is chaotic.

Of course, these problems are more or less sophistic, and we are mainly concerned with the sufficient conditions for chaotic motions.

8.3 Sufficient Conditions for Chaotic Sets

In the earlier definition of chaotic set Λ by Devaney, the following conditions are assumed to be valid:

(C_1) Λ is a non-trivial quasi-minimal set of f^t;

(C_2) the periodic points of f^t is dense in Λ;

(C_3) f^t has sensitive dependence on initial conditions in Λ.

In 1992, J. Banks and his coworkers proved an elementary but somewhat surprising result [5] about the Devaney's definition of chaotic set in terms of maps. For our purpose, we will prove it in terms of flows.

Theorem 8.1 *The conditions (C_1) and (C_2) imply the condition (C_3).*

Proof. First observe that there is a number $d_0 > 0$ such that for all $x \in \Lambda$ there exists a periodic orbit Γ_q which is of distance at least $d_0/2$ from x.

In fact, choose two arbitrary distinct periodic orbits Γ_{q_1} and Γ_{q_2}. Then we have a constant $d_0 = \rho[\Gamma_{q_1}, \Gamma_{q_2}] > 0$. It follows from triangle inequality

$$\rho[\Gamma_{q_1}, \Gamma_{q_2}] \leq \rho[\Gamma_{q_1}, x] + \rho[x, \Gamma_{q_2}]$$

that

$$\rho[x, \Gamma_{q_1}] \geq \frac{d_0}{2} \quad \text{or} \quad \rho[x, \Gamma_{q_2}] \geq \frac{d_0}{2},$$

for every point $x \in \Lambda$. This proves the above-mentioned conclusion.

Then we claim that f^t has sensitive dependence on initial conditions in Λ with respect to the sensitive constant $\varepsilon_0 = d_0/8$.

Now let x be an arbitrary point in Λ and let $N(x)$ be some neighborhood of x. Since the periodic orbits are dense, there exists a periodic orbit Γ_p, such that $p \in U = N(x) \cap B_{\varepsilon_0}(x)$, where $B_{\varepsilon_0}(x)$ is the open ball centered at x with radius ε_0. Let T denote the the least period of Γ_p, $(T > 0)$.

As we showed above, there exists a periodic orbit Γ_q with

$$\rho[x, \Gamma_q] \geq d_0/2 = 4\varepsilon_0.$$

Set

$$V = \bigcap_{0 \leq t \leq T} f^{-t}(B_{\varepsilon_0}(f^t(q))).$$

Clearly V is open and it is non-empty since $q \in V$. Consequently, since f^t is transitive, there exist a point y in U and a sufficiently large number $t_k > 0$, such that $f^{t_k}(y) \in V$.

Now, let $t_k = jT - \tau$ such that $0 \le \tau < T$. By construction, one has

$$f^{jT}(y) = f^{jT-t_k}(f^{t_k}(y)) \in f^{jT-t_k}(V) \subset B_{\varepsilon_0}(f^{jT-t_k}(q)).$$

Note that $f^{jT}(p) = p$, and so by the triangle inequality,

$$\rho[f^{jT}(p), f^{jT}(y)] = \rho[p, f^{jT}(y)]$$
$$\ge \rho[x, f^{jT-t_k}(q)] - \rho[f^{jT-t_k}(q), f^{jT}(y)] - \rho[p, x].$$

Consequently, since $p \in B_\delta(x)$ and $f^{jT}(y) \in B_\delta(f^{jT-t_k}(q))$, one has

$$\rho[f^{jT}(p), f^{jT}(y)] > 4\varepsilon_0 - \varepsilon_0 - \varepsilon_0 = 2\varepsilon_0.$$

Thus, using the triangle inequality again, either $\rho[f^{jT}(x), f^{jT}(y)] > \varepsilon_0$ or $\rho[f^{jT}(x), f^{jT}(p)] > \varepsilon_0$. In either case, we have found a point $\xi = p$ or y in $B_\delta(x)$, satisfying

$$\rho[f^{jT}(\xi), f^{jT}(x)] > \varepsilon_0, \qquad jT \gg 1.$$

This completes the proof. □

This theorem shows that the conditions proposed by Devaney in the earlier definition of chaotic set are not independent. Indeed, (C_1) and (C_2) constitute the sufficient condition for the sensitive dependence on initial conditions. Although (C_2) provides a colorful feature, it is not at the heart of chaos (see [108]). We will show later in the chaotic closed surfaces that the property of sensitive dependence on initial conditions is independent of the condition (C_2).

In 1996, E. Akin and his coworkers [1] generalized the result of Banks and his collaborators to the following result.

Theorem 8.2 *Assume f^t has a quasi-minimal set Q, which contains a dense set of proper minimal sets of almost periodic motions. Then f^t has the sensitive dependence on initial conditions on Q.*

On the other hand, we could find some other useful sufficient conditions for the sensitive dependence on initial conditions as follows.

Theorem 8.3 *Assume f^t has a quasi-minimal set Q, which contains at least a singular point q_0 with the property:*

($\mathcal{H}_1$) arbitrarily close to any point in Q, there is an orbit positively tending to the singular point q_0.

Then f^t has the sensitive dependence on initial conditions on Q.

Proof. Take a point $p_0 \in Q$, such that $p_0 \neq q_0$. Then we have the distance $d_0 = \rho[p_0, q_0] > 0$.

Using the assumption ($\mathcal{H}_1$), given any point $\xi \in Q$, for any constant $\delta > 0$, we have a point $\eta \in B_\delta(\xi)$, such that

$$\rho[f^t(\eta), q_0] < \frac{d_0}{4}, \quad \forall\, t > T,$$

where T is a sufficiently large constant.

On the other hand, since Q is a quasi-minimal set, there is a motion $z = f^t(y)$ with a dense orbit Γ_y in Q. It follows that there is a point $u \in \Gamma_y \cap B_\delta(\xi)$, such that

$$\rho[f^s(u), p_0] < \frac{d_0}{4}, \quad \text{for some large constant } s > T.$$

Then consider the point $f^s(\xi)$ with the following possible cases.

$$\begin{cases} 1.\ \rho[f^s(\xi), p_0] < \frac{d_0}{4}; \\ 2.\ \rho[f^s(\xi), p_0] \geq \frac{d_0}{4}. \end{cases}$$

In the first case, using the inequality

$$\rho[p_0, f^s(\xi)] + \rho[f^s(\xi), f^s(\eta)] + \rho[f^s(\eta), q_0] > \rho[p_0, q_0] = d_0,$$

we obtain

$$\rho[f^s(\xi), f^s(\eta)] > d_0 - \rho[p_0, f^s(\xi)] - \rho[f^s(\eta), q_0] > \frac{d_0}{2}. \qquad (8.1)$$

In the second case, using the inequality

$$\rho[p_0, f^s(\xi)] + \rho[f^s(\xi), f^s(u)] + \rho[f^s(u), q_0] > \rho[p_0, q_0] = d_0,$$

we obtain

$$\rho[f^s(\xi), f^s(u)] > d_0 - \rho[p_0, f^s(\xi)] - \rho[f^s(u), q_0] > \frac{d_0}{2}. \qquad (8.2)$$

It follows from (8.1) and (8.2) that there exists a point $x = u$ or $x = \eta$ with $\rho[x, \xi] < \delta$, satisfying

$$\rho[f^s(x), f^s(\xi)] > \frac{d_0}{2}, \quad \text{for some } s > 0.$$

We have thus proved that f^t has the sensitive dependence of initial conditions on Λ with respect to the sensitive constant $\varepsilon_0 = \frac{d_0}{2}$.
The proof of Theorem 8.3 is thus completed. □

As a direct consequence of Theorem 8.3, we have the following result.

Corollary 8.1 *Assume f^t has a quasi-minimal set Q, which contains a singular point q and an orbit Γ_p, such that Γ_p is negatively dense in Q and positively tends to q (i.e., $A_p = Q$ and $\Omega_p = q$). Then f^t has the sensitive dependence on initial conditions on Q.*

8.4 Chaotic Closed Surfaces

Let $\mathfrak{F}$ be a compact oriented two-dimensional C^1-differentiable manifold without boundary (i.e., it is an oriented C^1-differentiable closed surface), and let $g[\mathfrak{F}]$ be the genus of the closed surface $\mathfrak{F}$.

Assume f^t is the dynamical system defined by a C^1-differentiable vector field V on $\mathfrak{F}$.

When $g[\mathfrak{F}] = 0$, $\mathfrak{F}$ is topologically equivalent to a sphere in $\mathbb{R}^3$. Since the Poincaré-Bendixson theorem is valid on the sphere, there is no chaotic motion of f^t on M.

When $g[\mathfrak{F}] = 1$, $\mathfrak{F}$ is topologically equivalent to a torus in $\mathbb{R}^3$. If there is a singular point or a closed orbit in $\mathfrak{F}$, then $\mathfrak{F}$ is not a minimal set. If $\mathfrak{F}$ is a minimal set, then we have the following alternatives:

(A_1) All motions of f^t in $\mathfrak{F}$ are almost periodic;

(A_2) All motions of f^t in $\mathfrak{F}$ are not almost periodic.

It is easy to give examples of (A_1) (e.g., the Example 5.4), but it is not trivial to construct examples of (A_2).

Birkhoff first gave an example of (A_2). However, the example of Birkhoff was given by a discontinuous vector field. We will improve this example by constructing an analytic vector field in the next Chapter. It follows from Lemma 8.2 that all motions in M is Liapunov unstable, but it is a question if these motions are chaotic.

When $g[\mathfrak{F}] > 1$, then f^t has at least an equilibrium point in $\mathfrak{F}$.

In general, when $g[\mathfrak{F}] \geq 1$, assume the condition

($\mathbb{H}$): f^t *has at least an equilibrium point and a dense orbit in $\mathfrak{F}$, with a finite number of equilibrium points.*

The main purpose of this section is to prove the following theorem, which demonstrates an interesting fact that chaotic motions are extensively existent on closed surfaces.

Theorem 8.4 *When the condition* (ℍ) *is satisfied, then* f^t *is chaotic on the closed surface* $\mathfrak{F}$.

The proof of the theorem will be based on the following lemmas (see, [3] and [51]).

Lemma 8.3 *If f^t has a dense orbit on $\mathfrak{F}$, then there is no closed orbit.*

Proof. Since f^t has a dense orbit on $\mathfrak{F}$, assume without loss of generality that there is a positively dense semi-orbit Γ_p^+ on $\mathfrak{F}$.

Assume f^t has a closed orbit J. Since the orbit passing through a given point is unique, we assert that $\Gamma_p \cap J = \emptyset$.

Let $q \in J$ and let N_q be a line-segment centered at q, such that N_q is transversal to J. Since the semi-orbit Γ_p^+ is dense on $\mathfrak{F}$, the motion $z = f^t(p)$ must meet the line-segment N_p at infinitely many times. Let t_1 and t_2 be two consequent times with $0 < t_1 < t_2$, such that the motion $z = f^t(p)$ meets the line-segment N_p at $t = t_1$ and $t = t_2$, and does not intersect N_p during $t_1 < t < t_2$. Let $p_1 = f^{t_1}(p)$ and $p_2 = f^{t_2}(p)$. Denote by $N_p[p_1, p_2]$ the closed interval on N_p from the point p_1 to the point p_2, and denote by $\Gamma_p^+[p_2, p_1]$ the closed arc of Γ_p^+ from the point p_2 to the point p_1. It follows that

$$C = N_p[p_1, p_2] \cup \Gamma_p^+[p_2, p_1]$$

is a simple closed curve on $\mathfrak{F}$. Since $C \cap J = \emptyset$, the simple closed curves C and J bound a proper region G on $\mathfrak{F}$. It can be seen that the motion $z = f^t(p)$ will come in or go out the region G as $t > t_2$ forever. This is in conflict with the assumption that Γ_p^+ is dense on $\mathfrak{F}$.

The proof of Lemma 8.3 is completed. □

Roughly speaking, the classification of isolated planar singular points is very natural although the mathematical proofs are tedious (see [3] and [123], for example). Especially, the structure of the isolated singular point is simple if f^t has a dense orbit in $\mathfrak{F}$.

An isolated singular point q in $\mathfrak{F}$ is said to be *center-type*, if there exist closed orbits surrounding q in any small neighborhood of q.

Therefore, Lemma 8.3 implies that f^t has no center-type singular point if there is a dense orbit on $\mathfrak{F}$. In this case, in any neighborhood of the isolated singular point q, there is at least an orbit positively or negatively tending to q.

An orbit Γ is called ω-*parabolic* (or α-*parabolic*) if Γ tends positively (or negatively) to q.

An orbit Γ is called *elliptic* if Γ is both ω-parabolic and α-parabolic with respect to one singular point. Similarly to Lemma 8.3, we can prove that f^t has no elliptic orbit in $\mathfrak{F}$ if there is a dense orbit on $\mathfrak{F}$.

Let $\delta > 0$ be a constant, such that q is the unique singular point of f^t in the closed ball $\widehat{B}_\delta(q)$. It follows that there is an ω-parabolic orbit J^+ (or an α-parabolic orbit J^-) in $B_\delta(q)$, such that J^+ positively tends to q and negatively intersects first the boundary of $B_\delta(q)$ at a point p_- (J^- negatively tends to q and positively intersects first the boundary of $B_\delta(q)$ at a point p_+). The orbit-arc of J^+ from p_- to q is denoted by Υ^+, which is called an ω-*separatrix* of q. Similarly, the orbit-arc of J^- from p_+ to q is denoted by Υ^-, which is called an α-*separatrix* of q.

If f^t has a dense orbit and at most finitely many equilibrium points on $\mathfrak{F}$, then there are at most finitely many ω-separatrices and α-separatrices of q. Moreover, the ω-separatrices Υ_i^+ and the α-separatrices Υ_j^- separate each other in such a way

$$\Upsilon_1^+,\ \Upsilon_1^-,\ \Upsilon_2^+,\ \Upsilon_2^-,\ \cdots,\ \Upsilon_m^+,\ \Upsilon_m^-,$$

that Υ_i^+, Υ_i^- bound a sector Θ_i in $B_\delta(q)$ and Υ_i^-, Υ_{i+1}^+ bound a sector Φ_i in $B_\delta(q)$ ($1 \leq i \leq m$), where $\Upsilon_{m+1}^+ = \Upsilon_1^+$. In the interior of each sector, the orbits will leave the sector in the both positive and negative directions, they are called *hyperbolic orbits*. Both Θ_i and Φ_i are called *hyperbolic sectors* of q. The singular point q is called *saddle-type* of $2m$ sectors.

In summary, we have the following result.

Lemma 8.4 *Assume f^t satisfies the condition* (ℍ) *and denote all its equilibrium points by $\{q_1, \cdots, q_k\}$ on $\mathfrak{F}$. Then each q_i is a saddle-type point of $2m_i$ sectors, such that $2 \leq 2m_i \leq 2(2g[\mathfrak{F}] - 1)$, $(i = 1, \cdots, k)$.*

Proof. It is suffices to prove that $2m_i \leq 2(2g[\mathfrak{F}] - 1)$, $(i = 1, \cdots, k)$.

In fact, applying the Poincaré-Bendixson formula to q_j, we have

$$\text{Ind}(q_j) = \frac{2 - 2m_j}{2} = (1 - m_j).$$

It follows from the Poincaré-Hopf's formula of index that

$$\sum_{j=1}^m (1 - m_j) = 2 - 2g[\mathfrak{F}].$$

Since $(1 - m_j) \leq 0$, we have $(1 - m_i) \geq 2 - 2g[\mathfrak{F}]$, which implies

$$2m_i \leq 2(2g[\mathfrak{F}] - 1).$$

The proof of Lemma 8.4 is thus completed. □

Let J be an orbit of f^t. If both the ω-limit set $\Omega(J)$ and the α-limit set $A(J)$ of J are equilibrium points, then J is called a *separatrix*.

Assume there are s different separatrices $J_1, \cdots, J_s$ joining s equilibrium points $q_1, \cdots, q_s$, where the equilibrium points may be not different from each other, such that

$$A(J_i) = q_i \quad \text{and} \quad \Omega(J_i) = q_{i+1} \quad (1 \leq i \leq s)$$

where $q_{s+1} = q_1$. If the separatrices J_{i-1} and J_i bound a hyperbolic sector of q_i in the right-hand (or left-hand) side of $J_{i-1} \cup J_i$ $(i = 1, \cdots, s)$, then the family of separatrices $J_1, \cdots, J_s$ together with the equilibrium points $q_1, \cdots, q_s$ is said to be a *right-sided* (or *left-sided*) *contour*. A right-sided (or left-sided) contour is called a one-sided contour.

In general, a right-sided contour is not a left-sided contour, and vice versa. If a right-sided contour is also a left-sided contour, then each equilibrium point in the contour is a fake saddle (i.e., it has only two hyperbolic sectors).

Lemma 8.5 *If f^t satisfies* $(\mathbb{H})$, *then there is no one-sided contour on $\mathfrak{F}$.*

Proof. The proof is similar to that of Lemma 8.3. □

The following lemma (see [3]) plays an important role in the proof of theorem.

Lemma 8.6 *If f^t satisfies the condition* $(\mathbb{H})$, *then*

1) any non-trivial $P^{\pm}$-recurrent semi-orbit belong to exactly one quasi-minimal set;

2) the number of quasi-minimal sets is finite and does not exceed the genus $g[\mathfrak{F}]$;

3) two (different) quasi-minimal sets can intersect only in equilibrium points and separatrices going from an equilibrium point to an equilibrium point.

Now, we are in position to prove Theorem 8.4.

Let $\{q_1, \cdots, q_k\}$ be the set of equilibrium points of f^t. It follows from Lemma 8.4 that all of the equilibrium points are saddle-type. Let L_1 be a separatrix which positively tends to q_1 and negatively tends to q_j. It follows from Lemma 8.5 that $q_j = q_1$. Then take a separatrix L_j positively tending to q_j and negatively tending to q_i. Without loss of generality, let $L_2 = L_j$, $q_2 = q_j$ and $q_3 = q_i$. It follows again from Lemma 8.5 that $q_3 \neq q_1, q_2$. In this manner, since the number of equilibrium points is finite, Lemma 8.5 implies that there is an orbit L_n positively tending a saddle-type point q_n, such that L_n is P^--recurrent.

It follows that the closure of L_n is a non-trivial quasi-minimal set Q_n in $\mathfrak{F}$. It is noted that since f^t satisfies the condition (ℍ), $\mathfrak{F}$ is a non-trivial quasi-minimal set.

We want to prove that $Q_n = \mathfrak{F}$.

Assume the contrary. Then $Q_n = Q_n \cap \mathfrak{F}$ is a proper subset of $\mathfrak{F}$. It follows from the conclusion 3) of Lemma 8.6 that Q_n is a separatrix going from an equilibrium point to an equilibrium point. However, this contradicts that Q_n is a quasi-minimal set.

Therefore, we have $Q_n = \mathfrak{F}$. It follows that L_n is an orbit in $\mathfrak{F}$, such that it positively tends to an equilibrium point q_n and is negatively dense in $\mathfrak{F}$. Then, using Corollary 8.1, we conclude that f^t is chaotic on $\mathfrak{F}$.

The proof of Theorem 8.4 is thus completed. □

8.5 Applications

8.5.1 *Strange Attractor of the Lorenz Model*

We are inspired by the results in the Chapter 14 of the textbook by M. Hirsch, S. Smale and R. Devaney [75] because Theorem 8.3 is useful to the proof for the sensitive dependence on initial conditions in the geometrical attractor of Lorenz system. On the other hand, it is not easy to introduce the Lorenz geometrical model in details. Hence, we are satisfied to use the basic knowledge in [75]:

• Consider the Lorenz system

$$\begin{cases} \dot{x} = \sigma(y - x), \\ \dot{y} = rx - y - xz, \\ \dot{z} = xy - bz, \end{cases} \quad (8.3)$$

where $\sigma > 0$, $b > 0$ and $r > 0$ are specific parameters. It has been proved in the textbook that the Lorenz system has an attractor of zero-volume. In the ensuing 40 years, much progress has been made in the theory of chaotic systems. However, as we know, the analysis of the chaotic behavior of particular systems like the Lorenz attractor is usually extremely difficult. Most of the chaotic behavior that is readily understandable arises from geometric models rather than from the actual systems themselves. Indeed, this is the avenue pursed in literature.

• Before 30 years, mimicking the Lorenz system (8.3), Guckenheimer and Williams gave a dynamical system $\mathcal{L}$ defined piecewise by linear differential equations in a compact cube in the (x, y, z)-space. It was proved that $\mathcal{L}$ has a compact attractor $\mathcal{A}$, which is called the *geometrical model of Lorenz attractor*. As reported in [75], the geometric model $\mathcal{A}$ was only shown to be equivalent to the attractor of Lorenz system in the year 1999.

• In the Chapter 14 of [75], it is proved by using geometrical methods that $\mathcal{L}$ has sensitive dependence on initial conditions on $\mathcal{A}$. In addition, the following properties are verified (see the exercises there):

(P_1) $\mathcal{A}$ is topologically transitive;

(P_2) arbitrarily close to any point in $\mathcal{A}$, there is an orbit positively tending to the singular point at the origin of $\mathbb{R}^3$.

It follows that $\mathcal{A}$ is a quasi-minimal set of $\mathcal{L}$, such that $\mathcal{L}$ satisfies the above-mentioned condition $\mathbb{H}_1$ in $\mathcal{A}$. Hence, using the above Theorem 8.3 yields that $\mathcal{L}$ has sensitive dependence on initial conditions on $\mathcal{A}$. It follows that $\mathcal{A}$ is a strange attractor of the Lorenz model $\mathcal{L}$.

Our proof of the Lorenz strange attractor seems to be different from the geometric proof in [75]. But the former is based on the properties (P_1) and (P_2), which are still involved with geometrical considerations.

8.5.2 Schweitzer's Attractor

In 1950, Seifert proposed the following problem and conjectured the positive answer.

(P-1) *Let V be a non-singular C^1 (or an analytic vector field) on the three-dimensional sphere $\mathbb{S}^3$. Is there a periodic closed orbit in $\mathbb{S}^3$ for the vector field V?*

This problem is indeed equivalent to the question of the validity of the Poincaré-Bendixson annular region theorem in three-dimensional Euclidean space; that is,

(P-2) *Let $\mathcal{T}$ be a closed solid torus [2] in $\mathbb{R}^3$, and let V be a non-singular*

---
[2] A closed solid torus is the compact region in $\mathbb{R}^3$ bounded by a torus $\mathbb{T}^2$.

C^1 (or an analytic vector field) on $\mathcal{T}$, such that it always points inward on the boundary $\partial \mathcal{T} = \mathbb{T}^2$. Is there a periodic closed orbit in $\mathcal{T}$ for the vector field V?

"Prove or disprove it" has long been a difficult problem in the theory of ordinary differential equations. In 1974, Schweitzer constructed a C^1 counterexample to the Seifert's conjecture [113]. L. Markus highly praised this counterexample, and considered it as one of the major results in recent years.

To express the relationship between the problems (**P-1**) and (**P-2**), we prove the formula

$$\mathbb{S}^3 = \mathcal{T}_1 \cup \mathcal{T}_2, \tag{8.4}$$

where $\mathcal{T}_1$ and $\mathcal{T}_2$ are two solid toruses having a common boundary; that is,

$$\mathbb{T}^2 = \mathcal{T}_1 \cap \mathcal{T}_2 = \partial \mathcal{T}_1 = \partial \mathcal{T}_2.$$

In fact, let $\langle r, \theta, z \rangle$ be the cylindric coordinates of the point (x, y, z) in $\mathbb{R}^3$; that is, $x = r\cos\theta$, $y = r\sin\theta$, $z = z$. Given $\theta \in [0, 2\pi)$ and $\sigma \in [1, \infty)$, consider the circle

$$C_\sigma(\theta): \quad \left[r - \frac{\sigma + \sigma^{-1}}{2}\right]^2 + z^2 = \left[\frac{\sigma - \sigma^{-1}}{2}\right]^2,$$

in the half-plane

$$H^+(\theta): \quad r > 0, \quad z \in \mathbb{R}^1,$$

where θ is fixed. Note that $C_1(\theta) = (\cos\theta, \sin\theta, 0)$ is a point (i.e., $C_1(\theta)$ is a degenerate circle). It is obvious that the circle $C_\sigma(\theta)$ intersects the r-axis at $r = \sigma^{-1}$ and $r = \sigma$, and $C_{\sigma'} \cap C_{\sigma''} = \emptyset$ if $\sigma' \neq \sigma''$. The half-plane $H^+(\theta)$ is filled up by the circles $C_\sigma(\theta)$ $(1 \leq \sigma < \infty)$.

Consider the disk

$$D = \{(x, y, z) \in C_\sigma(0): \quad 1 \leq \sigma \leq 2\},$$

bounded by the circle $C_2(0)$ in the half-plane $H^+(0)$. Then the rotation of D about z-axis is a solid torus

$$\mathcal{T}_1 = \{(x, y, z) \in C_\sigma(\theta): \quad 1 \leq \sigma \leq 2; \quad 0 \leq \theta < 2\pi\} = D \times \mathbb{S}^1.$$

It is noticed that the solid torus $\mathcal{T}_1$ separates the (x, y)-plane into the regions

$$B_1: \quad 0 \leq \sqrt{x^2 + y^2} \leq 2^{-1},$$

and
$$B_2: \quad 2 \leq \sqrt{x^2+y^2} < \infty,$$
such that the regions
$$\overset{\circ}{B_1}: \quad 0 < \sqrt{x^2+y^2} \leq 2^{-1}$$
and
$$B_2: \quad 2 \leq \sqrt{x^2+y^2} < \infty$$
are connected by the circles $C_\sigma(\theta)$ from the point
$$(\sigma^{-1}\cos\theta, \ \sigma^{-1}\sin\theta, \ 0) \in \overset{\circ}{B_1}$$
to the point
$$(\sigma\cos\theta, \ \sigma\sin\theta, \ 0) \in B_2,$$
where $2 \leq \sigma < \infty$ and $0 \leq \theta < 2\pi$. Let Z be the set of points at the z-axis. Hence, we have
$$\mathbb{R}^3 = \bigcup_{2\leq\sigma<\infty, \ 0\leq\theta<2\pi} C_\sigma(\theta) \cup Z \cup T_1,$$
which together with
$$\mathbb{S}^3 = \mathbb{R}^3 \cup \{\infty\}$$
implies
$$\mathbb{S}^3 = \bigcup_{2\leq\sigma<\infty, \ 0\leq\theta<2\pi} C_\sigma(\theta) \cup (Z \cup \{\infty\}) \cup T_1 = T_2 \cup T_1,$$
where
$$T_2 = \bigcup_{2\leq\sigma<\infty, \ 0\leq\theta<2\pi} C_\sigma(\theta) \cup (Z \cup \{\infty\}).$$
It can be seen that $C_\sigma(\theta)$ are circles which is continuous for the parameters $2 \leq \sigma < \infty$, $0 \leq \theta < 2\pi$, and
$$\lim_{\sigma\to\infty} C_\sigma(\theta) = Z \cup \{\infty\} = \mathbb{S}^1.$$

Therefore, T_2 is topologically equivalent to the solid torus $B_1 \times \mathbb{S}^1$. The expression (8.4) is thus proved. □

Chapter 8. Chaotic Motions 213

It can be seen that if we are able to construct a nonsingular vector field V_1 on T_1, such that $V_1 \mid_{\partial T_1}$ points inward T_1 and V_1 has no closed orbit in T_1, then we can construct a similar nonsingular vector field V_2 on T_2, such that $V_2 \mid_{\partial T_2}$ points outward T_2 and V_2 has no closed orbit in T_2. Therefore, using the expression (8.4), we have a nonsingular vector field V on $\mathbb{S}_3$, such that V has no closed orbit in $\mathbb{S}^3$. In this way, the Seifert's conjecture can be stated with respect to the problem (**P-2**).

The following lemma is a key point of the Schweitzer's example.

Let N be a connected q-dimensional manifold (without boundary), and suppose that Z is a nonsingular C^r vector field on N $(r \geq 1)$ with a compact invariant set $F \subset N$. Let $I = [-1, 1]$. Let $X_0 = (0_N, c)$ and $Z_0 = (Z, 0)$ be the vector fields on $N \times I$, where $c \neq 0$ is a constant, and 0_N is the zero vector field on N.

Lemma 8.7 *There exists a homotopy of C^r vector fields $X_0 \simeq X_1$ on $N \times I$ with compact support [3] contained in $N \times (-1, 1)$, such that every closed orbit of X_1 is either*

(1) an arc joining $(p, -1)$ to $(p, 1)$ for some $p \in (N \setminus F)$; or

(2) a closed orbit of Z_0 contained in $F \times \{\pm\frac{1}{2}\}$.

Proof. Using a standard method, we construct a C^∞ function

$$\Psi : \quad N \times [0, 1] \to [0, 1],$$

such that Ψ has a compact support $K \subset N \times (0, 1)$, and $\Psi^{-1}(1) = F \times \{\frac{1}{2}\}$. Then define

$$X_1 \mid_{N \times [0,1]} = (1 - \Psi) \cdot X_0 + \Psi \cdot Z_0.$$

In order to extend X_1 to $N \times I$, construct the transformation $\Phi : (p, t) \to (p, -t)$ on $N \times I$ such that X_1 satisfies $d\Phi(X_1) = -X_1$. Clearly, X_1 and the homotopy $X_s = (1-s)X_0 + sX_1$ are well defined. They are nonsingular C^r vector fields, and X_s, X_0 agree outside the compact set $K \cup \Phi(K)$.

It remains to show that all closed orbits of X_1 must be of the form stated in the Lemma.

We first note that the set of points in $N \times I$ on which the second component of X_1 is zero (or $X_1 = \pm Z_0$) is exactly $F \times \{\pm\frac{1}{2}\}$. We can readily see that it is a compact invariant set for X_1. Hence any closed integral curve of X_1 intersecting with this set must be type (2).

---

[3] The support can be intuitively considered as a compact region where the function is nonzero in its interior and zero on its boundary and exterior.

Let $\alpha : J \to I$ be a closed orbit of X_1 which does not intersect with the set $F \times \{\pm\frac{1}{2}\}$. We now show that it must be of type (1). We note that α has the following two properties:

(a) the velocity vector $d(P_2 \circ \alpha)\dfrac{\partial}{\partial t} \neq 0$, where $P_2 : N \times I \to I$ is the projection;

(b) $\operatorname{Im}(P_2 \circ \alpha)$ is compact, where $\operatorname{Im}(\cdot)$ is the image set of $(\cdot)$.

Hence $\operatorname{Im}(P_2 \circ \alpha) = I$, and $\operatorname{Im}(\alpha)$ is an arc joining $(p, -1)$ to $(p', 1)$, where $p, p' \in N$. Since α crosses $N \times \{0\}$ at a unique point and α is invariant under Φ, we have $p = p'$. Since α does not intersect $F \times \{\pm\frac{1}{2}\}$, we have $d(P_1 \circ \alpha)\dfrac{\partial}{\partial t} = \Psi \cdot Z_0$, where $P_1 : N \times I \to N$ is the projection. Hence $\operatorname{Im}(P_1 \circ \alpha) \subset Z_0$ is contained in an integral curve in $N \times \{0\}$. We have thus see that $F \times I$ is invariant. The orbits inside it must be separated by $F \times \{\pm\frac{1}{2}\}$, and thus cannot connect $(p, -1)$ to $(p, 1)$. This shows that the orbit α described above must be outside $F \times I$ and thus we have $p \in (N \setminus F)$. This proves the lemma.

We now describe the construction of Schweitzer's example. Let X be the nonsingular C^1 vector field on the torus $\mathbb{T}^2$ such that X does not have any closed orbit. Moreover, X has the Cantorus F; that is, X is the Denjoy's field. Hence, $(\mathbb{T}^2 \setminus F)$ is a nonempty open set in $\mathbb{T}^2$. Inside it, we remove s closed disk and consider the remaining part as the manifold N (2-dimensional) of the previous lemma. Moreover, we will choose an appropriate vector field $Z = X\mid_N$.

We make an appropriate deformation on N as follows.

Then, using the lemma above, we obtain the figure for $B = N \times I$ in the following Figure 20.

Moreover, there exists a C^1 vector field X_1 on B such that there is no periodic closed orbit in B, and the direction of X_1 is the same as that of X_0 on the boundary of B. An orbit entering B at the point $(p, -1)$ either approaches the invariant set $F \times \{-\frac{1}{2}\}$, or leaves at the point $(p, 1)$. There are two points $p_2 = (p_*, -1)$ and $p_1 = (p_*, 1)$ such that the positive semi-orbit entering at p_2 and the negative semi-orbit leaving at p_1 always stay inside B, and so on.

Fig. 20.

Then in the solid $\mathcal{T}$, we construct a nonsingular C^∞ vector field Y such that at the boundary of $\mathcal{T}$ it points toward the interior of $\mathcal{T}$, and the central axis γ of $\mathcal{T}$ is the only closed orbit of Y. On γ, choose an arbitrary point p_0 and modify the vector field Y in some neighborhood $U(p_0)$ of p_0, so that all orbits inside some cylinder $W(p_0)$ (with p_0 as the center) are straight lines parallel to the vector $Y(p_0)$. Moreover, inside $W(p_0)$, γ corresponds to the central axis of $W(p_0)$.

We then embed $e : B \to W(p_0)$ and replace the original vector field $Y \mid_{e(B)}$ by the vector field induced by e from B to $e(B)$. The resulting vector field is still a nonsingular C^1 vector field (denoted again by Y), with $z_1 = e(p_1)$ and $z_2 = e(p_2)$. Then the orbits on the boundary of $e(B)$ will coalesce with the corresponding orbits of $W(p_0)$.

In this manner, we have constructed a new C^1 nonsingular vector field Y on $\mathcal{T}$, such that it points to its interior at its boundary $\partial \mathcal{T}$; and there is no closed orbit inside $\mathcal{T}$. This is the example of Schweitzer. □

It is obvious that the above vector field Y has a Denjoy's minimal set $\mathfrak{D}$ in the solid torus $\mathcal{T}$. Indeed, it is easy to modify the above construction a little, such that $\mathfrak{D}$ is a strange attractor of the dynamical system.

Finally, we remark that since a C^2 vector field on the torus $\mathbb{T}^2$ does not admit of the Denjoy's set (see [21]), the method of Schweitzer cannot improve the above C^1 vector field Y to a C^r vector field with $r \geq 2$. Therefore, the Schweitzer's example does not perfectly answer the Seifert's conjecture in case of analytic vector field on $\mathcal{T}$. Until 1996, G. Kuperberg and K. Kuperberg construct an analytic vector field on $\mathbb{S}^3$, such that there exist neither singular point nor closed orbit [83]. The Seifert's conjecture is thus completely disproved.

Chapter 9

Perturbation Method

9.1 Nonlinear Differential Equation of Second Order

In the theory of nonlinear oscillations, we frequently encounter the second order differential equation

$$x'' + f(x, x')x' + g(x) = p(t), \tag{9.1}$$

where x' and x'' denote, respectively, the first and second derivatives of x with respect to the variable t. For simplicity, we will assume $g(x) \in C^1(\mathbb{R}^1, \mathbb{R}^1)$, and $f(x, y) \in C^1(\mathbb{R}^2, \mathbb{R}^1)$; while $p(t) \in C(\mathbb{R}^1, \mathbb{R}^1)$ is a T-periodic function.

Equation (9.1) usually serves as a mathematical model for certain laws of motion. For example, for the oscillation of the unit mass, $f(x, x')$ is the damping coefficient, $g(x)$ is the spring restoring force and $p(t)$ represents the forcing term. Denote by $\omega = \dfrac{2\pi}{T}$ the forcing frequency.

9.1.1 *Linear System*

As a guideline for the study of the system (9.1), we first consider the familiar linear differential equation with constant coefficients and periodic forcing

$$x'' + cx' + kx = p(t), \tag{9.2}$$

where $c \geq 0$ and $k > 0$ are constants, and $p(t)$ is a T-periodic continuous function. It is well-known that if $x = u(t)$ and $x = v(t)$ form a fundamental system of solutions of the corresponding homogeneous equation

$$x'' + cx' + kx = 0, \tag{9.3}$$

9.1. Nonlinear Differential Equation of Second Order

then the general solution of (9.3) is

$$x = \alpha_0 u(t) + \beta_0 v(t),$$

where α_0 and β_0 are arbitrary constants. For simplicity, assume $x = u(t)$ and $x = v(t)$ are the standard solutions; that is, they satisfies, respectively, the initial conditions:

$$\begin{cases} u(0) = 1, & u'(0) = 0; \\ v(0) = 0, & v'(0) = 1. \end{cases}$$

It can be easily verified that the homogeneous equation does not have nontrivial T-periodic solution if and only if

$$\Delta(T) := \begin{vmatrix} u(T) - 1 & v(T) \\ u(T) & v(T) - 1 \end{vmatrix} \neq 0.$$

Using the method of variations for the constants α_0 and β_0, assume the general solution of (9.2) is

$$x = \alpha(t)u(t) + \beta(t)v(t),$$

where $\alpha(t)$ and $\beta(t)$ are functions to be determined. Then the technique of substitution (see [56], for example) yields

$$x = \alpha_0 u(t) + \beta_0 v(t) + \int_0^t G(t,s)p(s)\,ds \tag{9.4}$$

as the general solution of (9.2), where

$$G(t,s) = e^{cs}[u(s)v(t) - v(s)u(t)] \tag{9.5}$$

is called the *kernel function*. It follows from (9.6) that

$$x = x_0 u(t) + x_0' v(t) + \int_0^t G(t,s)p(s)\,ds \tag{9.6}$$

is the solution of (9.2) satisfying the initial condition

$$x(0) = x_0, \qquad x'(0) = x_0'. \tag{9.7}$$

It can be seen from (9.6) that the motion of (9.2) is the sum of a term $[x_0 u(t) + x_0' v(t)]$ caused by the initial perturbation and another term caused by the forcing perturbation $p(t)$. In other words, the influences of the initial perturbation and the forcing perturbation are separated from each other. This is a significant property of linear systems.

Theorem 9.1 *When the linear homogeneous equation (9.3) does not have a nontrivial T-periodic solution, equation (9.2) has one and only one T-periodic solution.*

Corollary 9.1 *If $c > 0$, then equation (9.2) has a unique harmonic (i.e., T-periodic) solution $x = x_0(t)$, such that any solution $x = x(t)$ of (9.2) tends to the harmonic solution; that is,*

$$\lim_{t \to \infty} [x(t) - x_0(t)] = 0, \qquad \lim_{t \to \infty} [x'(t) - x_0'(t)] = 0. \tag{9.8}$$

Corollary 9.2 *If $c = 0$ and $k \neq (2\pi/T)^2$, then equation (9.2) has a unique harmonic solution $x = x_0(t)$. In this case, this harmonic solution does not have the asymptotic property (9.8).*

The proofs of the above results are immediate.

Example 9.1 Consider the differential equation

$$x'' + cx' + kx = E \sin \omega t, \tag{9.9}$$

where $c \geq 0$, $k > 0$, $E > 0$ and $\omega > 0$ are constants.

As an important exercise, the main steps in solving the equation will be shown in details.

1) $c > 0$ (i.e., Damped Case):

(i) Suppose $D = c^2 - 4k = d^2 > 0$. In this case, the equation (9.3) has the characteristic roots

$$\lambda = \frac{-c + d}{2} < 0, \qquad \mu = \frac{-c - d}{2} < 0 \qquad (\lambda \neq \mu),$$

and the general solution

$$x = \alpha_0 e^{\lambda t} + \beta_0 e^{\mu t} + F \sin(\omega t + \phi),$$

where

$$\phi = \arctan \frac{c\omega}{\omega^2 - k} \qquad \text{and} \qquad F = \frac{E}{\sqrt{(k - \omega^2)^2 + (c\omega)^2}} \tag{9.10}$$

are respectively called the *phase-shift* and *amplitude*.

(ii) Suppose $D = c^2 - 4k = d^2 = 0$. In this case, the equation (9.3) has the characteristic roots $\lambda = \mu = \frac{-c}{2} < 0$ and the general solution

$$x = \alpha_0 e^{\lambda t} + \beta_0 t e^{\lambda t} + F \sin(\omega t + \phi),$$

where the phase-shift ϕ and the amplitude F are the same as above.

($\ddot{u}$) Suppose $D = c^2 - 4k = -d^2 < 0$. In this case, the equation (9.3) has the characteristic roots

$$\lambda = \frac{-c+id}{2}, \quad \mu = \frac{-c-id}{2} \quad (\lambda \neq \mu),$$

and the general solution

$$x = e^{-\lambda t}(\alpha_0 \cos \omega_0 t + \beta_0 \sin \omega_0 t + F \sin(\omega t + \phi)),$$

where $\lambda = c/2$ and $\omega_0 = \sqrt{D}$, and the phase-shift ϕ and the amplitude F are the same as above.

It can be seen in the damped case that equation (9.2) has a unique harmonic solution $x = x_0(t) = F\sin(\omega t + \phi)$, which has the asymptotic property (9.8). Using (9.10), we have

$$\lim_{\omega \to \infty} x_0(t) = 0 \quad \text{(uniformly in } t\text{)}. \tag{9.11}$$

In other words, the forced oscillation with high frequency has the effect of reducing vibrations.

2) $c = 0$ (i.e., Undamped Case): In this case, we call $\omega_0 = \sqrt{k}$ the *inherent frequency* of the oscillating spring.

(i) Suppose $\omega \neq \omega_0$. Then the general solution of (9.9) is

$$x = \alpha_0 \cos \omega_0 t + \beta_0 \sin \omega_0 t + \frac{E}{\omega_0^2 - \omega^2} \sin \omega t,$$

which contains at least a harmonic solution

$$x = x_0(t) = \frac{E}{\omega_0^2 - \omega^2} \sin \omega t.$$

It does not have the asymptotic property (9.8), but satisfies the condition (9.11).

($\ddot{u}$) Suppose $\omega = \omega_0$. Then the general solution of (9.9) is

$$x = \alpha_0 \cos \omega_0 t + \beta_0 \sin \omega_0 t - \frac{E}{2\omega_0} t \cos \omega_0 t.$$

It follows that every solution of (9.9) is unbounded and no periodic solution exists. This is commonly known as *linear resonance phenomenon*.

This example illustrates the fact that the existence of harmonic solutions is a more complicated problem for the undamped case (conservative systems) than the damped case (dissipative systems).

9.1.2 Almost Linear Systems

In the previous subsection, we investigate the harmonic solution of linear systems with constant coefficients by solving the equation directly. Now, consider an almost linear system as follows

$$x'' + [c + \varepsilon L(x, x')]x' + [k + \varepsilon M(x, x')]x = p(t), \tag{9.12}$$

where ε is a small parameter, $L(x, y)$ and $M(x, y)$ are continuous functions in $(x, y) \in \mathbb{R}^2$. Indeed, we will consider a more general equation

$$x'' + cx' + kx = p(t) + \varepsilon Q(t, x, x', \varepsilon) \tag{9.13}$$

where the perturbation $Q(t, x, x', \varepsilon)$ is continuous in (t, x, x', ε) and T-periodic with respect to t. Moreover, assume $Q(t, x, x', \varepsilon)$ is differentiable with respect to the variables x, x' and ε.

In order to study the harmonic solution of the almost linear equation (9.13), let us recall the theorem concerning the dependence of its solution on initial conditions and parameters.

Assume $x = x(t, x_0, y_0, \varepsilon)$ is the solution of (9.13) satisfying the initial condition

$$x(0) = x_0, \qquad x'(0) = y_0, \tag{9.14}$$

where the initial point (x_0, y_0) belongs to the disk

$$\mathfrak{D}_h = \{(x, y) \in \mathbb{R}^2 : \quad x^2 + y^2 \leq h^2\}$$

of any given radius $h > 0$. Then, for any given $T_1 > T$, $x = x(t, x_0, y_0, \varepsilon)$ exists on the interval $|t| \leq T_1$ for $(x_0, y_0) \in \mathfrak{D}_h$ and $|\varepsilon| < \varepsilon_0$ provided that ε_0 is small enough. Moreover, $x = x(t, x_0, y_0, \varepsilon)$ and $x' = x'_t(t, x_0, y_0, \varepsilon)$ are differentiable with respect to $(t, x_0, y_0, \varepsilon) \in [-T_1, T_1] \times \mathfrak{D}_h \times [-\varepsilon_0, \varepsilon_0]$.

Now, we state and prove the following main result in this subsection.

Theorem 9.2 *Suppose the corresponding linear homogeneous equation of (9.13) (i.e., (9.3)) has no nontrivial T-periodic solution, then there is a sufficiently small $\varepsilon_0 > 0$ such that, for $|\varepsilon| \leq \varepsilon_0$, equation (9.13) has a harmonic solution $x = x(t, \varepsilon)$ which depends continuously on (t, ε) and has the property that*

$$\lim_{\varepsilon \to 0} x(t, \varepsilon) = x_0(t), \tag{9.15}$$

where $x = x_0(t)$ is the unique harmonic solution of (9.2).

Proof. It follows from the previous subsection that the linear equation (9.2) has a unique T-periodic solution $x = x_0(t)$. Let $\xi = x_0(0)$ and $\eta = x'(0)$; and consider the closed disk

$$\mathfrak{D}_\rho = \{(x,y) \in \mathbb{R}^2 : (x-\xi)^2 + (y-\eta)^2 \leq \rho^2\} \subset \mathfrak{D}_h,$$

of radius $\rho > 0$ centered at (ξ, η), where $\mathfrak{D}_h$ is defined-above for sufficiently large $h > 0$.

Assume $x = x(t, x_0, y_0, \varepsilon)$ is the solution of (9.13) defined as above. It follows that $x = x(t, x_0, y_0, \varepsilon)$ and $x' = x'_t(t, x_0, y_0, \varepsilon)$ are continuously differentiable with respect to $(t, x_0, y_0, \varepsilon) \in [0, T] \times \mathfrak{D}_\rho \times [-\varepsilon_0, \varepsilon_0]$. By the variations of constants method, we readily deduce that $x(t) = x(t, x_0, y_0, \varepsilon)$ satisfies the integral equation

$$x(t) = x_0 u(t) + y_0 v(t)$$

$$+ \int_0^t G(t,s)[p(s) + \varepsilon Q(s, x(s), x'(s), \varepsilon)]\, ds, \tag{9.16}$$

where $u(t)$ and $v(t)$ are standard solutions of (9.3), and $G(t,s)$ is the kernel function mentioned-above.

It is known that $x = x(t)$ is a T-periodic solution if and only if it satisfies the periodic boundary condition: $x(0) = x(T)$, $x'(0) = x'(T)$; that is,

$$\Phi(x_0, y_0, \varepsilon) = 0, \qquad \Psi(x_0, y_0, \varepsilon) = 0, \tag{9.17}$$

where

$$\Phi(x_0, y_0, \varepsilon) := [u(T) - 1]x_0 + v(T)y_0 + A$$

$$+ \varepsilon \int_0^T G(T, s) Q(s, x(s), x'(s), \varepsilon)\, ds,$$

$$\Psi(x_0, y_0, \varepsilon) := u(T)x_0 + [v(T) - 1]y_0 + B$$

$$+ \varepsilon \int_0^T G(T, s) Q(s, x(s), x'(s), \varepsilon)\, ds,$$

with constants

$$A = \int_0^T G(T, s) p(s)\, ds \quad \text{and} \quad B = \int_0^T G'_t(T, s)\, ds.$$

Since $x = x_0(t) = x(t, \xi, \eta, 0)$ is a harmonic solution of (9.2), satisfying the initial condition

$$x_0(0) = \xi, \qquad x_0'(0) = \eta,$$

we have

$$\Phi(\xi, \eta, 0) = 0, \qquad \Psi(\xi, \eta, 0) = 0.$$

On the other hand, it is clear that

$$\left.\frac{\partial(\Phi, \Psi)}{\partial(x_0, y_0)}\right|_{\varepsilon=0} = \Delta(T) \neq 0.$$

Using the implicit function theorem, we assert that equation (9.17) has a unique solution $(x_0, y_0) = (x_0(\varepsilon), y_0(\varepsilon))$ which is continuously differentiable in $\varepsilon \in [\varepsilon_1, \varepsilon_1] \subset [\varepsilon_0, \varepsilon_0]$.

It follows that

$$x = x(t, \varepsilon) := x(t, x_0(\varepsilon), y_0(\varepsilon), \varepsilon)$$

is the T-periodic solution of (9.13) having the property

$$\lim_{\varepsilon \to 0} x(t, \varepsilon) = x(t, 0) = x_0(t).$$

The proof of Theorem 9.2 is complete. □

When $\Delta(T) = 0$, the equation (9.3) is called *critical*.

In the critical case, we can not, in general, guarantee the existence of T-periodic solution for the equation (9.13). The major difficulty arises due to the possibility of "resonance phenomenon".

In the subsequent, we will analyze the conditions for the existence of harmonic solution in the critical case for the equation (9.13).

In the critical case, we can deduce that $c = 0$ and $k = \left(\dfrac{2m\pi}{T}\right)^2$, where m is some positive integer. Hence, we have

$$u(t) = \cos\frac{2m\pi t}{T}, \qquad v(t) = \frac{T}{2m\pi} \sin\frac{2m\pi t}{T}.$$

Note that $x = u(t)$ and $x = v(t)$ are two T-periodic solutions of (9.3). On the other hand, equation (9.13) has a T-periodic solution which depends continuously on ε if and only if equation (9.17) has a solution $(x_0, y_0) =$

$(x_0(\varepsilon), y_0(\varepsilon))$ which depends continuously on ε. In particular, when $\varepsilon = 0$, equation (9.13) becomes

$$\begin{cases} [u(T) - 1]x_0 + v(T)y_0 + A = A = 0, \\ u'(T)x_0 + [v'(T) - 1]y_0 + B = B = 0, \end{cases}$$

which includes the orthogonality condition

$$\int_0^T p(s) \begin{pmatrix} \cos \frac{2m\pi}{T} s \\ \sin \frac{2m\pi}{T} s \end{pmatrix} ds = \begin{pmatrix} 0 \\ 0 \end{pmatrix}. \tag{9.18}$$

In other words, the orthogonality condition (9.18) is a necessary condition for the almost linear differential equation (9.13) to have a T-periodic solution which depends continuously on ε. Hence, we assume in the following that the orthogonality condition (9.18) is valid. It follows that equation (9.17) can be written as

$$\varepsilon \Phi_1(x_0, y_0, \varepsilon) = 0, \qquad \varepsilon \Psi_1(x_0, y_0, \varepsilon) = 0,$$

where

$$\Phi_1(x_0, y_0, \varepsilon) = \int_0^T G(T, s) Q(s, x(s, x_0, y_0, \varepsilon), x'(s, x_0, y_0, \varepsilon), \varepsilon) \, ds,$$

and

$$\Psi_1(x_0, y_0, \varepsilon) = \int_0^T G'_t(T, s) Q(s, x(s, x_0, y_0, \varepsilon), x'(s, x_0, y_0, \varepsilon), \varepsilon) \, ds.$$

Now, consider the implicit equation

$$\Phi_1(x_0, y_0, \varepsilon) = 0, \qquad \Psi_1(x_0, y_0, \varepsilon) = 0. \tag{9.19}$$

Clearly, any continuous solution $(x_0, y_0) = (x_0(\varepsilon), y_0(\varepsilon))$ of (9.19) is also a continuous solution of (9.17). We have thus obtain the following result.

Theorem 9.3 *Suppose equation (9.13) is in the critical case (i.e., $c = 0$ and $k = (2m\pi/T)^2$ for some positive integer m). Furthermore, suppose there are constants ξ and η, satisfying*

$$\Phi_1(\xi, \eta, 0) = 0, \qquad \Psi_1(\xi, \eta, 0) = 0$$

and

$$\left| \frac{\partial(\Phi_1, \Psi_1)}{\partial(x_0, y_0)} \right| \neq 0, \qquad \text{when } (x_0, y_0, \varepsilon) = (\xi, \eta, 0).$$

Then equation (9.13) has a T-periodic solution $x = x(t, \varepsilon)$ which depends continuously on ε and satisfies

$$\lim_{\varepsilon \to 0} x(t, \varepsilon) = \xi \cdot \cos \frac{2m\pi}{T} t + \eta \cdot \frac{T}{2m\pi} \sin \frac{2m\pi}{T} t$$

$$+ \frac{T}{2m\pi} \int_0^t p(s) \sin \frac{2m\pi}{T}(t - s)\, ds.$$

Example 9.2 As an exercise, consider the existence of harmonic solution for the critical differential equation

$$x'' + x = E_0 \sin t + \varepsilon x^3,$$

where ε is a small parameter.

On the other hand, it follows from Theorem 9.2 that in non-critical case ($c > 0$), equation (9.13) has the harmonic solution $x = x(t, \varepsilon)$. In the mechanical intuition, it is seems obvious that $x = x(t, \varepsilon)$ has the asymptotic property similar to that of (9.8). We leave its mathematical proof as an exercise for readers (see [123]).

9.2 Method of Averaging

The method of averaging is a very useful appropriate procedure to analyze non-autonomous differential equations. However, it is quite complicated to establish the background for the theory. It is shown here that we can use a relatively simple procedure to prove a fundamental theorem concerning the method of averaging applied to periodic differential equations.

Consider a system of differential equations

$$\frac{dx}{dt} = \varepsilon f(t, x, y, \varepsilon), \qquad \frac{dy}{dt} = \varepsilon g(t, x, y, \varepsilon), \tag{9.20}$$

where ε ($|\varepsilon| \leq \varepsilon_0$) is a small parameter, the functions f and g are continuous in $(t, x, y, \varepsilon) \in \mathbb{R}^3 \times [-\varepsilon_0, \varepsilon_0]$ and continuously differentiable with respect to $(x, y) \in \mathbb{R}^2$. Moreover, assume they are $T(\varepsilon)$-periodic with respect to t where $T(\varepsilon) > 0$ is continuous in $\varepsilon \in [-\varepsilon_0, \varepsilon_0]$.

For given $(x, y, \varepsilon) \in \mathbb{R}^2 \times [-\varepsilon_0, \varepsilon_0]$, calculate the mean values

$$F(x, y, \varepsilon) := \frac{1}{T(\varepsilon)} \int_0^{T(\varepsilon)} f(t, x, y, \varepsilon)\, dt$$

and

$$G(x,y,\varepsilon) := \frac{1}{T(\varepsilon)} \int_0^{T(\varepsilon)} g(t,x,y,\varepsilon)\, dt.$$

Then consider the *averaging system*

$$\frac{dx}{dt} = \varepsilon F(x,y,\varepsilon), \qquad \frac{dy}{dt} = \varepsilon G(x,y,\varepsilon) \qquad (9.21)$$

for the system (9.20).

Let ξ and η be constants satisfying

$$F(\xi,\eta,0) = 0, \qquad G(\xi,\eta,0) = 0, \qquad (9.22)$$

and assume the Jacobian determinant

$$J_0 = \left.\frac{\partial(F,G)}{\partial(x,y)}\right|_{(x,y,\varepsilon)=(\xi,\eta,0)}.$$

Theorem 9.4 *Suppose that $J_0 \neq 0$. Then there exists a positive $\varepsilon_1 < \varepsilon_0$ such that when $|\varepsilon| \leq \varepsilon_1$, system (9.20) has a harmonic solution*

$$(x,y) = (x(t,\varepsilon), y(t,\varepsilon)),$$

which depends continuously on (t,ε) and satisfies the asymptotic condition

$$\lim_{\varepsilon \to 0} (x(t,\varepsilon), y(t,\varepsilon)) = (\xi,\eta).$$

Proof. Let

$$x = x(t) = x(t,x_0,y_0,\varepsilon), \qquad y = y(t) = y(t,x_0,y_0,\varepsilon), \qquad (9.23)$$

be the solution of (9.20) satisfying the initial condition

$$x(0) = x_0, \qquad y(0) = y_0.$$

From the theorem concerning the dependence of solutions on the initial conditions and parameters, we know that the solution (9.23) is continuous in (t,x_0,y_0,ε) for

$$-2T_0 \leq t \leq 2T_0, \qquad (x_0-\xi)^2 + (y_0-\eta)^2 \leq 1, \qquad -\varepsilon_1 \leq \varepsilon \leq \varepsilon_1,$$

where $\varepsilon_1 > 0$ is a sufficiently small constant. Moreover, the solution (9.23) is continuously differentiable with respect to (x_0,y_0). Note that

$$\lim_{\varepsilon \to 0} (x(t,x_0,y_0,\varepsilon), y(t,x_0,y_0,\varepsilon)) = (x_0,y_0). \qquad (9.24)$$

On the other hand, we have

$$\begin{cases} x(t) = x_0 + \varepsilon \int_0^t f(t, x(t), y(t), \varepsilon)\, dt, \\ y(t) = y_0 + \varepsilon \int_0^t g(t, x(t), y(t), \varepsilon)\, dt, \end{cases}$$

which together with the periodic boundary condition

$$x(0) = x(T), \qquad y(0) = y(T),$$

yields that

$$\Phi(x_0, y_0, \varepsilon) = 0, \qquad \Psi(x_0, y_0, \varepsilon) = 0, \tag{9.25}$$

where

$$\begin{cases} \Phi(x_0, y_0, \varepsilon) = \int_0^{T(\varepsilon)} f(t, x(t), y(t), \varepsilon)\, dt, \\ \Psi(x_0, y_0, \varepsilon) = \int_0^{T(\varepsilon)} g(t, x(t), y(t), \varepsilon)\, dt. \end{cases}$$

Using (9.24), we readily obtain

$$\Phi(x_0, y_0, 0) = T(0)F(x_0, y_0, 0), \qquad \Psi(x_0, y_0, 0) = T(0)G(x_0, y_0, 0).$$

It can be seen that (9.22) is equivalent to

$$\Phi(\xi, \eta, 0) = 0, \qquad \Psi(\xi, \eta, 0) = 0,$$

with the Jacobian determinant

$$\left.\frac{\partial(\Phi, \Psi)}{\partial(x_0, y_0)}\right|_{(x_0, y_0, \varepsilon) = (\xi, \eta, 0)} = T_0^2 J_0 \neq 0, \qquad \text{where } T_0 = T(0).$$

Therefore, for sufficiently small parameter $|\varepsilon| < \varepsilon_1$, the implicit equation (9.22) has a continuous solution

$$x_0 = x_0(\varepsilon), \qquad y_0 = y_0(\varepsilon),$$

satisfying

$$x_0(0) = \xi, \qquad y_0(0) = \eta.$$

Finally, we obtain the desired harmonic solution of system (9.20); that is,
$$\begin{cases} x = x(t,\varepsilon) = x(t, x_0(\varepsilon), y_0(\varepsilon), \varepsilon), \\ y = y(t,\varepsilon) = y(t, x_0(\varepsilon), y_0(\varepsilon), \varepsilon). \end{cases}$$

Theorem 9.4 is thus proved. □

Example 9.3 For an application of Theorem 9.4, consider the Doffing's equation

$$\frac{d^2u}{dt^2} + \varepsilon c_1 \frac{du}{dt} + u + \varepsilon d_1 u^3 = \varepsilon B_1 \cos\omega t, \qquad (9.26)$$

where $c_1 \geq 0$, $d_1 \neq 0$, $B_1 \neq 0$ and $\omega > 0$ are constants, and $\varepsilon \geq 0$ is a small parameter. Note that equation (9.26) is $\frac{2\pi}{\omega}$-periodic.

When $\varepsilon = 0$, equation (9.3) represents free oscillations without damping with inherent frequency $\omega_0 = 1$. Suppose that the forced frequency ω satisfies the condition $\omega^2 = 1 + \varepsilon\beta$ where β is a constant. In other words, the forced frequency ω is very close to the inherent frequency $\omega_0 = 1$ when the parameter ε is small enough.

In this case, *does equation (9.26) have a harmonic solution ? If there is one, how can we determine it approximately ?*

Let (9.26) be put in its equivalent system

$$u' = v, \qquad v' = -u - \varepsilon c_1 v - \varepsilon d_1 u^3 + \varepsilon B_1 \cos\omega t.$$

Performing the van der Pol transformation

$$u = x\sin\omega t + y\cos\omega t, \qquad v = \omega(x\cos\omega t - y\sin\omega t),$$

we obtain

$$\begin{cases} x' = \frac{\varepsilon}{\omega}(\beta u - d_1 u^3 - c_1 v + B_1 \cos\omega t)\cos\omega t, \\ y' = -\frac{\varepsilon}{\omega}(\beta u - d_1 u^3 - c_1 v + B_1 \cos\omega t)\sin\omega t, \end{cases} \qquad (9.27)$$

where $\beta = (\omega^2 - 1)/\varepsilon$ and $\varepsilon > 0$. Now, we want to find the averaging equation for equation (9.27). For simplicity, denote $(x,y) = (r\sin\theta, r\cos\theta)$, then

$$\begin{cases} u = x\sin\omega t + y\cos\omega t = r\cos(\omega t - \theta), \\ v = \omega(x\cos\omega t - y\sin\omega t) = -\omega r\sin(\omega t - \theta). \end{cases}$$

Chapter 9. Perturbation Method

We then readily obtain the averaging system

$$\begin{cases} x' = \dfrac{\varepsilon}{2\omega}\left(\beta y - \dfrac{3d_1}{4}r^2 y - c_1\omega x + B_1\right), \\ y' = \dfrac{-\varepsilon}{2\omega}\left(\beta x - \dfrac{3d_1}{4}r^2 x + c_1\omega y\right), \end{cases} \quad (9.28)$$

where $r^2 = x^2 + y^2$ and $\omega^2 = 1 + \varepsilon\beta$. Corresponding to (9.22), we obtain

$$\begin{cases} F(x,y,0) = \dfrac{1}{2}\left(\beta y - \dfrac{3d_1}{4}r^2 y - c_1 x + B_1\right) = 0, \\ G(x,y,0) = -\dfrac{1}{2}\left(\beta x - \dfrac{3d_1}{4}r^2 x - cy\right) = 0, \end{cases}$$

which is equivalent to

$$x = \frac{c_1}{B_1}r^2, \quad y = \frac{1}{B_1}\left(\frac{3d_1}{4}r^2 - \beta\right)r^2, \quad (9.29)$$

where r satisfies the equation

$$c_1^2 r^2 + \left(\beta - \frac{3d_1}{4}r^2\right)^2 r^2 = B_1^2. \quad (9.30)$$

Thus from Theorem 9.4, as long as the Jacobian determinant

$$J_0 = \left.\frac{\partial(F,G)}{\partial(x,y)}\right|_{\varepsilon=0} = \frac{1}{4}\left[(c_1^2 + \beta^2) + \frac{45}{16}d_1^2 x^2 y^2 - \frac{9d_1\beta}{4}r^2\right] \neq 0,$$

equations (9.290) and (9.30) determine constants x and y which form an approximate solution for equation (9.27) provided that $|\varepsilon|$ is sufficiently small. Hence, $u = r\cos(\omega - \theta)$ is an approximate harmonic solution for equation (9.26). This proves that if the Jacobian determinant $J_0 \neq 0$, then (9.26) has a harmonic solution whose amplitude is approximately given by the positive root of the algebraic equation (9.30) provided that $|\varepsilon|$ is sufficiently small.

Let $c = \varepsilon c_1$, $d = \varepsilon d_1$ and $B = \varepsilon B_1$. Note that $\omega^2 = 1 + \varepsilon\beta$. Then (9.30) can be written as

$$\omega^2 = 1 + \frac{3d}{4}r^2 \pm \sqrt{\frac{B^2}{r^2} - c^2}. \quad (9.31)$$

We usually find $r = r(\omega) > 0$ from (9.31), which determines a curve $\mathfrak{R}$ on the (r,ω)-plane, called the *frequency response curve*. It can be seen that the frequency response curve $\mathfrak{R}$ gives the following information:

A point $(r,\omega) \in \mathfrak{R}$ corresponds to a $\dfrac{2\pi}{\omega}$-periodic (i.e., harmonic) solution of equation (9.26) with amplitude near r for sufficiently small ε.

Of course, formula (9.31) is valid only for ω in a small neighborhood at the inherent frequency $\omega = \omega_0 = 1$ since $\omega^2 = 1 + \varepsilon\beta$ for a small parameter ε (i.e., $\omega \approx 1$). In other words, the above result is valid if the forced frequency ω is not much higher compared with the inherent frequency ω_0. Therefore, it is an interesting problem to demonstrate the existence of harmonic solution when the forced frequency is much higher (i.e., $\omega \gg 1$). We will do it in next section.

9.3 High Frequency Forced Oscillations

9.3.1 Small Amplitude Harmonic Solutions

Consider the equation

$$\ddot{x} + c\dot{x} + kx = f(x,\dot{x}) + E\sin\omega t \qquad \left(\dot{} = \frac{d}{dt}\right), \qquad (9.32)$$

where the forced frequency is higher. Assume $f(x,y)$ is a differentiable function, satisfying the condition

$$f(0,0) = f'_x(0,0) = f'_y(0,0) = 0.$$

For example, a polynomial

$$f(x,y) = \sum_{i+j=2}^{N} c_{ij} x^i y^j,$$

of order $N \geq 2$ with constant coefficients c_{ij} satisfies such a condition.

Theorem 9.5 *If the forced frequency ω is sufficiently large, then equation (9.32) has a harmonic solution $x = x(t,\omega)$ which is continuous in (t,ω), and satisfies*

$$|x(t,\omega)| \leq \frac{M_0}{\omega}, \qquad |\dot{x}(t,\omega)| \leq \frac{M_0}{\omega}, \qquad (9.33)$$

where M_0 is a constant independent of ω.

Proof. Equation (9.32) is equivalent to

$$\dot{x} = y, \qquad \dot{y} = -cy - kx + f(x,y) + E\sin\omega t. \qquad (9.34)$$

Let $\tau = \omega t$, $u = \omega x$, $v = \omega y$ and $\varepsilon = 1/\omega$, then equation (9.34) becomes

$$\frac{du}{d\tau} = \varepsilon v, \qquad \frac{dv}{d\tau} = -\varepsilon k u - \varepsilon c v + f(\varepsilon u, \varepsilon v) + E \sin \tau. \tag{9.35}$$

Note that ε is a small parameter since ω is a large parameter, and the system (9.35) belongs to the critical case when $\varepsilon = 0$. We will follow the basic idea of small parameter expansion technique.

We first consider the homogeneous linear system

$$\frac{du}{d\tau} = \varepsilon v, \qquad \frac{dv}{d\tau} = -\varepsilon k u - \varepsilon c v. \tag{9.36}$$

Let

$$U(\tau, \varepsilon) = \begin{pmatrix} \tilde{u}_1(\tau, \varepsilon) & \tilde{u}_2(\tau, \varepsilon) \\ \tilde{v}_1(\tau, \varepsilon) & \tilde{v}_2(\tau, \varepsilon) \end{pmatrix}$$

be the fundamental matrix of solutions for the linear homogeneous system (9.36), satisfying the initial condition

$$U(0, \varepsilon) = I = \begin{pmatrix} 1 & 0 \\ 0 & 1 \end{pmatrix}.$$

Using the method of constant variation, we obtain the solution

$$\begin{pmatrix} u(\tau) \\ v(\tau) \end{pmatrix} = U(\tau, \varepsilon) \begin{pmatrix} u_0 \\ v_0 \end{pmatrix}$$

$$+ \int_0^\tau U(\tau, \varepsilon) U^{-1}(s, \varepsilon) \begin{pmatrix} 0 \\ f(\varepsilon u(s), \varepsilon v(s)) + E \sin s \end{pmatrix} ds$$

of (9.35) satisfying the initial condition

$$\begin{pmatrix} u(0) \\ v(0) \end{pmatrix} = \begin{pmatrix} u_0 \\ v_0 \end{pmatrix}.$$

It is 2π-periodic if and only if

$$\Phi(u_0, v_0, \varepsilon) = 0, \qquad \Psi(u_0, v_0, \varepsilon) = 0, \tag{9.37}$$

where

$$\begin{pmatrix} \Phi \\ \Psi \end{pmatrix} = (U(2\pi, \varepsilon) - I) \begin{pmatrix} u_0 \\ v_0 \end{pmatrix}$$

$$+ \int_0^{2\pi} U(2\pi, \varepsilon) U^{-1}(s, \varepsilon) \begin{pmatrix} 0 \\ f(\varepsilon u(s), \varepsilon v(s)) + E \sin s \end{pmatrix} ds.$$

It follows from $f(\varepsilon u, \varepsilon v) = \mathcal{O}(\varepsilon^2)$ that

$$\Phi(u_0, v_0, \varepsilon) = \varepsilon \Phi_1(u_0, v_0, \varepsilon), \qquad \Psi(u_0, v_0, \varepsilon) = \varepsilon \Psi_1(u_0, v_0, \varepsilon),$$

where

$$\begin{cases} \Phi_1(u_0, v_0, \varepsilon) = 2\pi v_0 + \displaystyle\int_0^{2\pi} (2\pi - s) E \sin s \, ds + \mathcal{O}(\varepsilon) \\ \Psi_1(u_0, v_0, \varepsilon) = -2\pi(k u_0 + c v_0) + \displaystyle\int_0^{2\pi} (2\pi - s) E \sin s \, ds + \mathcal{O}(\varepsilon) \end{cases}$$

Then (9.37) is equivalent to

$$\Phi_1(u_0, v_0, \varepsilon) = 0, \qquad \Psi_1(u_0, v_0, \varepsilon) = 0. \tag{9.38}$$

Clearly, the equations $\Phi_1(u_0, v_0, 0) = 0$ and $\Psi_1(u_0, v_0, 0) = 0$ uniquely determine

$$u_0 = \xi = -\frac{(1+c)E}{2k\pi} \int_0^{2\pi} s \sin s \, ds, \qquad v_0 = \eta = \frac{E}{2\pi} \int_0^{2\pi} s \sin s \, ds.$$

On the other hand, we have the Jacobian determinant

$$\frac{\partial(\Phi_1, \Psi_1)}{\partial(u_0, v_0)}(\xi, \eta, 0) = \begin{vmatrix} 0 & 2\pi \\ -2\pi k & -2\pi c \end{vmatrix} = 4k\pi^2 \neq 0.$$

Therefore, the implicit function theorem implies that there exists $\varepsilon_0 > 0$ such that (9.38)) has a continuous solution

$$u_0 = u_0(\varepsilon), \qquad v_0 = v_0(\varepsilon) \qquad \text{(satisfying } u_0(0) = \xi, \quad v_0(0) = \eta\text{)}.$$

It follows that (9.35) has a 2π-periodic solution

$$\begin{cases} u = \tilde{u}(\tau, \varepsilon) = u(\tau, u_0(\varepsilon), v_0(\varepsilon), \varepsilon), \\ v = \tilde{v}(\tau, \varepsilon) = v(\tau, u_0(\varepsilon), v_0(\varepsilon), \varepsilon), \end{cases}$$

which is continuous in $(\tau, \varepsilon) \in [0, 2\pi] \times [-\varepsilon_0, \varepsilon_0]$. Let

$$M_0 = \max_{(0 \le \tau \le 2\pi,\ |\varepsilon| \le \varepsilon_0)} \{|\tilde{u}(s, \varepsilon)|, |\tilde{v}(s, \varepsilon)|\}.$$

Then

$$x = x(t, \omega) = \frac{1}{\omega} \tilde{u}\left(\omega t, \frac{1}{\omega}\right)$$

is the desired $2\pi/\omega$-solution of (9.32) having the property (9.33).
Theorem 9.5 is thus proved. $\square$

9.3.2 Large Amplitude Harmonic Solutions

Consider the periodically perturbed equation

$$x'' + f(x)x' + g(x) = p(\omega t), \tag{9.39}$$

where

$$f(x) = \sum_{i=0}^{m} a_{2i+1} x^{2i+1}, \qquad g(x) = \sum_{i=0}^{n} b_{2i+1} x^{2i+1}$$

are polynomials with constant coefficients, and $p(\omega t)$ is a $2\pi/\omega$-periodic perturbation. Assume $f(x)$ and $g(x)$ satisfy the conditions

$$\begin{cases} b_{2n+1} = \beta^2 > 0, & n \ge 2(m+1), & \text{if } f(x) \not\equiv 0, \\ b_{2n+1} = \beta^2 > 0, & n > 0, & \text{if } f(x) \equiv 0. \end{cases}$$

For the equation (9.39), perform the transformation

$$t = \frac{s}{\omega}, \qquad x = \omega^{\frac{1}{n}} u, \qquad \omega = \frac{1}{\varepsilon^n},$$

we obtain the equivalent 2π-periodic system

$$\frac{du}{ds} = v, \qquad \frac{dv}{ds} = -\beta^2 u^{2n+1} + \varepsilon F(s, u, v, \varepsilon), \tag{9.40}$$

where

$$F(s, u, v, \varepsilon) = \varepsilon^{2n} p(s) - \sum_{i=0}^{n-1} b_{2i+1} u^{2i+1} \varepsilon^{2(n-i)-1}$$

$$- \sum_{j=0}^{m} a_{2j+1} u^{2j+1} v \varepsilon^{n-(2j+1)}.$$

Denote by
$$u = u_\varepsilon(s, u_0, v_0), \qquad v = v_\varepsilon(s, u_0, v_0),$$
the solution of system (9.40) passing through the initial point (u_0, v_0). Then
$$H_\varepsilon : \quad (u_0, v_0) \quad \mapsto \quad (u_\varepsilon(2\pi, u_0, v_0), v_\varepsilon(2\pi, u_0, v_0)),$$
defines the Poincaré map of (9.40). Since (9.40) is an analytic system with respect to the variables (u, v), its Poincaré map H_ε is analytic in (u_0, v_0).

Then consider the auxiliary equation
$$\frac{du}{ds} = v, \qquad \frac{dv}{ds} = -\beta^2 u^{2n+1}. \tag{9.41}$$

Clearly, system (9.41) has a first integral
$$\frac{1}{2}v^2 + \frac{\beta^2}{2n+2} u^{2n+2} = c, \tag{9.42}$$
where $c \geq 0$ is an integral constant. For given $c > 0$, the integral (9.42) defines a closed orbit Γ_c of (9.41), which is starlike with respect to the origin o. Let $\wp(c) > 0$ be the minimal period of Γ_c. It is not hard to check that $\wp(c)$ is monotonically decreasing and satisfies
$$\lim_{c \to 0} \wp(c) = \infty \quad \text{and} \quad \lim_{c \to \infty} \wp(c) = 0.$$
There is a unique $c_0 > 0$ such that $\wp(c_0) = 2\pi$. Let
$$c_1 = c_0 - \delta_0 > 0, \qquad c_2 = c_0 + \delta_0,$$
where $\delta_0 > 0$ is a small constant. Then we have a starlike annulus $\mathcal{A}_0$ with the inner boundary Γ_{c_1} and the outer boundary Γ_{c_2}.

For brevity, let
$$(u(s), v(s)) = (u_0(s, \xi, \eta), v_0(s, \xi, \eta)).$$
Then, we have
$$H_0 : \quad (\xi, \eta) \quad \mapsto \quad (u(2\pi), v(2\pi)).$$
In what follows, we will use polar coordinates
$$(\xi, \eta) = \langle \rho, \sigma \rangle, \qquad (u(s), v(s)) = \langle r(s), \theta(s) \rangle,$$
where
$$r(s) = r(s, \rho, \sigma), \qquad \theta(s) = \theta(s, \rho, \sigma),$$

are continuously differentiable about (s, ρ, σ) for $\rho = \sqrt{\xi^2 + \eta^2} > 0$.
It follows from (9.41) that

$$\frac{d\theta}{ds} = -\sin^2\theta - \beta^2 r^{2n} \cos^{2n+2}\theta < 0, \tag{9.43}$$

which implies that the motion of (9.41) moves clockwise on the closed orbit Γ_c. Let

$$\Theta(\rho, \sigma) := \theta(2\pi, \rho, \sigma) + 2\pi - \sigma.$$

It follows from the property of $\wp(c)$ that

$$\begin{cases} \Theta(\rho, \sigma) > 0 & \text{for } \langle \rho, \sigma \rangle \in \Gamma_{c_1}; \\ \Theta(\rho, \sigma) < 0 & \text{for } \langle \rho, \sigma \rangle \in \Gamma_{c_2}. \end{cases} \tag{9.44}$$

In a similar manner, let

$$(u_\varepsilon(s), v_\varepsilon(s)) = (u_\varepsilon(s, \xi, \eta), v_\varepsilon(s, \xi, \eta)).$$

Then, we have

$$H_\varepsilon: \quad (\xi, \eta) \quad \mapsto \quad (u_\varepsilon(2\pi), v_\varepsilon(2\pi)).$$

We will use polar coordinates

$$(\xi, \eta) = \langle \rho, \sigma \rangle, \quad (u_\varepsilon(s), v_\varepsilon(s)) = \langle r_\varepsilon(s), \theta_\varepsilon(s) \rangle,$$

where

$$r_\varepsilon(s) = r(s, \rho, \sigma, \varepsilon), \quad \theta_\varepsilon(s) = \theta(s, \rho, \sigma),$$

are continuously differentiable about (s, ρ, σ) for $\rho = \sqrt{\xi^2 + \eta^2} > 0$.
Let

$$\Theta_\varepsilon(\rho, \sigma) := \theta_\varepsilon(2\pi, \rho, \sigma) + 2\pi - \sigma.$$

It follows from (9.44) together with the continuity property that there is a sufficiently $\varepsilon_0 > 0$, such that the following result is valid.

Lemma 9.1 *If ε_0 is sufficiently small, we have*

$$\begin{cases} \Theta_\varepsilon(\rho, \sigma) > 0 & \text{for } \langle \rho, \sigma \rangle \in \Gamma_{c_1}; \\ \Theta_\varepsilon(\rho, \sigma) < 0 & \text{for } \langle \rho, \sigma \rangle \in \Gamma_{c_2}, \end{cases} \quad \text{for } |\varepsilon| < \varepsilon_0.$$

It implies that H_ε is a twist analytic map on $\mathcal{A}_0$ for $|\varepsilon| < \varepsilon_0$. Therefore, we have clearly the following result.

Lemma 9.2 H_ε has a radially transformation closed curve Δ_ε in $\mathcal{A}_0$ for $|\varepsilon| < \varepsilon_0$, such that $\Delta_\varepsilon \approx \Gamma_{c_0}$ when δ_0 is sufficiently small.

Now, differentiating equation (9.43) with respect to ρ, we obtain

$$\frac{d}{ds}\frac{\partial \theta}{\partial \rho} = A_1 \frac{\partial \theta}{\partial \rho} + B_1 \frac{\partial r}{\partial \rho}, \tag{9.45}$$

where

$$\begin{cases} A_1 = -\sin 2\theta + (2n+2)\beta^2 r^{2n} \cos^{2n+1}\theta \sin\theta, \\ B_1 = -2n\beta^2 r^{2n-1} \cos^{2n+2}\theta \leq 0. \end{cases}$$

On the other hand, putting (9.42) in the polar coordinates

$$\frac{1}{2}r^2 \sin^2\theta + \frac{\beta^2}{2n+2} r^{2n+2} \cos^{2n+2}\theta = c,$$

and differentiating it with respect to ρ, we obtain

$$E_1 \frac{\partial r}{\partial \rho} = F_1 + G_1 \frac{\partial \theta}{\partial \rho}, \tag{9.46}$$

where

$$\begin{cases} E_1 = r\sin^2\theta + \beta^2 r^{2n+1} \cos^{2n+2}\theta > 0, \\ F_1 = \rho\sin^2\theta + \beta^2 \rho^{2n+1} \cos^{2n+2}\theta > 0, \\ G_1 = -r^2 \sin\theta \cos\theta + \beta^2 r^{2n+2} \cos^{2n+1}\theta \sin\theta. \end{cases}$$

Eliminating $\dfrac{\partial r}{\partial \rho}$ from (9.45) and (9.46), we obtain

$$\frac{d}{ds}\frac{\partial \theta}{\partial \rho} = A \frac{\partial \theta}{\partial \rho} + B, \tag{9.47}$$

with

$$A = \frac{A_1 E_1 + B_1 G_1}{E_1}, \qquad B = \frac{B_1 F_1}{E_1} \leq 0.$$

Note that

$$\theta(0, \rho, \sigma) = \sigma, \qquad \frac{\partial \theta(0, \rho, \sigma)}{\partial \rho} = 0.$$

Hence, from (9.47), we obtain

$$\frac{\partial \Theta(\rho,\sigma)}{\partial \rho} = e^{\int_0^{2\pi} A\,d\tau} \int_0^{2\pi} Be^{-\int_0^s A\,dt}\,ds.$$

Since $B < 0$ for almost all θ, we obtain

$$\frac{\partial \Theta(\rho,\sigma)}{\partial \rho} = e^{\int_0^{2\pi} A\,d\tau} \int_0^{2\pi} Be^{-\int_0^s A\,dt}\,ds < 0, \qquad \text{for } \rho > 0.$$

Then, using the continuity property yields the following result.

$$\frac{\partial \Theta_\varepsilon(\rho,\sigma)}{\partial \rho} < 0, \qquad \text{for } \rho > 0, \tag{9.48}$$

which implies the following result.

Lemma 9.3 *The radially transformation curve Δ_ε intersects any radial-ray at one and only one point and, therefore, it is starlike with respect to the origin $\mathbf{o}$.*

Lemma 9.4 *Suppose that the periodic function $p(s)$ is odd. Then the Poincaré map H_ε of system (9.40) has at least two fixed points in $\mathcal{A}_0$ for $|\varepsilon| < \varepsilon_0$ whenever $\varepsilon_0 > 0$ is sufficiently small.*

Proof. Let

$$P_1 = (\alpha, 0) \in \Delta_\varepsilon, \qquad \alpha_1 < \alpha < \alpha_2.$$

Note that when $\varepsilon = 0$, P_1 is a fixed point of H_0 and Δ_ε is a radially transformed curve for the map H_ε. It follows that

$$H_\varepsilon(P_1) = P_2 \tag{9.49}$$

where

$$P_2 := (\tilde{\alpha}, 0), \qquad \alpha_1 < \tilde{\alpha} < \alpha_2,$$

for $|\varepsilon| < \varepsilon_0$, provided that ε_0 is small enough. That is,

$$u_\varepsilon(2\pi, \alpha, 0) = \tilde{\alpha}, \qquad v_\varepsilon(2\pi, \alpha, 0) = 0,$$

It follows from (9.49) that $H_\varepsilon^{-1}(P_2) = P_1$ which yields

$$u_\varepsilon(-2\pi, \tilde{\alpha}, 0) = \alpha, \qquad v_\varepsilon(-2\pi, \tilde{\alpha}, 0) = 0. \tag{9.50}$$

On the other hand, performing the transformation

$$y = -u, \qquad z = v \tag{9.51}$$

for (9.40), we obtain

$$\frac{dy}{ds} = -z, \qquad \frac{dz}{ds} = \beta^2 y^{2n+1} + \varepsilon F(s, -y, z, \varepsilon). \tag{9.52}$$

Let

$$y = y_\varepsilon(s, \xi, \eta), \qquad z = z_\varepsilon(s, \xi, \eta),$$

denote the solution of (9.52) satisfying the initial condition $(y(0), z(0)) = (\xi, \eta)$. From (9.51), we obtain

$$y_\varepsilon(s, \xi, \eta) = -u_\varepsilon(s, -\xi, \eta), \qquad z_\varepsilon(s, \xi, \eta) = v_\varepsilon(s, -\xi, \eta). \tag{9.53}$$

Furthermore, let $s = -t$, then (9.52) becomes

$$\frac{dy}{dt} = z, \qquad \frac{dz}{dt} = -\beta^2 y^{2n+1} + \varepsilon F(t, y, z, \varepsilon), \tag{9.54}$$

where we used the identity

$$F(-t, -y, z, \varepsilon) = -F(t, y, z, \varepsilon).$$

It follows that

$$y = y_\varepsilon(-t, \xi, \eta), \qquad z = z_\varepsilon(-t, \xi, \eta),$$

is the solution of (9.54) satisfying the initial condition $(y(0), z(0)) = (\xi, \eta)$. Since (9.40) and (9.54) are identical in the sense that

$$t = s, \quad y = u, \quad z = v.$$

It follows that

$$\begin{cases} y_\varepsilon(-t, \xi, \eta) = u_\varepsilon(t, \xi, \eta), \\ z_\varepsilon(-t, \xi, \eta) = v_\varepsilon(t, \xi, \eta). \end{cases}$$

Using (9.53) again, we obtain

$$\begin{cases} u_\varepsilon(s, \xi, \eta) = -u_\varepsilon(-s, -\xi, \eta), \\ v_\varepsilon(s, \xi, \eta) = v_\varepsilon(-s, -\xi, \eta). \end{cases}$$

Then, letting $s = 2\pi$, $-\xi = \tilde{\alpha}$ and $\eta = 0$, we have

$$\begin{cases} u_\varepsilon(2\pi, -\tilde{\alpha}, 0) = -u_\varepsilon(-2\pi, \tilde{\alpha}, 0), \\ v_\varepsilon(2\pi, -\tilde{\alpha}, 0) = v_\varepsilon(-2\pi, \tilde{\alpha}, 0), \end{cases} \qquad (9.55)$$

which together with (9.50) implies

$$u_\varepsilon(2\pi, -\tilde{\alpha}, 0) = -\alpha, \qquad v_\varepsilon(2\pi, -\tilde{\alpha}, 0) = 0.$$

It follows that

$$H_\varepsilon(Q_1) = Q_2, \qquad (9.56)$$

where $Q_1 = (-\tilde{\alpha}, 0)$ and $Q_2 = (-\alpha, 0)$. It follows from Lemma 9.3 that Q_1 is a point of Δ_ε.

1) Assume $\tilde{\alpha} = \alpha$. It follows from (9.56) and (9.49) that P_1 and Q_1 are two distinct fixed points of H_ε.

2) Assume $\tilde{\alpha} > \alpha$. Then $P_1 \in \delta_\varepsilon$ implies that $P_2 = H_\varepsilon(P_1)$ belongs to the outside of Δ_ε; $Q_1 \in \Delta_\varepsilon$ implies that $Q_2 = H_\varepsilon(Q_1)$ belongs to the inside of Δ_ε. Therefore, H_ε is a bend-twist map on $\mathcal{A}_0$, and it has at least two distinct fixed points in $\mathcal{A}_0$.

3) Assume $\tilde{\alpha} < \alpha$. We conclude in a similar manner that H_ε has at least two distinct fixed points in $\mathcal{A}_0$.

We have thus proved Lemma 9.4. □

In a similar manner, we can prove the following result.

Lemma 9.5 *Suppose that the periodic function $p(s)$ is even and $f(x) \equiv 0$. Then the Poincaré map H_ε of system (9.40) has at least two fixed points in $\mathcal{A}_0$ for $|\varepsilon| < \varepsilon_0$ whenever $\varepsilon_0 > 0$ is sufficiently small.*

Finally, as a direct consequence of Lemma 9.4 and Lemma 9.5, we can prove the following result.

Theorem 9.6 *Suppose that the periodic function $p(\omega t)$ is odd (or even and $f(x) \equiv 0$) and the forced frequency ω is sufficiently large. Then the differential equation (9.39) has two distinct harmonic solutions $x = x_i(t, \omega)$ having amplitude larger than $M_0 \sqrt[n]{\omega}$ for $i = 1, 2$, where M_0 is some positive constant.*

Proof. In fact, it follows from Lemma 9.4 (or Lemma 9.5) that H_ε has at least two distinct fixed points $(\xi_i, \eta_i) \in \mathcal{A}_0$. Then equation (9.39) has two distinct harmonic (i.e., $2\pi/\omega$-periodic) solution

$$x = x_i(t, \omega) = \omega^{\frac{1}{n}} u_\varepsilon(\omega t, \xi_i, \eta_i), \qquad i = 1, 2.$$

Then we have

$$\max_{(0\leq t\leq 2\pi/\omega,\ \omega\geq\omega_0)} |x_i(t,\omega)| = M_0 \sqrt[n]{\omega}, \qquad i=1,\,2, \qquad (9.57)$$

where

$$M_0 = \max_{(0\leq s\leq 2\pi,\ |\varepsilon|\leq\varepsilon_0)} |u_\varepsilon(s,\xi,\eta)| > 0.$$

It can be seen from (9.57) that the amplitude of $x_i(t,\omega)$ is larger than $M_0 \sqrt[n]{\omega}$ ($i=1,\,2$), which tends to ∞ as $\omega \to \infty$.

We have thus proved Theorem 9.6. $\qquad\square$

Chapter 10

Duffing Equations of Second Order

10.1 Periodic Oscillations

10.1.1 Linear and Nonlinear Oscillations

Consider the scalar Duffing equation

$$\ddot{x} + c\dot{x} + g(x) = p(t), \qquad x \in \mathbb{R}^1, \tag{10.1}$$

where $\dot{x}$ and $\ddot{x}$ denote, respectively, the first and second derivatives of x with respect to t, the constant $c \geq 0$ is the coefficient of friction, and the function $g(x)$ is the restoring force satisfying the local Lipschitz condition. Furthermore, let $g(x)$ satisfy the *sign condition*

$$(S)_a: \qquad xg(x) > 0, \qquad \text{for } |x| > a,$$

where $a \geq 0$ is some constant. When $a = 0$, the sign condition $(S)_0$ is said to be *strict*.

In the sequel, the forced term $p(t)$ is assumed to be a piecewise continuous periodic function with minimal period $T > 0$.

People are much interested in the scalar Duffing equation (10.1) since it is the simple model admitting of abundant complicated phenomena in mathematical physics.

Assume $g(x)$ satisfies the condition

$$xg(x) \geq bx^2 \quad (|x| > a),$$

with some positive constants a and b.

If $c > 0$, the Duffing equation (10.1) has the following property.

It is dissipative and all of its solutions are thus bounded. In fact, there is a compact attractor of the dissipative Duffing equation. However, people still know very little about the attractor despite the abundant numerical computations.

If $c = 0$, the Duffing equation

$$\ddot{x} + g(x) = p(t), \qquad x \in \mathbb{R}^1, \qquad (10.2)$$

is conservative. No doubt, the bounded-ness problem of solution of the conservative equation (10.2) is one of the most troublesome problems in the theory of differential equations [91]. Indeed, it is not an easy task to prove the existence of (even one) bounded solution.

Theorem 4.10 says that if there is a periodic solution of (10.2), then the Poincaré map of the equation has a positively stable continuum and a negatively stable continuum connecting the initial point of the periodic solution to the infinity. It leads to a group of bounded solutions. Anyway, to prove the existence of periodic solution is an achievement of nonlinear analysis.

In literature, there is a great amount of contributions to the Duffing equation. However, for the professional consideration, we are limited in the familiar content of some works of the author and the collaborator F. Zanolin (see the references of this book).

To introduce the basic terminologies, consider the following example.

Example 10.1 Consider the linear Duffing equation

$$\ddot{x} + Kx = p(t), \qquad x \in \mathbb{R}^1, \qquad (10.3)$$

where $K > 0$ is the coefficient of elasticity. Using the method of variation of constant, we obtain the general solution

$$x = c_1 \cos \sqrt{K} t + c_2 \sin \sqrt{K} t + \frac{1}{\sqrt{K}} \int_0^t p(s) \sin \sqrt{K}(t - s) \, ds, \quad (10.4)$$

where c_1 and c_2 are arbitrary constants.

It is obvious that the general solution (10.4) is bounded if and only if

$$\frac{1}{\sqrt{K}} \int_0^t p(s) \sin \sqrt{K}(t - s) \, ds$$

is bounded; or equivalently, both the functions

$$\int_0^t p(s) \cos \sqrt{K} s \, ds \qquad \text{and} \qquad \int_0^t p(s) \sin \sqrt{K} s \, ds$$

are bounded. The characteristic values of the elasticity coefficient K:

$$\lambda_j = \left(\frac{2j\pi}{T}\right)^2, \quad \text{where } j \geq 0 \text{ is an integer,}$$

are called the *resonance points* of (10.3). We have following conclusions:

(1) If K is not a resonance point (i.e., $K \neq \lambda_j$ for all integers $j \geq 0$), all the solutions of (10.3) are bounded;
(2) If K is a resonance point (say, $K = \lambda_n$ for some $n \in \mathbb{Z}^1$), then all the solutions of (10.3) are unbounded whenever $|a_n| + |b_n| \neq 0$; or T-periodic whenever $|a_n| + |b_n| = 0$, where the constants

$$a_n = \frac{2}{T}\int_0^T p(t)\cos\frac{2n\pi t}{T}\,dt \quad \text{and} \quad b_n = \frac{2}{T}\int_0^T p(t)\sin\frac{2n\pi t}{T}\,dt.$$

Note that periodic solution cannot coexist with unbounded solution for the linear Duffing equation. Beside the condition in the conclusion (1), the existence of periodic solution implies the bounded-ness of solution. These conclusions involve heavily the *resonance spectrum*

$$\Lambda = \left\{\lambda_j = \left(\frac{2j\pi}{T}\right)^2 : \text{ for integers } j \geq 0\right\}$$

and the harmonic components a_n and b_n ($n \geq 0$).

For nonlinear Duffing equation, the considerations are complicated. Shortly speaking [51], the non-resonance (i.e., the existence of bounded solution) is dependent on the relative positions of the resonance spectrum Λ and the *set of elasticity*

$$\Upsilon = \left\{y \in \mathbb{R}^1 : \liminf_{|x|\to\infty}\frac{g(x)}{x} \leq y \leq \limsup_{|x|\to\infty}\frac{g(x)}{x}\right\}.$$

10.1.2 Classification of Periodic Oscillations

We first consider the general differential equation of second order

$$\ddot{x} + F(t, x, \dot{x}) = 0, \tag{10.5}$$

or its equivalent system

$$\dot{x} = y, \quad \dot{y} = -F(t, x, y), \quad (x, y) \in \mathbb{R}^2, \tag{10.6}$$

where the function $F(t, x, y)$ is smooth enough in $(t, x, y) \in \mathbb{R}^3$, and periodic in t with minimal period $T > 0$.

Now, let
$$x = u(t), \quad y = v(t) \tag{10.7}$$
be a periodic solution of (10.6) with minimal period $T_0 > 0$. To clarify the geometric meaning of the solution (10.7), let us consider an equivalent autonomous system of (10.6)
$$\begin{cases} \dot{x} = y, \\ \dot{y} = -F(\tau, x, y), \quad (\tau, x, y) \in \mathbb{S}_T \times \mathbb{R}^1 \times \mathbb{R}^1, \\ \dot{\tau} = 1, \end{cases} \tag{10.8}$$
where $\mathbb{S}_T = \mathbb{R}^1/\mathrm{mod}(T)$. Therefore, the three-dimensional manifold $E^3 = \mathbb{S}_T \times \mathbb{R}^2$ is the phase space of (10.8). Obviously,
$$\tau = t, \quad x = u(t), \quad y = v(t) \tag{10.9}$$
is a solution of (10.8) associated to the solution (10.7). Let Γ be the orbit of the solution (10.9) in the space E^3. It follows that the orbit Γ transversally intersects the section
$$S_0 = \{ (\tau, x, y) \in E^3 : \quad \tau = 0 \}$$
of E^3 at the points
$$(kT, x(kT), y(kT)) = (0, x(kT), y(kT)) \quad (k \in \mathbb{Z}^1). \tag{10.10}$$
If $\frac{T_0}{T}$ is an irreducible fraction $\frac{p}{q}$, then (10.10) is a finite set
$$\{(0, x(kqT_0/p), y(kqT_0/p)) : \quad 0 \le k \le p-1 \},$$
which implies that Γ is a closed orbit in the phase space E^3.

If $\frac{T_0}{T}$ is an irrational number μ, then (10.10) is an infinite set
$$\{ (0, x(k\mu T_0), y(k\mu T_0)) \},$$
which implies that Γ is a nontrivial recurrent orbit in E^3.

Definition 10.1
(1) (10.7) is called a *harmonic solution* if $\frac{T_0}{T} = 1$;
(2) (10.7) is called a *subharmonic solution* of order m if $\frac{T_0}{T} = m$ is an integer > 1;

(3) (10.7) is called a *superharmonic solution* of order n if $\frac{T_0}{T} = \frac{1}{n}$ with an integer $n > 1$;

(4) (10.7) is called a *hyperharmonic solution* of order $\frac{m}{n}$ if $\frac{T_0}{T} = \frac{m}{n}$ is an irreducible fraction with integers $m > 1$ and $n > 1$;

(5) (10.7) is called a *quasi-periodic solution* if $\frac{T_0}{T} = \mu$ is an irrational number.

The geometric meanings of the solutions in the definition are clarified in the above discussions. It is remarked that the 'periodicity' of the solution (10.7) is determined not only by the period T_0 of the solution itself but also by the period T of the system.

Example 10.2 Consider the periodic system

$$\dot{x} = y, \quad \dot{y} = -x - (x^2 + y^2 - 1)\sin\omega t \quad (\omega > 0)$$

with minimal period $T = 2\pi/\omega$. There is a particular solution

$$\Gamma_0: \quad x = \sin t, \quad y = \cos t,$$

of minimal period $T_0 = 2\pi$. Note that $\frac{T_0}{T} = \omega$. Then we conclude that

(1) Γ_0 is harmonic if $\omega = 1$;

(2) Γ_0 is subharmonic of order m if ω is an integer $m > 1$;

(3) Γ_0 is superharmonic of order n if $\omega = \frac{1}{n}$ with an integer $n > 1$;

(4) Γ_0 is hyperharmonic of order $\frac{m}{n}$ if $\omega = \frac{m}{n}$ is an irreducible fraction for integers $m > 1$ and $n > 1$;

(5) Γ_0 is quasi-periodic if ω is an irrational number.

Lemma 10.1 *The periodic solution of (10.5) is harmonic or subharmonic if the variables (t, x, y) of $F(t, x, y)$ are separated in the form:*

$$F(t, x, y) = G(x, y) + h(t),$$

where $h(t)$ is periodic in t with minimal period $T > 0$.

Proof. Let $x = \phi(t)$ be a periodic solution of (10.5) with minimal period $T_0 > 0$. Then we have

$$h(t) = -[\ddot{\phi}(t) + G(\phi(t), \dot{\phi}(t))], \quad t \in \mathbb{R}^1,$$

which yields

$$h(t + T_0) = -[\ddot{\phi}(t + T_0) + G(\phi(t + T_0), \dot{\phi}(t + T_0))], \quad t \in \mathbb{R}^1.$$

Since $x = \phi(t)$ is T_0-periodic, we have

$$\ddot\phi(t+T_0) + G(\phi(t+T_0), \dot\phi(t+T_0)) = \ddot\phi(t) + G(\phi(t), \dot\phi(t)).$$

It follows that $h(t+T_0) = h(t)$ for all $t \in \mathbb{R}^1$. On the other hand, since $h(t)$ is periodic with minimal period $T > 0$, we have $T_0 = mT$ for some positive integer $m \geq 1$. Therefore, $x = \phi(t)$ is harmonic if $m = 1$ or subharmonic if $m > 1$.

Lemma 10.1 is thus proved. □

Corollary 10.1 *The periodic solution of Duffing equation is harmonic or subharmonic.*

10.1.3 Classification of Duffing Equations

In literature, the Duffing equations are usually classified as follows.

Definition 10.2

(1) If $g(x)$ satisfies the super-linear condition

$$(S_p): \qquad \lim_{|x|\to\infty} \frac{g(x)}{x} = +\infty,$$

then (10.1) and (10.2) are said to *super-linear*;

(2) If $g(x)$ satisfies the sub-linear condition

$$(S_b): \qquad \lim_{|x|\to\infty} \frac{g(x)}{x} = +0,$$

then (10.1) and (10.2) are said to *sub-linear*;

(3) If $g(x)$ satisfies the semi-linear condition

$$(S_m): \qquad 0 < \liminf_{|x|\to\infty} \frac{g(x)}{x} \leq \limsup_{|x|\to\infty} \frac{g(x)}{x} < \infty,$$

then (10.1) and (10.2) are said to *semi-linear*.

However, the Definition does not include the particular cases:

$$(\check{S}_p): \qquad \limsup_{|x|\to\infty} \frac{g(x)}{x} = +\infty$$

and

$$(\check{S}_b): \qquad \liminf_{|x|\to\infty} \frac{g(x)}{x} = +0.$$

Conditions (S_p) and (S_m) imply the condition

$$(g_{\pm\infty}): \quad g(x) \to \pm\infty, \quad \text{as } x \to \pm\infty.$$

But the condition (S_b) does not imply the condition $(g_{\pm\infty})$, which will be tacitly assumed for sub-linear Duffing equations.

Denote by

$$G(x) = \int_0^x g(x)\,dx$$

the primitive of $g(x)$. Then assume the following conditions

$$(S_p^*): \quad \lim_{|x|\to\infty} \frac{2G(x)}{x^2} = \infty;$$

$$(S_b^*): \quad \lim_{|x|\to\infty} \frac{2G(x)}{x^2} = 0;$$

$$(S_m^*): \quad 0 < \liminf_{|x|\to\infty} \frac{2G(x)}{x^2} \le \limsup_{|x|\to\infty} \frac{2G(x)}{x^2} < \infty,$$

respectively.

In addition, if $g(x)$ is continuously differentiable, assume the following conditions

$$(S_p'): \quad \lim_{|x|\to\infty} g'(x) = \infty;$$

$$(S_b'): \quad \lim_{|x|\to\infty} g'(x) = 0;$$

$$(S_m'): \quad 0 < \liminf_{|x|\to\infty} g'(x) \le \limsup_{|x|\to\infty} g'(x) < \infty,$$

respectively.

Obviously, we have the following implications:

$$(S_p') \Rightarrow (S_p) \Rightarrow (S_p^*);$$

$$(S_b') \Rightarrow (S_b) \Rightarrow (S_b^*);$$

$$(S_m') \Rightarrow (S_m) \Rightarrow (S_m^*).$$

The Duffing equation is usually considered under these conditions. In 1982, Wang Duo studied a mixed type of Duffing equation, where $g(x)$ is super-linear for $x > 0$ and sub-linear for $x < 0$ (see [117]). In 1986, Fučik considered the mixed type of semi-linear Duffing equation

$$\ddot{x} + ax^+ + bx^- + h(x) = p(t),$$

where $a > 0$ and $b > 0$ are constants,

$$x^+ = \max\{x, 0\}, \qquad x^- = \{x, 0\},$$

and the continuous function $h(x)$ satisfies the condition

$$\lim_{|x| \to \infty} \frac{h(x)}{x} = 0.$$

Therefore, if $a \neq b$, that is,

$$\lim_{x \to +\infty} \frac{g(x)}{x} \neq \lim_{x \to -\infty} \frac{g(x)}{x},$$

the Fučik equation is asymmetric.

10.1.4 Solution of Conservative Duffing Equation

It is known that all the solutions of the dissipative Duffing equation (10.1) approach positively to a compact attractor in the phase plane.

Now, we are concerned with the bound of solution for conservative Duffing equation (10.2). Consider its equivalent system

$$\dot{x} = y, \quad \dot{y} = -g(x) + p(t). \tag{10.11}$$

Let

$$x = x(t, x_0, y_0), \qquad y = y(t, x_0, y_0) \tag{10.12}$$

be the solution of (10.11) satisfying the initial condition $(x(0), y(0)) = (x_0, y_0)$. It can be seen that the solution (10.12) is existent and unique on the interval $-\infty < t < \infty$, and continuous with respect to initial condition on the finite interval $|t| \leq T$.

Furthermore, the Poincaré map

$$\Phi : \quad (x_0, y_0) \quad \mapsto \quad (x(T, x_0, y_0), y(T, x_0, y_0))$$

of (10.11) is a homeomorphism of the phase plane. Let D_d be the closed disk in the plane with radius $d > 0$ centered at the origin. Then

$$K = \{(x,y) = (x(t,x_0,y_0), y = y(t,x_0,y_0)), \quad \text{for } |t| \leq T,\ (x_0,y_0) \in D_d\}$$

is a compact set in $\mathbb{R}^2$. Hence, there is a sufficiently large constant $a > d$, such that

$$K \subset D_a,$$

which leads to the following result.

Lemma 10.2 *Given a constant $b > 0$, there is a constant $a > b$, such that*

$$x^2(t,x_0,y_0) + y^2(t,x_0,y_0) > b^2, \quad \text{for } |t| \leq T,$$

whenever $x_0^2 + y_0^2 > a^2$.

Now, we write the solution (10.12) in polar coordinates

$$x(t) = r(t)\cos\theta(t), \quad y(t) = r(t)\sin\theta(t)$$

with the corresponding initial values

$$x_0 = r_0\cos\theta_0, \quad y_0 = r_0\sin\theta_0.$$

It follows from (10.11) that

$$\begin{cases} \dot{r} = [r\cos\theta - g(r\cos\theta) + p(t)]\sin\theta, \\ \dot{\theta} = -\sin^2\theta - \dfrac{1}{r}[g(r\cos\theta) - p(t)]\cos\theta. \end{cases} \quad (10.13)$$

It follows from Lemma 10.2 that the solution (10.12) can be written in the form

$$r = r(t) = r(t,r_0,\theta_0) > b, \quad \theta = \theta(t) = \theta(t,r_0,\theta_0) \quad (|t| \leq T)$$

if it satisfies the initial condition $r(0) = r_0 > a,\ \theta(0) = \theta_0$.

Lemma 10.3 *There is a sufficiently large constant $a > 0$, such that*

$$\dot{\theta}(t) < 0, \quad \text{whenever } r(t) > a.$$

Proof. It follows from the condition $(g_{\pm\infty})$ that there is constant $N > 0$, such that

$$\frac{g(x) - p(t)}{x}\cos^2\theta > 0, \quad \text{for } |x| \geq N.$$

Therefore, (10.13) yields

$$\dot\theta = -\left[\sin^2\theta + \frac{g(x)-p(t)}{x}\cos^2\theta\right] < 0,$$

whenever $|x| \geq N$.

If $|x| \leq N$, we can choose a sufficient large constant $H > 0$, such that

$$\sin^2\theta > \frac{1}{2} \quad \text{and} \quad \left|\frac{g(x)-p(t)}{r}\cos\theta\right| < \frac{1}{4},$$

whenever $r \geq H$. It yields that

$$\dot\theta(t) = -\left[\sin^2\theta + \frac{g(x)-p(t)}{r}\cos\theta\right] < 0,$$

whenever $|r(t)| \geq H$.

Therefore, in both cases, we have

$$\dot\theta(t) < 0, \quad \text{as } r(t) \geq a = \sqrt{N^2 + H^2}.$$

Lemma 10.3 is thus proved. □

The geometric meaning of Lemma 10.3 states that the motion of (10.2) moves outside the disk D_a in the clockwise direction.

Finally, we have the Poincaré map in polar coordinates

$$\Phi: \quad \langle r_0, \theta_0 \rangle \mapsto \langle R(r_0,\theta_0), \theta_0 + \Theta(r_0,\theta_0)\rangle,$$

where

$$R(r_0,\theta_0) = r(T, r_0, \theta_0), \quad \Theta(r_0,\theta_0) = \theta(T, r_0, \theta_0) - \theta_0.$$

It follows from Lemma 10.3 that

$$\Theta(r_0,\theta_0) < 0, \quad \text{if } r_0 \text{ is large enough.} \tag{10.14}$$

10.2 Time-Map

10.2.1 Closed Orbit of Autonomous Duffing Equation

Consider the autonomous Duffing equation

$$\ddot x + g(x) = 0, \tag{10.15}$$

or its equivalent system

$$\dot{x} = y, \quad \dot{y} = -g(x), \tag{10.16}$$

where $g(x)$ satisfies the local Lipschitz condition and the condition $(g_{\pm\infty})$. It is noted that the condition $(g_{\pm\infty})$ implies the sign condition $(S)_a$ with some constants $a \geq 0$.

Choosing the primitive function

$$G(x) = \int_0^t g(x)\,dx,$$

we obtain the first integral of (10.15)

$$\frac{1}{2}y^2 + G(x) = c, \tag{10.17}$$

containing an arbitrary constant c. The condition $(g_{\pm\infty})$ yields

$$\lim_{|x|\to\infty} G(x) = +\infty.$$

Hence, we have the lower bound

$$B_0 = \inf_{x\in\mathbb{R}^1} G(x) > -\infty.$$

In general, the constant may be negative. However, $B_0 = 0$ whenever $g(x)$ satisfies the strict sign condition $(S)_0$.

It follows from $(g_{\pm\infty})$ that $G'(x) = g(x) \neq 0$ (as $|x| \gg 1$). For sufficiently large constant $c \geq c^*$, using the existence theorem of inverse function, we get two differentiable branches $x = k(c) < 0$ and $x = h(c) < 0$ of the inverse function satisfying

$$G(k(c)) = c, \qquad G(h(c)) = c.$$

Then, using

$$y^2 = 2[c - G(x)] > 0, \qquad k(c) < x < h(c),$$

we get the semi-orbits

$$y = +\sqrt{2[c - G(x)]}, \qquad k(c) \leq x \leq h(c),$$

and

$$y = -\sqrt{2[c - G(x)]}, \qquad k(c) \leq x \leq h(c),$$

which form a closed orbit Γ_c intersecting the x-axis orthogonally at the points $(k(c), 0)$ and $(h(c), 0)$. Obviously, the closed orbit Γ_c surrounds the origin clockwise. Hence, the first integral (10.17) determines a family of closed orbits Γ_c for $c > c^*$.

If the strict sign condition $(S)_0$ is satisfied, we have $c^* = 0$. Then the origin $\mathbf{o}$ is a center of the system.

If the sign condition $(S)_a$ holds only for some constant $a > 0$, we have $c^* > 0$. Then there are another types of singular points surrounded by Γ_c.

Example 10.3 Consider the autonomous Duffing system

$$\dot{x} = y, \quad \dot{y} = -(x + x^3),$$

where $g(x) = x + x^3$ satisfies the sign condition $(S)_a$ for $a = 0$. The first integral

$$y^2 + x^2 + \frac{1}{2}x^4 = c$$

determines a family of closed orbits Γ_c ($c > 0$) contractible to the center $\mathbf{o}$ as $c \to 0$.

Example 10.4 Consider the autonomous Duffing system

$$\dot{x} = y, \quad \dot{y} = x - x^3,$$

where $g(x) = -x + x^3$ satisfies the sign condition $(S)_a$ for $a = 1$. The system has a saddle $\mathbf{o} = (0, 0)$ and two centers $(-1, 0)$ and $(1, 0)$. The first integral

$$y^2 - x^2 + \frac{1}{2}x^4 = c$$

determines a family of closed orbits Γ_c (when sufficiently large $c > c^*$) in the outside of the contour of type ∞ across the saddle $\mathbf{o}$.

10.2.2 *Characteristic Frequency*

Let

$$x = x(t, \xi), \quad y = y(t, \xi), \tag{10.18}$$

be the motion of (10.15) satisfying the initial condition $x(0) = \xi$, $y(0) = 0$ for $\xi > 0$ large enough. Then the corresponding orbit Γ_ξ of motion (10.18) is a closed orbit with minimal period $\tau(\xi)$. It follows from

$$\frac{1}{2}y^2(t,\xi) + G(x(t,\xi)) = G(\xi)$$

that

$$\tau(\xi) = \sqrt{2}\int_k^h \frac{dx}{\sqrt{G(\xi) - G(x)}}, \qquad (10.19)$$

where $k = k(\xi) < 0$ and $h = h(\xi) > 0$ are defined above for $\xi > \xi_0 > 0$ ($G(k) = G(h)$). The function $\tau(\xi)$ given by the integration (10.19) is called the *time-map* of the Duffing equation (10.15) or (10.2), which denotes the minimal period of the orbit Γ_ξ.

Example 10.5 Consider the free oscillation of a spring

$$\ddot{x} + Kx = 0, \qquad \text{with elasity coefficient } K > 0.$$

In this case, the time-map

$$\tau_0(\xi) = \frac{2\pi}{\sqrt{K}}$$

is a constant, and the frequency of oscillation is $\sqrt{K} = \dfrac{2\pi}{\tau_0(\xi)}$.

The function

$$\omega(\xi) = \frac{2\pi}{\tau(\xi)}$$

is called the *characteristic frequency* of the Duffing equation (10.15) or (10.2), which has a significant meaning in physics.

Example 10.6 The super-linear Duffing equation

$$\ddot{x} + 2x^3 = 0$$

has the time-map

$$\tau(\xi) = 2\int_{-\xi}^{\xi} \frac{dx}{\sqrt{\xi^4 - x^4}} = \frac{I_0}{\xi} \qquad (\xi > 0)$$

where

$$I_0 = 4\int_0^1 \frac{ds}{\sqrt{1 - s^4}} > 0$$

is a constant. Then we obtain the characteristic frequency

$$\omega(\xi) = \frac{2\pi\xi}{I_0}.$$

Example 10.7 The sub-linear Duffing equation

$$\ddot{x} + \frac{2}{3}\sqrt[3]{x} = 0$$

has the time-map

$$\tau(\xi) = 2 \int_{-\xi}^{\xi} \frac{dx}{\sqrt{\sqrt[3]{\xi^4} - \sqrt[3]{x^4}}} = 4 J_0 \sqrt[3]{\xi} \qquad (\xi > 0)$$

where

$$J_0 = \int_0^1 \frac{ds}{\sqrt{1 - \sqrt[3]{s^4}}} > 0$$

is a constant. Then we obtain the characteristic frequency

$$\omega(\xi) = \frac{\pi}{2 J_0 \sqrt[3]{\xi}} \qquad (\xi > 0).$$

Theorem 10.1 *Let $\tau(\xi)$ and $\omega(\xi)$ be the time-map and the characteristic frequency of the Duffing equation, respectively. Then we have the following conclusions:*

1) *The super-linear condition (S_p) implies*

$$(\tau_0): \qquad \lim_{\xi \to \infty} \tau(\xi) = 0,$$

which is equivalent to

$$(\omega_\infty): \qquad \lim_{\xi \to \infty} \omega(\xi) = \infty;$$

2) *The sub-linear condition (S_b) implies*

$$(\tau_\infty): \qquad \lim_{\xi \to \infty} \tau(\xi) = +\infty,$$

which is equivalent to

$$(\omega_0): \qquad \lim_{\xi \to \infty} \omega(\xi) = 0;$$

Chapter 10.   Duffing Equations of Second Order

3) The semi-linear condition (S_m) implies

$$(T_*): \quad 0 < \liminf_{\xi \to \infty} \tau(\xi) \le \limsup_{\xi \to \infty} \tau(\xi) < \infty,$$

which is equivalent to

$$(\omega_*): \quad 0 < \liminf_{\xi \to \infty} \omega(\xi) \le \limsup_{\xi \to \infty} \omega(\xi) < \infty.$$

Proof. Using the polar coordinates, we get

$$\dot\theta = - \left[\sin^2\theta + \frac{g(r\cos\theta)}{r\cos\theta} \cos^2\theta \right]$$

from the Duffing equation (10.15). It follows that

$$\tau(\xi) = \int_0^{2\pi} \frac{d\theta}{\sin^2\theta + \frac{g(r\cos\theta)}{r\cos\theta}\cos^2\theta}.$$

Let $A > 0$ and $B > 0$ be sufficiently large constants, satisfying $A/B < \varepsilon \ll 1$. Then consider, respectively, the following regions:

$$D_1 = \{\, (x,y) \in \mathbb{R}^2 : \quad |x| \le A, \ y > B \,\};$$

$$D_2 = \{\, (x,y) \in \mathbb{R}^2 : \quad x \ge A, \ |y| < \infty \,\};$$

$$D_3 = \{\, (x,y) \in \mathbb{R}^2 : \quad |x| \le A, \ y < -B \,\};$$

$$D_4 = \{\, (x,y) \in \mathbb{R}^2 : \quad x \le -A, \ |y| < \infty \,\}.$$

Finally, let

$$D = D_1 \cup D_2 \cup D_3 \cup D_4.$$

It follows from Lemma 10.3 that for sufficiently large $\xi > 0$, the closed orbit Γ_ξ will stays in the region D and rotates clockwise there. In one turn of the rotation, assume the motion stays in D_1 during the time-interval $[t_1, t_2]$, and in D_2 during the time-interval $[t_2, t_3]$, then in D_3 during the time-interval $[t_3, t_4]$, and finally D_4 during the time-interval $[t_4, t_5]$. It follows that

$$\tau(\xi) = (t_2 - t_1) + (t_3 - t_2) + (t_4 - t_3) + (t_5 - t_4).$$

Using $dx = y\,dt$, we get

$$t_2 - t_1 = \int_{t_1}^{t_2} dt = \int_{-A}^{+A} \frac{1}{y}\,dx < \frac{2A}{B}$$

and

$$t_4 - t_3 = \int_{t_3}^{t_4} dt = \int_{+A}^{-A} \frac{1}{y}\,dx < \frac{2A}{B}.$$

Then, according the conditions (S_p), (S_b) and (S_m), we will estimate the time-intervals $[t_2, t_3]$ and $[t_5, t_4]$ as follows:

(1) First assume (S_p) is valid.

Then we can take the constant $B > 0$ large enough, such that

$$\frac{g(x)}{x} > K, \qquad \text{as } |x| \geq B,$$

where $K > 0$ is a sufficiently large constant. It follows that

$$t_3 - t_2 = \int_{\theta(t_3)}^{\theta(t_2)} \frac{d\theta}{\sin^2\theta + \frac{g(x)}{x}\cos^2\theta} \leq \int_{-\frac{\pi}{2}}^{+\frac{\pi}{2}} \frac{d\theta}{\sin^2\theta + K\cos^2\theta}$$

$$= \frac{1}{\sqrt{K}} \arctan \frac{\tan\theta}{\sqrt{K}} \bigg|_{-\frac{\pi}{2}}^{+\frac{\pi}{2}} = \frac{\pi}{\sqrt{K}} \ll 1.$$

Similarly, we have

$$t_5 - t_4 \leq \frac{\pi}{\sqrt{K}} \ll 1.$$

Therefore, we get

$$\tau(\xi) < \frac{4A}{B} + \frac{2\pi}{\sqrt{K}} \ll 1, \qquad \text{whenever } B \gg 1.$$

Conclusion 1) of Theorem 10.1 is thus proved.

(2) Nextly, assume (S_b) is valid.

Then, for given positive constant $\delta \ll 1$, choose constant $B \gg 1$, such that

$$0 < \frac{g(x)}{x} < \delta, \qquad \text{whenever } |x| \geq B.$$

It follows that

$$t_3 - t_2 = \int_{\theta(t_3)}^{\theta(t_2)} \frac{d\theta}{\sin^2\theta + \frac{g(x)}{x}\cos^2\theta}$$

$$\geq \int_{-\frac{\pi}{2}}^{+\frac{\pi}{2}} \frac{d\theta}{\sin^2\theta + \cos^2\theta} = \frac{\pi}{\sqrt{\delta}}.$$

Similarly, we get

$$t_5 - t_4 \geq \frac{\pi}{\sqrt{\delta}}.$$

Then we have

$$\tau(\xi) > \frac{2\pi}{\sqrt{\delta}},$$

which proves the conclusion 2) of the theorem.

(3) Finally, assume (S_m) is valid.

Then there are constants $\alpha > 0$, $\beta > 0$ and $a_0 > 0$, such that

$$\alpha < \frac{g(x)}{x} < \beta, \qquad \text{whenever } |x| \geq a_0. \qquad (10.20)$$

Besides, let us consider the auxiliary functions

$$h(x) = \begin{cases} \dfrac{g(a_0)}{a_0} x, & 0 \leq x \leq a_0, \\ g(a_0), & a_0 < x < \infty, \\ \dfrac{g(-a_0)}{-a_0} x, & -a_0 \leq x \leq 0, \\ g(-a_0), & -\infty < x < -a_0, \end{cases}$$

and

$$E(x) = \begin{cases} g(x), & 0 \leq x \leq a_0, \\ g(a_0), & a_0 < x < \infty, \\ g(x), & -a_0 \leq x \leq 0, \\ g(-a_0), & -\infty < x < -a_0. \end{cases}$$

Let
$$\hat{g}(x) = g(x) - E(x) + h(x).$$

It is clear that both $h(x)$ and $E(x)$ are bounded for $x \in \mathbb{R}^1$, and $\hat{g}(x)$ satisfies the condition
$$0 < \alpha \le \frac{\hat{g}(x)}{x} \le \beta < \infty, \qquad x \in \mathbb{R}^1, \tag{10.21}$$

The equation (10.2) can be put in the form
$$\dot{x} = y, \quad \dot{y} = -\hat{g}(x) + Q(x), \tag{10.22}$$

where the function $Q(x) = -E(x) + h(x)$ is bounded in $x \in \mathbb{R}^1$. If $a_0 = 0$ in the condition (10.20), then we have $\hat{g}(x) = g(x)$ and $Q(x) = 0$.

Obviously, the closed orbit Γ_ξ of (10.16) is identical to the closed orbit of (10.22) for sufficiently large initial condition $\xi > 0$.

Using polar coordinates, we write Equation (10.22) into the form
$$\begin{cases} \dot{r} = \left[\sin\theta \cos\theta + \dfrac{Q(x) - \hat{g}(x)}{r} \sin\theta \right] r, \\ \dot{\theta} = - \left[\sin^2\theta + \dfrac{\hat{g}(x)}{x} \cos^2\theta \right] - \dfrac{Q(x)}{r} \cos\theta. \end{cases}$$

Note that for given $\varepsilon > 0$, there is a constant $a > 0$, such that
$$\left| \frac{Q(x)}{r} \cos\theta \right| < \varepsilon = \varepsilon(\cos^2\theta + \sin^2\theta), \qquad \text{whenever } r \ge a,$$

which together with (10.20) implies
$$\dot{\theta} > -[(1+\varepsilon)\sin^2\theta + (\beta+\varepsilon)\cos^2\theta]$$

and
$$\dot{\theta} < -[(1-\varepsilon)\sin^2\theta + (\alpha-\varepsilon)\cos^2\theta]$$

whenever $r > \max\{a, a_0\}$. Hence, we obtain
$$\tau(\xi) > \int_0^{2\pi} \frac{d\theta}{[(1+\varepsilon)\sin^2\theta + (\beta+\varepsilon)\cos^2\theta]} = \frac{2\pi}{\sqrt{(1+\varepsilon)(\beta+\varepsilon)}}$$

and
$$\tau(\xi) < \int_0^{2\pi} \frac{d\theta}{[(1-\varepsilon)\sin^2\theta + (\alpha-\varepsilon)\cos^2\theta]} = \frac{2\pi}{\sqrt{(1-\varepsilon)(\alpha-\varepsilon)}}$$

Chapter 10. Duffing Equations of Second Order

for sufficiently large $\xi > 0$, which lead to the conclusion 3) of the theorem. In summary, the proof of Theorem 10.1 is completed. □

10.2.3 Increment of Time-Map

Let
$$\tau^* = \limsup_{\xi \to \infty} \tau(\xi) \quad \text{and} \quad \tau_* = \liminf_{\xi \to \infty} \tau(\xi),$$

and define the *increment* of the time-map $\tau(\xi)$ by
$$\Delta \tau = \tau^* - \tau_*.$$

Assume $\Delta \tau = 0$ if $\tau^* = \tau_* = \infty$. Then we have $\Delta \tau \geq 0$, and $\Delta \tau = 0$ if and only if $\tau^* = \tau_*$.

Similarly, we can define the increment of the characteristic frequency
$$\Delta \omega = \omega^* - \omega_*$$

where
$$\omega^* = \limsup_{\xi \to \infty} \omega(\xi) \qquad \omega_* = \liminf_{\xi \to \infty} \omega(\xi).$$

It is noticed that all the examples of Duffing equation given above have null increment of time-map (i.e., $\Delta \tau = 0$). Now, we give an example of Duffing equation having $\Delta \tau \neq 0$.

Example 10.8 Consider the Duffing equation
$$\frac{d^2 u}{dt^2} + \tilde{g}(u) = 0, \tag{10.23}$$

where
$$\tilde{g}(u) = \left(m_0^2 + \frac{1}{10} \cos \log(1 + u^2) - \frac{u^2}{10(1 + u^2)} \sin \log(1 + u^2) \right) u,$$

where $m_0 > 0$ is a constant. The Duffing equation is equivalent to the system
$$\dot{u} = v, \quad \dot{v} = -\tilde{g}(u),$$

with the first integral
$$\frac{1}{2} v^2 + \tilde{G}(u) = c(= \tilde{G}(\xi)),$$

where
$$\tilde{G}(u) = \int_0^u \tilde{g}(u)\,du = \frac{1}{2}u^2\left[m_0^2 + \frac{1}{10}\cos\log(1+u^2)\right]$$
is an even function of u. Hence, we obtain the corresponding time-map
$$\tilde{\tau}(\xi) = 2\sqrt{2}\int_0^\xi \frac{du}{\sqrt{\tilde{G}(\xi) - \tilde{G}(u)}} = \frac{4}{\sqrt{Q(\xi)}}\int_0^1 \frac{ds}{\sqrt{1 - s^2 Q(\xi s)/Q(\xi)}},$$
where
$$Q(x) = m_0^2 + \frac{1}{10}\cos\log(1+x^2).$$
Let
$$\alpha_k = \sqrt{e^{2k\pi} - 1}, \quad \beta_k = \sqrt{e^{2k+1\pi} - 1};$$
and
$$a_k = \tilde{G}(\alpha_k), \quad b_k = \tilde{G}(\beta_k);$$
for $k = 1, 2, \cdots$. Then we have
$$a_k < b_k, \quad \text{for } k > 0,$$
with
$$\lim_{k\to\infty} a_k = \lim_{k\to\infty} b_k = \infty.$$
Using the inequalities
$$0 < \frac{Q(\alpha_k s)}{\alpha_k} \leq 1 \quad \text{and} \quad \frac{Q(\alpha_k s)}{\alpha_k} \geq 1 \quad (0 \leq \xi \leq 1),$$
we obtain
$$\tilde{\tau}(a_k) = \frac{4}{\sqrt{Q(\alpha_k)}}\int_0^1 \frac{ds}{\sqrt{1 - s^2 Q(\alpha_k s)/Q(\alpha_k)}}$$
$$\leq \frac{4\sqrt{10}}{\sqrt{10m_0^2 + 1}}\int_0^1 \frac{ds}{\sqrt{1 - s^2}} \leq \frac{2\pi}{m_0} - \sigma$$
and
$$\tilde{\tau}(b_k) = \frac{4}{\sqrt{Q(\beta_k)}}\int_0^1 \frac{ds}{\sqrt{1 - s^2 Q(\beta_k s)/Q(\beta_k)}}$$

$$\geq \frac{4\sqrt{10}}{\sqrt{10m_0^2+1}} \int_0^1 \frac{ds}{\sqrt{1-s^2}} \geq \frac{2\pi}{m_0} + \sigma$$

where

$$\sigma = \frac{2\pi}{m_0} \min\left\{1 - \frac{\sqrt{10}m_0}{\sqrt{10m_0^2+1}}, \frac{\sqrt{10}m_0}{\sqrt{10m_0^2-1}} - 1\right\}.$$

Hence, the time-map of Duffing equation (10.8) has increment

$$\Delta\tilde{\tau} \geq 2\sigma > 0.$$

10.3 Duffing Equation of Super-Linear Type

10.3.1 Super-linear Duffing Equation

Assume the Duffing equation

$$\ddot{x} + g(x) = p(t), \tag{10.24}$$

namely,

$$\dot{x} = y, \quad \dot{y} = -g(x) + p(t),$$

where $g(x)$ satisfies the local Lipschitz condition, and $p(t)$ is piecewise continuous and periodic (with minimal period $T > 0$).

To prove the existence of periodic solution, the routine method in the degree theory needs to know first a prior bound of the periodic solutions. However, such a prior bound of super-linear Duffing equation does not exist since it will be shown that the most Duffing equations possess an infinity of periodic solutions extending to the infinity.

In 1975, Fučik and Lovicar [67] proved the existence of periodic solution of the super-linear Duffing equation via the existence of the fixed point of the corresponding Poincaré map. Then the work was generalized by Struwe [112] and Shekhter [111], respectively, to the equation

$$\ddot{x} + g(x) = p(t, x, \dot{x}),$$

where the function $p(t, x, \dot{x})$ may be unbounded with respect to $\dot{x}$. In 1990, Capietto, Mawhin and Zanolin [17] generalized the existence theorems of periodic solution in [111] and [112] further to systems of differential equations by using a continuation approach.

In 1982, W. Ding proved that the super-linear Duffing equation has an infinity of T-periodic solutions $x = \varphi_i(t)$ extending to the infinity; that is,

$$\lim_{i \to \infty} \left(\sup_{t \in [0,T]} (|\varphi_i(t)| + |\varphi'(t)|) \right) = \infty,$$

where [63] he generalized the earlier works of Cesari, Harvey, Morris and Micheletti (see [19], [71], [94] and [112]) by successfully applying the generalized Poncaré-Birkhoff's twist theorem to the Duffing equation.

In a similar manner, for any given integer $m > 1$, we proved the result [58] that the super-linear Duffing equation has an infinity of mT-periodic solutions $x = \varphi_i(t)$ extending to the infinity. But it remained to be an open question whether or not mT is the minimal period of these periodic solutions.

Using the method of variation, A. Bahri and H. Berestycki [4] proved the existence of multiple periodic solutions of Duffing equation

$$\frac{d^2 x}{dt^2} + g(x) = p(t), \qquad x \in \mathbb{R}^n,$$

in higher dimensional space $\mathbb{R}^n$, where $p(t) \in L^2_{loc}[\mathbb{R}^1, \mathbb{R}^n]$ and $g(x) = G'(x)$ satisfies the condition

$$0 < G(x) \leq \kappa \langle g(x), x \rangle, \qquad \forall\, |x| \geq R > 0, \qquad (10.25)$$

with a constant κ ($0 < \kappa < 1/2$). In the special case: $n = 1$, the condition (10.25) implies

$$\lim_{|x| \to \infty} \left(\frac{g(x) \operatorname{sign}(x)}{|x|^{\frac{1}{\kappa}}} \right) = \lim_{|x| \to \infty} \left(\frac{(\frac{1}{\kappa} + 1) G(x)}{|x|^{\frac{1}{\kappa}+1}} \right) = \infty,$$

which is stronger than the conditions (S_p) and (S_p^*).

It is proved in the paper [60] that the equation (10.24) has abundant harmonic solutions as well as abundant subharmonic solutions of any order under some weak conditions, which will be called later to be the *nearly super-linear conditions*.

10.3.2 Super-Quadratic Potential

The Duffing equation (10.1) is said to have the *super-quadratic potential* if it satisfies the condition (S_p^*). Since the condition (S_p) implies the condition

(S_p^*), the super-linear Duffing equation has the super-quadratic potential. But the converse is not true. For example, the Duffing equation

$$\ddot{x} + x + x^3(1 + \cos x^4) = p(t) \tag{10.26}$$

is not super-linear, but it has a super-quadratic potential

$$G(x) = \frac{2x^2 + x^4 + \sin x^4}{4}.$$

In fact, we have the following result.

Proposition 10.1 *The implications*

$$(S_p) \quad \Rightarrow \quad (g_{\pm\infty}) + (\tau_0) \quad \Rightarrow \quad (S_p^*) \tag{10.27}$$

are valid and non-invertible.

Proof. Note that the first implication has been already proved before. Hence, it suffices to prove the second implication.

Now, let the condition $(g_{\pm\infty}) + (\tau_0)$ be valid. It is obvious that the condition $(g_{\pm\infty})$ implies

$$\lim_{|x|\to\infty} G(x) = +\infty.$$

On the other hand, there is a constant $c_0 > 0$, such that

$$G(x) > G(c_0) > 0, \qquad \text{whenever } x > c_0.$$

Assume $d < 0$ is the constant satisfying $G(d) = G(h)$. Then, letting $h > c_0$, we have

$$\frac{1}{2}\tau(h) = \int_d^h \frac{dx}{\sqrt{G(h) - G(x)}} > \int_0^h \frac{dx}{\sqrt{G(h) - G(x)}}$$

$$\geq \int_{c_0}^h \frac{dx}{\sqrt{G(h) - G(x)}} \geq \frac{h - c_0}{\sqrt{G(h)}},$$

which together with the condition (τ_0) implies

$$\lim_{h\to\infty} \frac{G(h)}{h^2} = +\infty.$$

Similarly, we can prove

$$\lim_{d\to-\infty} \frac{G(d)}{d^2} = +\infty.$$

10.3. Duffing Equation of Super-Linear Type

We have thus proved the implication

$$(g_{\pm\infty}) + (\tau_0) \quad \Rightarrow \quad (S_p^*).$$

It follows that the implications in (11.11) hold true.

Finally, the non-invertibility of the implications in (11.11) will be shown by the following examples.

(A) We first consider the Duffing equation (10.26). We have known that it is not super-linear. Now, consider the corresponding time-map

$$\tau(h) = \int_0^h \frac{4\sqrt{2}dx}{\sqrt{(h^4 + 2h^2 + \sin h^4) - (x^4 + 2x^2 + \sin x^4)}}$$

$$= \frac{4\sqrt{2}}{h} \int_0^1 \frac{du}{\sqrt{(1 + \frac{2}{h^2} + \frac{\sin h^4}{h^4}) - (u^4 + \frac{2u^2}{h^2} + \frac{\sin(hu)^4}{h^4})}}.$$

Since the limit

$$\lim_{h \to \infty} \int_0^1 \frac{du}{\sqrt{(1 + \frac{2}{h^2} + \frac{\sin h^4}{h^4}) - (u^4 + \frac{2u^2}{h^2} + \frac{\sin(hu)^4}{h^4})}}$$

is equal to the constant

$$\sigma_0 = \int_0^1 \frac{du}{\sqrt{1 - u^4}} > 0,$$

we obtain

$$\lim_{h \to \infty} \tau(h) = 0 \quad \text{(i.e., the condition } (\tau_0) \text{ is valid)}.$$

Hence, (10.26) is an example demonstrating that the condition $(g_{\pm\infty}) + (\tau_0)$ does not imply the super-linear condition (S_p).

(B) Then we consider the Duffing equation

$$\ddot{x} + \breve{g}(x) = p(t), \tag{10.28}$$

where $\breve{g}(x)$ is defined as follows.

First, choose a C^∞-differentiable function $\eta(x)$ satisfying

$$\begin{cases} \eta(x) = 0, & -\infty < x \leq 0; \\ \eta'(x) \geq 0, & 0 \leq x \leq 1; \\ \eta(x) = 1, & 1 \leq x < \infty. \end{cases}$$

Chapter 10. Duffing Equations of Second Order

Next, for any integer $n \geq 0$, define the constants

$$a_n = \frac{4}{3}(4^n - 1), \quad b_n = 4a_n + 1, \quad c_n = b_n + 1, \quad d_n = c_n + 1.$$

Notice that

$$a_{n+1} = d_n + 1, \quad b_n > 4a_n \quad (n \geq 0).$$

Then define a C^∞-differentiable even function

$$h(x) = \begin{cases} K, & x \in [a_n, b_n); \\ K + a_{n+2}^2 \eta(x - b_n), & x \in [b_n, c_n); \\ K + a_{n+2}^2, & x \in [c_n, d_n); \\ K + a_{n+2}^2 [1 - \eta(x - d_n)], & x \in [d_n, a_{n+1}), \end{cases}$$

where $n \geq 0$, and $K > 0$ is a constant.

Finally, let

$$\breve{g}(x) = xh(x), \quad \breve{G}(x) = \int_0^x \breve{g}(x)\, dx \quad (x \in \mathbb{R}^1).$$

It can be seen that $\breve{g}(x)$ is C^∞-differentiable and satisfies

$$\frac{\breve{g}(x)}{x} \geq K > 0 \quad (x \neq 0).$$

On the other hand, for any $x \in [a_{n+1}, a_{n+2})$, we have

$$\frac{\breve{G}(x)}{x^2} > \frac{1}{a_{n+2}^2} \left(\int_{c_n}^{d_n} \breve{g}(x)\, dx \right) = \frac{(K + a_{n+2}^2)(d_n^2 - c_n^2)}{2d_{n+2}^2}$$

$$= \frac{1}{2}\left(1 + \frac{K}{a_{n+2}^2}\right)(c_n + d_n) > \frac{c_n + d_n}{2}.$$

Hence $\breve{G}(x)$ satisfies the condition (S_p^*) since it is an even function. It follows that the Duffing equation (10.28) has the super-quadratic potential. However, its time-map $\breve{\tau}(\xi)$ satisfies

$$\frac{1}{2\sqrt{2}} \breve{\tau}(b_n) = \int_0^{b_n} \frac{dx}{\sqrt{\breve{G}(b_n) - \breve{G}(x)}} = b_n \int_0^1 \frac{dt}{\sqrt{\breve{G}(b_n) - \breve{G}(b_n t)}}$$

$$> b_n \int_{\frac{1}{4}}^{1} \frac{dt}{\sqrt{\check{G}(b_n) - \check{G}(b_n t)}}$$

$$= b_n \int_{\frac{1}{4}}^{1} \frac{dt}{\sqrt{(b_n - b_n t) \int_0^1 \check{g}(b_n \chi(t, \lambda))\, d\lambda}}$$

(where $\chi(t, \lambda) = (1 - \lambda)t + \lambda$)

$$= \int_{\frac{1}{4}}^{1} \frac{dt}{\sqrt{(1-t) \int_0^1 h(b_n \chi(t,\lambda)) \chi(t,\lambda)\, d\lambda}}$$

$$> \int_{\frac{1}{4}}^{1} \frac{dt}{\sqrt{K(1-t)}} = \sqrt{\frac{3}{K}}.$$

(In the above calculation, we have used the fact that $a_n < b_n \chi(t, \lambda) < b_n$ when $\frac{1}{4} < t < 1$ and $0 \leq \lambda \leq 1$.) Hence, we obtain

$$\limsup_{h \to \infty} \check{\tau}(h) \geq \sqrt{\frac{24}{K}} > 0,$$

which proves that the condition (S_p^*) does not imply the condition (τ_0).

It follows that the implications in (11.11) are not invertible. □

The Duffing equation (10.24) is called *nearly super-linear* if it satisfies the condition $(g_{\pm\infty}) + (\tau_0)$.

It is obvious that the super-linear Duffing equation (10.24) is nearly super-linear. The Duffing equation (10.28) satisfies the condition (S_p^*), but it is not nearly super-linear since it does not satisfy the condition (τ_0).

10.3.3 Elementary Lemmas

Let $z = z(t, z_0)$ be the solution of the Duffing equation (10.24), satisfying the initial condition $z(0) = z_0$, and denote the polar coordinates of $z(t, z_0)$ by $\langle r(t, z_0), \theta(t, z_0) \rangle$.

Lemma 10.4 *Let (10.24) be a nearly super-linear Duffing equation. Let m be a given positive integer. Then, for any given integer $n > 0$, there is a constant $R_n > 0$, such that*

$$\Delta_m(z_0) := \theta(mT, z_0) - \theta(0, z_0) < -2n\pi, \tag{10.29}$$

Chapter 10. Duffing Equations of Second Order

whenever $|z_0| \geq R_n$.

Proof. It is known that for given a constant $c_0 > 0$, there is a constant $d_0 > 0$, such that

$$\begin{cases} r(t, z_0) > c_0, \\ \theta(t, z_0) < 0, \end{cases} \quad (0 \leq t \leq mT), \tag{10.30}$$

whenever $|z_0| > d_0$. In the following, assume the motion $z = z(t, z_0)$ ($0 \leq t \leq mT$) satisfies the condition (10.30).

Then the motion $z = z(t, z_0)$ ($0 \leq t \leq mT$) will move clockwise in the region

$$\mathfrak{D} = \{(x, y) \in \mathbb{R}^2 : \quad |(x, y)| = \sqrt{x^2 + y^2} > c_0\}.$$

Now, let us cut $\mathfrak{D}$ into the following six sub-regions

$$\mathfrak{D}_1 = \{(x, y) \in \mathfrak{D} : \quad |x| \leq c_0, \quad y > 0\};$$
$$\mathfrak{D}_2 = \{(x, y) \in \mathfrak{D} : \quad x \geq c_0, \quad y \geq 0\};$$
$$\mathfrak{D}_3 = \{(x, y) \in \mathfrak{D} : \quad x \geq c_0, \quad y \leq 0\};$$
$$\mathfrak{D}_4 = \{(x, y) \in \mathfrak{D} : \quad |x| \leq c_0, \quad y < 0\};$$
$$\mathfrak{D}_5 = \{(x, y) \in \mathfrak{D} : \quad x \leq -c_0, \quad y \leq 0\};$$
$$\mathfrak{D}_6 = \{(x, y) \in \mathfrak{D} : \quad x \leq -c_0, \quad y \geq 0\}.$$

Without loss of generalization, assume $z_0 \in \mathfrak{D}_1$. Then the motion $z = z(t, z_0)$ ($0 \leq t \leq mT$) will rotates as follows

$$\mathfrak{D}_1 \to \mathfrak{D}_2 \to \mathfrak{D}_3 \to \mathfrak{D}_4 \to \mathfrak{D}_5 \to \mathfrak{D}_6 \to \mathfrak{D}_1.$$

Given k, assume $[t_1, t_2] \subset [0, mT]$, such that

$$z(t, z_0) \in \mathfrak{D}_k, \quad t_1 \leq t \leq t_2.$$

First, assume $z(t) \in \mathfrak{D}_1$ for $t_1 \leq t \leq t_2$. It follows that $x(t)$ is monotonically increasing and $y(t) \geq \sqrt{d_0^2 - c_0^2}$. Then, we have

$$t_2 - t_1 = \int_{t_1}^{t_2} dt = \int_{x(t_1)}^{x(t_2)} \frac{dx}{y} \leq \int_{-c_0}^{+c_0} \frac{dx}{y} \leq \frac{2c_0}{\sqrt{d_0^2 - c_0^2}} < \varepsilon,$$

provided that d_0 is sufficiently large.

Similarly, if $z(t) \in \mathfrak{D}_4$ for $t_1 \leq t \leq t_2$, then we have

$$t_2 - t_1 < \varepsilon, \quad \text{whenever } d_0 \gg 1.$$

Next, assume $z(t) \in \mathfrak{D}_2$ for $t_1 \leq t \leq t_2$. It follows that $x(t)$ is also monotonically increasing and $y(t) \geq 0$. Consider the auxiliary function

$$w(t) = G(x(t)) - Mx(t) + \frac{1}{2}y^2(t),$$

where

$$M = \sup_{0 \leq t \leq T} |p(t)|.$$

Then we have

$$\begin{aligned}\dot{w}(t) &= (g(x(t)) - M)\dot{x}(t) + y(t)\dot{y}(t) \\ &= (g(x(t)) - M)y(t) + y(t)(-g(x(t)) + p(t)) \\ &= -y(t)(M - p(t)) \leq 0 \quad (t_1 \leq t \leq t_2),\end{aligned}$$

which implies

$$G(x(t)) - Mx(t) + \frac{1}{2}y^2(t) \geq G(x(t_2)) - Mx(t_2),$$

for $t_1 \leq t \leq t_2$. It follows that

$$\frac{1}{2}y^2(t) \geq [G(x(t_2)) - G(x(t))] - M[x(t_2) - x(t)]$$

$$= \frac{G(x(t_2)) - G(x(t))}{2} + \frac{x(t_2) - x(t)}{2}\left[\frac{G(x(t_2)) - G(x(t))}{x(t_2) - x(t)} - 2M\right]$$

$$= \frac{G(x(t_2)) - G(x(t))}{2} + \frac{x(t_2) - x(t)}{2}[g(\xi) - 2M],$$

where $\xi = \xi(t) \in [x(t), x(t_2)]$ is produced by the Lagrange mean-value theorem. Since $x(t) \geq c_0$ ($t_1 \leq t \leq t_2$), we can take c_0 sufficiently large, such that $g(\xi) > 2M$. Therefore, we obtain

$$\dot{x}(t) = y(t) \geq \sqrt{G(x(t_2)) - G(x(t))}, \quad t_1 \leq t \leq t_2,$$

which yields

$$0 < t_2 - t_1 \leq \int_{t_1}^{t_2} \frac{y(t)dt}{\sqrt{G(x(t_2)) - G(x(t))}}$$

$$= \int_{x(t_1)}^{x(t_2)} \frac{dx}{\sqrt{G(x(t_2)) - G(x)}}$$

$$\leq \int_0^{x(t_2)} \frac{dx}{\sqrt{G(x(t_2)) - G(x)}} \leq \frac{\tau(x(t_2))}{\sqrt{2}}.$$

Since $x(t_2) \geq c_0 \gg 1$, the condition (τ_0) implies that $\tau(x(t_2)) \ll 1$. Hence, we have

$$0 < t_2 - t_1 < \varepsilon, \qquad \text{whenever } c_0 \gg 1.$$

Similarly, for $k = 3, 5, 6$, we can prove

$$0 < t_2 - t_1 < \varepsilon, \qquad \text{for } z(t, z_0) \in \mathfrak{D}_k \text{ as } t \in [t_1, t_2],$$

whenever $c_0 \gg 1$. Let $d_0 > 0$ be sufficiently large, such that $z(t, z_0) \in \mathfrak{D}$ whenever $|z_0| > d_0$ and $t \in [s_1, s_2]$, together with

$$\theta(s_2, z_0) - \theta(s_1, z_0) = -2\pi.$$

It follows that

$$0 < s_2 - s_1 < 7\varepsilon.$$

It means that the motion $z = z(t, z_0)$ rotates clockwise a turn in $\mathfrak{D}$ in a period less than 7ε. Therefore, during the period mT, it can rotates sufficiently large number of (clockwise) turns provided $|z_0| > d_0 \gg 1$.

The proof of Lemma 10.4 is thus completed. □

10.3.4 Multiplicity of Harmonic Solutions

Theorem 10.2 *If the Duffing equation (10.24) is nearly super-linear, then it has infinitely many harmonic solutions $x = x_j(t)$, satisfying*

$$\lim_{j \to \infty} \left(\sup_{0 \leq t \leq T} \sqrt{|x_j(t)|^2 + |\dot{x}_j(t)|^2} \right) = \infty.$$

Proof. Let $m = 1$ in Lemma 10.4, and consider

$$\Delta_1(z_0) = \theta(T, z_0) - \theta(0, z_0), \qquad |z_0| > d_0.$$

It is obvious that for any given $a_1 > d_0$, there is a large integer $K_1 > 0$, such that

$$\Delta_1(z_0) = \theta(T, z_0) - \theta(0, z_0) > -2K_1\pi, \qquad \text{for } |z_0| = a_1. \quad (10.31)$$

On the other hand, using Lemma 10.4, we have a constant $b_1 > a_1$, such that

$$\Delta_1(z_0) = \theta(T, z_0) - \theta(0, z_0) < -2K_1\pi, \qquad \text{for } |z_0| = b_1. \quad (10.32)$$

Then consider the annulus

$$\mathcal{A}_1: \qquad 0 < a_1 \le |z| \le b_1.$$

Now, let

$$\Phi: \qquad \langle \rho_0, \phi_0 \rangle \mapsto \langle \rho_1, \phi_1 \rangle \qquad (\rho_0 > c_0 > 0)$$

be the Poincaré map of the Duffing equation (10.24) for $\rho_0 > d_0$. Since the Duffing equation (10.24) is conservative, Φ is area-preserving. It follows from (10.31) and (10.32) that Φ is a twist area-preserving map on the annulus $\mathcal{A}_1$. Therefore, using the generalized Poincaré-Birkhoff twist theorem, we assert that Φ has two fixed-points

$$\zeta_{1i} = \langle r_i, \theta_i \rangle \in \mathcal{A}_1 \qquad (i = 1, 2).$$

It follows that

$$z = z(t, \zeta_{1i}) \qquad (i = 1, 2)$$

are two T-periodic solution of (10.24), satisfying the initial conditions $z(0, \zeta_{1i}) = \zeta_{1i}$ ($i = 1, 2$), respectively.

On the other hand, since the period of the periodic solution of (10.24) must be a period of $p(t)$ (that is kT with $k \ge 1$), then T is the minimal period. Therefore, $z = z(t, \zeta_{1i})$ are harmonic solutions of the Duffing equation (10.24).

In a similar manner, we can prove that there is a sequence

$$(a_1 < b_1 <) a_2 < b_2 < \cdots < a_k < b_k < \cdots (\to \infty),$$

such that the area-preserving map Φ are twisted on the annuli

$$\mathcal{A}_k: \qquad a_k \le |z| \le b_k, \qquad k = 2, \cdots,$$

respectively. It follows from the generalized Poincaré-Birkhoff twist theorem that Φ has fixed points

$$\zeta_{k1}, \zeta_{k2} \in \mathcal{A}_k, \qquad k = 2, \cdots,$$

respectively. Then we obtain the harmonic solutions

$$z = z(t, \zeta_{ki}) \qquad (i = 1, 2)$$

for $k = 2, \cdots$. It follows from

$$\limsup_{0 \le t \le T} |z(t, \zeta_{ki})| \ge a_k$$

that these harmonic solutions extend to infinity.

Theorem 10.2 is thus proved. $\qquad\square$

10.3.5 Multiplicity of Sub-Harmonic Solutions

Theorem 10.3 *If the Duffing equation (10.24) is nearly super-linear, then for any given integer $m \ge 2$, it has infinitely many harmonic solutions $x = x_j(t)$ of order m, satisfying*

$$\lim_{j \to \infty} \left(\sup_{0 \le t \le mT} \sqrt{|x_j(t)|^2 + |\dot{x}_j(t)|^2} \right) = \infty.$$

Proof. Now, let $m > 1$ in in Lemma 10.4, and consider

$$\Delta_m(z_0) = \theta(mT, z_0) - \theta_0, \qquad |z_0| > d_0,$$

where $\theta_0 = \theta(0, z_0)$.

On the other hand, since the period of the periodic solution of (10.24) must be a period of $p(t)$ (that is kT with $k \ge 1$), then T is the minimal period. Therefore, $z = z(t, \zeta_{1i})$ are harmonic solutions of the Duffing equation (10.24).

In a similar manner as in the proof of Theorem 10.2, we can prove that for the given integer $m > 1$, there is a sequence

$$0 < a_1 < b_1 < a_2 < b_2 < \cdots < a_k < b_k < \cdots (\to \infty),$$

such that the area-preserving map Φ^m are twisted on the annuli

$$\mathcal{A}_s: \qquad a_s \le |z| \le b_s, \qquad s = 2, \cdots,$$

respectively. Without destroying the generality, assume

$$\begin{cases} \theta(mT, z_0) - \theta_0 < -2q_s\pi, & |z_0| = a_s, \\ \theta(mT, z_0) - \theta_0 > -2q_s\pi, & |z_0| = b_s, \end{cases}$$

where q_s is a positive prime integer. Then, using the generalized Poincaré-Birkhoff twist theorem, we assert that Φ^m has fixed points

$$\zeta_{s1},\ \zeta_{s2} \in \mathcal{A}_s, \qquad s = 1, 2, \cdots,$$

satisfying

$$\theta(mT, \zeta_{s1}) - \theta_0 = \theta(mT, \zeta_{s2}) - \theta_0 = -2q_s\pi, \qquad (s = 1, 2, \cdots), \qquad (10.33)$$

respectively. It follows that

$$z = z(t, \zeta_{si}) \qquad (i = 1, 2)$$

are mT-periodic solutions of (10.24) for $s = 1, 2, \cdots$.

We want to prove that mT is the minimal period.

Assume the contrary. Let $z = z(t, \zeta_{si})$ has a minimal period nT with $0 < nT < mT$. It follows that ζ_{s1} and ζ_{s2} are n-periodic point of Φ with minimal period n. Let $m = pn + q$ with $0 \le q < n$. Therefore, we have

$$\zeta_{si} = \Phi^m(\zeta_{si}) = \Phi^{q+pn}(\zeta_{si}) = \Phi^q \circ \Phi^{np}(\zeta_{si}) = \Phi^q(\zeta_{si}),$$

which implies that ζ_{si} is a periodic point of Φ with period q. If $q > 0$, then ζ_{si} is a q-periodic point. Therefore, it is in conflict with the fact that $n(> q)$ is the minimal period. It follows that $q = 0$ (i.e., $m = pn$). It follows that $p > 1$.

On the other hand, note that $\theta'(t) < 0$. Since ζ_{si} is an n-periodic point of Φ, we have

$$\theta((l+1)nT, \zeta_{s1}) - \theta(lnT, \zeta_{s1}) = -2b_s\pi$$

where $b_s \ge 1$ is some integer. Hence, we have

$$\theta(mT, \zeta_{si}) - \theta(0, \zeta_{si}) = -2pb_s\pi,$$

which together with (10.33) implies that $q_s = pb_s$. However, it contradicts that q_s is a prime integer. This proves that mT is the minimal period of the solutions $z = z(t, \zeta_{si})$ $(i = 1, 2)$, and hence they are subharmonic solutions of order $m > 1$.

It follows from

$$\limsup_{0 \le t \le mT} |z(t, \zeta_{ki}| \ge |\zeta_{ki}| \ge a_k \quad \to \quad \infty \qquad (\text{as } k \to \infty)$$

that these subharmonic solutions extend to infinity.

We have thus proved Theorem 10.3. $\qquad\square$

Finally, we introduce the following propositions for the application of Theorems 10.2 and 10.3.

Proposition 10.2 Let the condition $(g_{\pm\infty})$ be valid. For any constant $L > 0$, if there are constants $R \geq d_0$ and $\sigma > 0$, such that the function

$$\phi_L(s) = G(s) - Ls^2, \qquad s \in \mathbb{R}^1,$$

satisfies the inequality

$$\phi_L(s_1) \leq \phi_L(s_2), \qquad (s_1, s_2) \in \mathfrak{S}, \qquad (10.34)$$

where

$$\mathfrak{S} = \{(s_1, s_2): \quad R \leq s_1 \leq s_2 - \sigma \quad \text{and} \quad -R \geq s_1 \geq s_2 + \sigma\},$$

then the condition (τ_0) is valid.

Proof. Let

$$\tau^+(c) = \sqrt{2} \int_0^c \frac{d\xi}{\sqrt{G(c) - G(\xi)}}, \qquad \tau^-(d) = \sqrt{2} \int_d^0 \frac{d\xi}{\sqrt{G(d) - G(\xi)}},$$

where

$$c > 0, \qquad d < 0 \qquad (G(d) = G(c)).$$

Then we have

$$\tau(c) = \tau^+(c) + \tau^-(d) \qquad \text{(for sufficiently large constant } c\text{).}$$

It follows from (10.34) that

$$G(c) - G(s) \geq L(c^2 - s^2), \qquad \text{whenever } R \leq s \leq c - \sigma,$$

which yields

$$\frac{\sqrt{2}}{2} \tau^+(c) \leq \int_0^R \frac{d\xi}{\sqrt{G(c) - G(\xi)}}$$

$$+ \int_R^{c-\sigma} \frac{d\xi}{\sqrt{L(c^2 - \xi^2)}} + \int_{c-R}^c \frac{d\xi}{\sqrt{G(c) - G(\xi)}}$$

$$\leq \frac{R}{\sqrt{G(c) - G(R)}} + \frac{1}{\sqrt{L}} \left(\arcsin \frac{c - R}{c} - \arcsin \frac{R}{c} \right)$$

$$+\frac{R}{\sqrt{\min_{c-R \leq s \leq c} g(s)}}.$$

Using the condition $(g_{\pm\infty})$ yields

$$\limsup_{c \to \infty} \tau^+(c) \leq \sqrt{\frac{1}{2L}}\pi.$$

Letting $L \to \infty$, then we obtain

$$\lim_{c \to \infty} \tau^+(c) = 0.$$

Similarly, we have

$$\lim_{d \to -\infty} \tau^-(d) = 0.$$

The proof of the proposition is thus completed. □

Proposition 10.3 *Let the condition $(g_{\pm\infty})$ be valid. If there is a constant $M > 0$, such that*

$$\lim_{|x| \to \infty} \frac{G(x+M) - G(x)}{x^2} = \infty, \tag{10.35}$$

then the condition (τ_0) is valid.

Proof. It is obvious that for any given constant $L > 0$, there exists a sufficiently large constant $R > 0$, such that

$$G(s_2) - G(s_1) \geq G(s_2) - G(s_2 - M) \geq 2L(s_2 - M)^2 \geq Ls_2^2 \geq L(s_2^2 - s_1^2),$$

whenever

$$R \leq s_1 \leq s_2 - M \quad \text{or} \quad -R \geq s_1 \geq s_2 + M.$$

Then, letting

$$\sigma = M \quad \text{and} \quad \phi_L(s) = G(s) - Ls^2,$$

then we complete the proof via Proposition 10.2. □

It is known that the condition (S_p^*) does not suffice to guarantee the condition (τ_0) (see the Duffing equation (10.28)). Roughly speaking, the condition (10.35) means that $G(x)$ is larger than $|x|^k$ for some $k > 3$ (e.g., the Duffing equation (10.26)), and so it is a stronger condition than (S_p^*).

Chapter 10. Duffing Equations of Second Order

Proposition 10.4 *Assume $g(x)$ satisfies the condition*

$$\frac{g(x)}{x} \geq K > 0 \qquad (|x| \geq d > 0),$$

with some constants K and d, and the primitive $G(x)$ of $g(x)$ satisfies

$$G(tx) \leq \sigma(t)G(x), \qquad (|x| \geq d, \;\; 0 \leq \sigma(t) < 1 \text{ for } 0 \leq t \leq 1),$$

then the condition (τ_0) is valid.

Proof. The proof is simple and is thus omitted. □

Before ending this section, we remark that despite the above Theorems 10.2 and 10.3, it remains an open question:

Does the condition (S_p^*) imply the existence of infinitely many harmonic and subharmonic solutions of the Duffing equation?

10.4 Duffing Equation of Sub-Linear Type

10.4.1 *Preliminaries*

Consider the Duffing equation

$$\ddot{x} + g(x) = p(t), \tag{10.36}$$

or equivalently,

$$\dot{x} = y, \quad \dot{y} = -g(x) + p(t),$$

where $g(x)$ satisfies the local Lipschitz condition in $x \in \mathbb{R}^1$, and $p(t)$ is a piecewise continuous function in $t \in \mathbb{R}^1$ with minimal period $T > 0$. Assume (10.36) is a sub-linear Duffing equation (i.e., it satisfies the conditions (S_b) and $(g_{\pm \infty})$).

In 1960, Z. Opial proved first that the sub-linear Duffing equation has at least a harmonic solution [101]. It was proved later in 1989 that the sub-linear Duffing equation has an infinity of subharmonic solutions extending to infinity [42]. More exactly, the result can be stated as follows.

Proposition 10.5 *If (10.36) is a sub-liner Duffing equation, satisfying the condition*

$$g'(x) > 0, \qquad x \in \mathbb{R}^1, \tag{10.37}$$

then there is a positive integer $\kappa \geq 2$, such that for any integer $m \geq \kappa$, the equation (10.36) has at least n_m subharmonic solutions

$$x = x_{mj}, \quad j = 1, \cdots, n_m,$$

of order m, where the integers n_m satisfy:

$$2 \leq n_\kappa \leq \cdots \leq n_m \leq \cdots \quad (\to \infty),$$

and the subharmonic solutions extend to the infinity; that is,

$$\lim_{m \to \infty} \inf_{1 \leq j \leq n_m} \sup_{0 \leq t \leq mT} \sqrt{x_{mj}^2(t) + \dot{x}_{mj}^2(t)} = \infty.$$

Subsequently, Z. Yang in his master thesis (1989) proved this result again under favor of a technique to eliminate the condition (10.37). The research is further generalized by Ding and Zanolin [62] to the equation

$$\ddot{x} + h(t, x) = p(t),$$

under the condition $h(t, x) \leq g(x)$, where $g(x)$ satisfies the conditions $(g_{\pm\infty})$ and (τ_∞). However, we are satisfied with the equation (10.36) for the brevity of this book. The arguments in this section are more or less similar to those in the above section, but the results are quite different.

Assume the condition $(g_{\pm\infty})$. Then the equation (10.36) has the time-map

$$\tau(h) = \sqrt{2} \int_d^h \frac{dx}{\sqrt{G(h) - G(x)}} \quad (h > c_0 > 0),$$

where $d = d(h) < 0$ satisfies $G(d) = G(h)$. Then it is easy to prove the following implication

$$(g_{\pm\infty}) + (S_b) \quad \Rightarrow \quad (g_{\pm\infty}) + (\tau_\infty).$$

For the Duffing equation of sub-linear type, it is sensible sometime to consider the one-sided conditions:

$$(S_b^+): \quad \lim_{x \to +\infty} \frac{g(x)}{x} = 0;$$

$$(S_b^-): \quad \lim_{x \to -\infty} \frac{g(x)}{x} = 0;$$

$$(S_b^{*+}): \quad \lim_{x \to +\infty} \frac{G(x)}{x^2} = 0;$$

$$(S_b^{*-}): \quad \lim_{x \to -\infty} \frac{G(x)}{x^2} = 0.$$

Moreover, let

$$\tau^+(h) = \sqrt{2} \int_0^h \frac{dx}{\sqrt{G(h) - G(x)}} \qquad (h > c_0 > 0),$$

and

$$\tau^-(h) = \sqrt{2} \int_d^0 \frac{dx}{\sqrt{G(h) - G(x)}} \qquad (d < 0, \ G(d) = G(h)).$$

Then we have

$$\tau(h) = \tau^-(h) + \tau^+(h).$$

Correspondingly, assume

$$(\tau_\infty^+): \quad \lim_{h \to +\infty} \tau^+(h) = \infty,$$

and

$$(\tau_\infty^-): \quad \lim_{h \to +\infty} \tau^-(h) = \infty.$$

Proposition 10.6 *The following implications hold:*

$$(g_{\pm\infty}) + (S_b^+) \quad \Rightarrow \quad (g_{\pm\infty}) + (S_b^{*+}) \quad \Rightarrow \quad (g_{\pm\infty}) + (\tau_\infty^+)$$

and

$$(g_{\pm\infty}) + (S_b^-) \quad \Rightarrow \quad (g_{\pm\infty}) + (S_b^{*-}) \quad \Rightarrow \quad (g_{\pm\infty}) + (\tau_\infty^-),$$

and they are non-invertible.

Proof. The proof is similar to that of Proposition 10.1. □

Let

$$z = z(t, z_0) = \langle r(t, z_0), \theta(t, z_0) \rangle$$

be the solution of (10.36) satisfying the initial condition $z(0) = z_0$. As we know before, if the condition $(g_{\pm\infty})$ is valid, then for any given integer $m > 0$ and any arbitrarily large constant $c_0 > 0$, there is a constant $d_0 > 0$ such that

$$r(t, z_0) > c_0, \quad \dot\theta(t, z_0) < 0 \quad (0 \le t \le mT),$$

whenever $|z_0| \ge d_0$.

Lemma 10.5 *If the condition* $(g_{\pm\infty}) + (\tau_\infty^+)$ *or* $(g_{\pm\infty}) + (\tau_\infty^-)$ *is valid and the solution* $z = z(t, z_0)$ *satisfies the condition*

$$r(t, z_0) \geq R_0 > 0, \qquad 0 < t_1 < t_2,$$

such that

$$\theta(t_2, z_0) - \theta(t_1, z_0) = -2\pi,$$

then $(t_2 - t_1)$ *is arbitrarily large provided* R_0 *is large enough.*[1]

Proof. Let $\mathfrak{D}$ and $\mathfrak{D}_i$ ($1 \leq i \leq 6$) be the regions defined as before. Without loss of generalization, assume $z = z(t, z_0) \in \mathfrak{D}$ if $t \in [t_1, t_2]$.

Let (τ_∞^+) be true. Assume the motion $z = (x(t), y(t)) = z(t, z_0)$ passes through the region $\mathfrak{D}_2$ during the time interval $[s_1, s_2] \subset [t_1, t_2]$. Hence, we have

$$x(t) > c_0, \quad \dot{x}(t) = y(t) > 0 \qquad (s_1 < t < s_2)$$

with

$$x_1 = x(s_1) = c_0, \quad y(s_2) = 0, \quad x_2 = x(s_2) > R_0.$$

It follows from $\max |p(t)| \leq M$ ($|t| \leq T$) that

$$y(t)\dot{y}(t) = y(t)[-g(x(t)) + p(t)] \leq -[g(x(t)) + M]\dot{x}(t),$$

which together with an integration on the interval $[t, s_2] \subset [s_1, s_2]$ yields

$$\frac{1}{2} y^2(t) \leq G(x_2) - G(x(t)) + M(x_2 - x(t))$$

$$= [G(x_2) - G(x(t))]\left(1 + \frac{M}{g(\xi)}\right),$$

where ξ is some point in $[x_1, x_2]$. Using the condition $(g_{\pm\infty})$, we obtain

$$1 < \left(1 + \frac{M}{g(\xi)}\right) < 2.$$

It follows that

$$\dot{x}(t) = y(t) = 2\sqrt{G(x_2) - G(x(t))}, \qquad s_1 \leq t \leq s_2.$$

---
[1] Note that $(t_2 - t_1) > 0$ is the period during which the motion $z = z(t, z_0)$ rotates a clockwise turn around the origin.

Hence, we have

$$s_2 - s_1 \geq \int_{c_0}^{x_2} \frac{dx}{2\sqrt{G(x_2) - G(x)}} = \frac{\tau^+(x_2) - \tau^+(c_0)}{2\sqrt{2}}.$$

Note that c_0 is a fixed constant and $x_2 > R_0$, where $R_0 > 0$ is an arbitrarily large constant whenever $|z_0|$ is large enough. Therefore, using the condition (τ_∞^+), we conclude that $(s_2 - s_1)$ is sufficiently large whenever $|z_0|$ is large enough.

It is clear that the same conclusion can be proved in a similar manner if the condition (τ_∞^-) is assumed to be true.

We have thus proved Lemma 10.5. □

10.4.2 Harmonic Solution

Theorem 10.4 *If the condition* $(g_{\pm\infty}) + (\tau_\infty^+)$ *or* $(g_{\pm\infty}) + (\tau_\infty^-)$ *is valid, then the Duffing equation has at least a harmonic solution.*

Proof. Using Lemma 10.5 yields that there is a large constant $\rho > 0$, such that when $|z_0| = \rho$, the motion $z = x(t, z_0)$ will move clockwise through a small angle, say $-\frac{\pi}{6} < (\theta(T, z_0) - \theta(0, z_0)) < 0$, during the period $0 \leq t \leq T$. On the other hand, let us consider the Poincaré map Φ of Duffing equation (10.36) on the disk

$$\mathfrak{D}_\rho = \{z \in \mathbb{R}^2 : \quad |z| \leq \rho\}.$$

Let $\mathcal{C}_\rho = \partial \mathfrak{D}_\rho$ and $\xi \in \mathcal{C}_\rho$. It follows from $-\frac{\pi}{6} < (\theta(T, z_0) - \theta(0, z_0)) < 0$ that $\Phi(\xi) \notin L_\xi$, where L_ξ is the ray issuing from the origin and passing through the point ξ. Thus the Poincaré map Φ satisfies the condition of the Poincaré-Bohl fixed-point theorem. Hence there is at least a fixed-point ζ_0 in the disk $\mathfrak{D}_\rho$. It follows that $z = z(t, \zeta_0)$ is a harmonic solution of the Duffing equation (10.36).

Theorem 10.4 is thus proved. □

Now, let $z = z_0(t) = (x_0(t), y_0(t))$ be a harmonic solution of the Duffing equation (10.36). Denote its orbit by Γ_0, which is of course a closed orbit. Note that Γ_0 may be not simple since the differential equation (10.36) is not autonomous. Let

$$x = u + x_0(t), \quad y = v + y_0(t).$$

Substituting it in (10.36), we obtain

$$\dot{u} = v, \quad \dot{v} = -[g(u + x_0(t)) - g(x_0(t))], \tag{10.38}$$

which has a trivial solution $w = w_0(t) = (u_0(t), v_0(t)) = (0,0)$, associated with the harmonic solution $z = z_0(t)$. Let $w = w(t, w_0)$ be the solution of (10.38) satisfying the initial condition $w(0) = w_0 \neq 0$. It follows that $w = w(t, w_0) \neq 0$ for $t \in \mathbb{R}^1$. Hence, it can be represented by polar coordinates

$$\Lambda: \quad u = u(t) = \varrho(t) \cos \varphi(t), \quad v = v(t) = \varrho(t) \sin \varphi(t),$$

where $\varrho(t) > 0$ and $\varphi(t)$ are continuous functions in $t \in \mathbb{R}^1$.

Lemma 10.6 *If $t_1 > 0$ is a constant, such that*

$$\varphi(t_1) - \varphi(0) < -2N\pi, \tag{10.39}$$

then for any $t_2 > t_1$ we have

$$\varphi(t_2) - \varphi(0) < -2N\pi + \pi. \tag{10.40}$$

Proof. Note that the orbit Λ cannot intersect the origin. It follows from $\dot{u}(t) = v(t)$ that the orbit Λ moves in the clockwise direction when it intersects the v-axis. It follows that if the orbit Λ intersects the positive (or the negative) v-axis at the time $t = \alpha$ and intersects subsequently the negative (or the positive) v-axis at the time $t = \beta > \alpha$, then we have

$$\varphi(\beta) - \varphi(\alpha) = -\pi.$$

On the other hand, if the orbit Λ stays in the right half-plane (or in the left half-plane) during the time interval $\mu \leq t \leq \nu$, the increase of the angle

$$\varphi(\nu) - \varphi(\mu) < \pi.$$

It follows that

$$\varphi(t_2) - \varphi(0) = [\varphi(t_1) - \varphi(0)] + [\varphi(t_2) - \varphi(t_1)] < -2N\pi + \pi.$$

Lemma 10.6 is thus proved. □

Corollary 10.2 *Letting $mT > t_1$, we obtain*

$$\varphi(mT) - \varphi(0) < -2N\pi + \pi. \tag{10.41}$$

Note that the orbit of the harmonic solution $z = z_0(t) = (x_0(t), y_0(t))$ is a closed orbit Γ_0 in the (x, y)-plane. Let Υ_ζ be the orbit of the solution $z = z(t) = z(t, \zeta)$ satisfying the initial condition $z(0) = \zeta$ in the (x, y)-plane. Consider the moving points

$$P(t) = z(t) \in \Upsilon_\zeta \quad \text{and} \quad Q(t) = z_0(t) \in \Gamma_0 \quad (t \in \mathbb{R}^1).$$

Let Δ_{OPQ} be the triangle withe vertices O, $P(t)$ and $Q(t)$. It is noted that the vector $\overrightarrow{QP}$ has the argument $\varphi(t, \zeta)$ and the vector $\overrightarrow{OP}$ has the argument $\theta(t, \zeta)$. Therefore, we obtain

$$\theta(t, \zeta) = \varphi(t, \zeta) + \angle OPQ,$$

where $|\angle OPQ| < \frac{\pi}{6}$ provided $|\zeta|$ is large enough. It follows from (10.41) that

$$\theta(mT, \zeta) - \varphi(0, \zeta) < -\left(2N - \frac{3}{2}\right)\pi \tag{10.42}$$

provided $|\zeta|$ is sufficiently large.

10.4.3 Subharmonic Solutions of Higher Order

Theorem 10.5 *If the condition $(g_{\pm\infty}) + (\tau_\infty^+)$ or $(g_{\pm\infty}) + (\tau_\infty^-)$ is valid, then there is a positive integer $m_0 \geq 2$, such that for any integer $m \geq m_0$, the Duffing equation (10.36) has at least a subharmonic solution $z = z_m(t)$ of order m. Furthermore, all the subharmonic solutions of order m are bounded in a compact region $\mathfrak{B}_m$ in the phase plane.*

Proof. The condition $(g_{\pm\infty})$ yields that there is a constant $c_0 > 0$, such that the solution of (10.36)

$$\Gamma: \quad z = z(t, z_0) = \langle r(t, z_0), \theta(t, z_0)\rangle$$

satisfies

$$\dot\theta(t) < 0, \quad \text{whenever } r(t, z_0) \geq c_0. \tag{10.43}$$

Moreover, assume

$$c_0 > \sup_{0 \leq t \leq T} \sqrt{x_0^2(t) + y_0^2(t)}$$

where $z = z_0(t) = (x_0(t), y_0(t))$ is the subharmonic solution mentioned above.

Then, for arbitrarily large $t_1 > 0$, there is a constant $d_0 > c_0$, such that when $0 \leq t \leq t_1$ we have

$$r(t, z_0) > c_0, \quad \dot{\theta}(t) < 0, \quad \text{when } |z_0| \geq d_0,$$

which yields

$$\theta(t_1, z_0) - \theta(0, z_0) < -2N\pi, \tag{10.44}$$

where $N \geq 0$ is some integer.

Claim: The integer N in (10.44) can be taken sufficiently large if t_1 is large enough.

In fact, assume the contrary. Then there is a constant $K > 0$, such that

$$\theta(t_1, z_0) - \theta(0, z_0) > -K, \quad \text{for some } t_1 \gg 1. \tag{10.45}$$

We will proceed the arguments in the following cases:

Case 1: $r(t_1, z_0) \to \infty$ as $t_1 \to \infty$.

It follows that

$$\lim_{t \to \infty} \theta(t, z_0) = \theta^* > -\infty.$$

Therefore, the orbit Γ has the asymptotical ray $\theta = \theta^*$. If $x(t) \to 0$ as $t \to \infty$, then $y(t) \to \pm\infty$. It follows that

$$\lim_{|y| \to \infty} \frac{dy}{dx} = \lim_{|y| \to \infty} \frac{-g(x) + p(t)}{y} = 0,$$

which means that when Γ approaches asymptotically to the y-axis its slope tends to zero. This conclusion is obviously absurd. Therefore, we have $\theta^* \neq \frac{(2n\pm 1)\pi}{2}$. Hence, $x \to \infty$ as $t \to \infty$. It follows from

$$\dot{\theta}(t) = -\sin^2 \theta(t) - \frac{g(x) - p(t)}{x} \cos^2 \theta(t)$$

and the condition (S_b) that

$$\theta^* = k\pi, \quad \text{for some integer } k.$$

Therefore, we have

$$(x(t), y(t)) \to (+\infty, +0) \quad \text{or} \quad (-\infty, -0), \quad \text{as } t \to \infty,$$

which together with $(g_{\pm\infty})$ implies

$$\lim_{|x| \to \infty} \frac{dy}{dx} = \lim_{|x| \to \infty} \frac{-g(x) + p(t)}{y} = -\infty.$$

This is contradicts the fact that Γ approaches asymptotically to the x-axis. Therefore Case 1 cannot appear.

Case 2: There exists arbitrarily large constants $t_j > 0$, such that $r(t_j, z_0) = c_0$, where c_0 is the constant given above.

Let $d_0 > 0$ be a sufficiently large constant. Using (10.45), we conclude that each orbit Γ^* starting from the circle $\mathcal{C}$ ($|z| = d_0$) has an asymptotic ray. It will similarly lead a contradiction. Hence, Case 2 cannot appear too.

Therefore, the above claim is proved.

Then, take a sufficiently large constant $d_0 > c_0$, such that the solutions $z = z(t, z_0)$ starting from the circle $|z_0| = d_0$ has the following property:

(P_1) For a given prime $q \geq 2$, there is a constant $t_1 > 0$, such that

$$|z(t, z_0)| > c_0, \qquad 0 \leq t < t_1,$$

with

$$|z(t_1, z_0)| = c_0 \quad \text{and} \quad \theta(t_1, z_0) - \theta(0, z_0) < -(2q+2)\pi;$$

or

(P_2) For a given prime $q \geq 2$ and for arbitrarily large $t_1 > 0$, the solution $z = z(t, z_0)$ satisfies

$$|z(t, z_0)| > c_0, \qquad 0 \leq t \leq t_1,$$

and

$$\theta(t_1, z_0) - \theta(0, z_0) < -(2q+2)\pi.$$

Then, let

$$E = \{t_1 > 0: \quad (P_1) \text{ is valid}\}.$$

Let

$$t_1^* = \begin{cases} \sup_{t_1 \in E} \{t_1\}, & \text{if } E \neq \emptyset; \\ 0, & \text{if } E = \emptyset. \end{cases}$$

Taking the integers

$$m_0 = \max\{2, [t_1^*/T] + 1\} \quad \text{and} \quad m \geq m_0,$$

it follows that

$$\theta(mT, z_0) - \theta(0, z_0) < -\left(2q + \frac{1}{2}\right)\pi, \qquad \text{where } |z_0| = d_0. \qquad (10.46)$$

On the other hand, it follows from Lemma 10.5 that there is a sufficiently large constant $b_m (> d_0)$, such that

$$-\frac{\pi}{6} < \theta(mT, z_0) - \theta(0, z_0) < 0, \qquad |z_0| = b_m. \qquad (10.47)$$

It follows that the iteration Φ^m of the Poincaré map is area-preserving and twisted on the annulus

$$\mathcal{A}_m : \qquad d_0^2 \leq x^2 + y^2 \leq b_m^2.$$

Then, using the generalized Poincaré-Birkhoff twist theorem, we obtain two fixed-points ζ_i ($i = 1, 2$) of Φ^m, satisfying

$$\theta(mT, \zeta_i) - \theta(0, \zeta_i) = -2q\pi. \qquad (10.48)$$

It is obvious that

$$z = z(t, \zeta_i), \qquad i = 1, 2,$$

are mT-periodic solutions. Similar to the proof of subharmonic solutions for the nearly super-linear Duffing equation, using (10.48) implies that mT is the minimal period. Hence, $z = z(t, \zeta_i)$ ($i = 1, 2$) are subharmonic solutions of order m.

Finally, it can be seen that

$$-\frac{\pi}{6} < \theta(mT, z_0) - \theta(0, z_0) < 0, \qquad |z_0| \geq R_m$$

holds for sufficiently large constant $R_m > 0$. Therefore, the subharmonic solutions of order m are bounded in the region

$$\mathcal{B}_m = \{\, (x, y) \in \mathbb{R}^2 : \quad \sqrt{x^2 + y^2} \leq R_m \,\}.$$

The proof of Theorem 10.5 is thus completed. □

10.4.4 Another Sub-Linear Condition

Corresponding to the conditions ($g_{\pm\infty}$) and (S_b), we can indeed assume another type of conditions. For example, consider the conditions.

$$(\hat{g}_E): \qquad \text{sign}\,(x) \cdot g(x) > E > 0, \qquad \text{for } |x| > a_0 \geq 0,$$

Chapter 10. Duffing Equations of Second Order

where E is a positive constant; and

$$(\hat{S}_b): \quad \liminf_{|x| \to \infty} \frac{g(x)}{x} = 0.$$

These conditions are obviously weaker than $(g_{\pm\infty})$ and (S_b).

Theorem 10.6 *Let the conditions $(\hat{g}_E)$ and $(\hat{S}_b)$ be valid, with*

$$E > \sup_{0 \le t \le T} |p(t)|. \tag{10.49}$$

Furthermore, let the condition

$$(\hat{S}_b^*): \quad \left|\frac{xg'(x)}{g(x)}\right| < M, \quad |x| > 0,$$

be satisfied for some constant $M > 0$. Then the Duffing equation (10.36) has at least a harmonic solution (see [54]).

Proof. Let $t = (2Ts)$ and $x = (2T)^2 u$. It follows from (10.36) that

$$\frac{d^2 u}{ds^2} + \breve{g}(u) = \breve{p}(s),$$

where $\breve{p}(s) = p(2Ts)$ is periodic in $s \in \mathbb{R}^1$ with minimal period $1/2$, and $\breve{g}(u) = g((2T)^2 u)$ satisfies the conditions $(\hat{g}_E)$, $(\hat{S}_b)$ and $(\hat{S}_b^*)$ with respect to $u \in \mathbb{R}^1$.

Then, without loss of generality, assume (10.36) has minimal period $1/2$. It is obvious that

$$\limsup_{|x| \to \infty} \frac{g(x)}{x} = 0 \tag{10.50}$$

implies (S_b).

It can be seen from the proof of Theorem 10.4 that Duffing equation (10.36) has at least a harmonic solution if (S_b) and (10.49) are valid. It follows that Theorem 10.6 holds true when (10.50) is valid.

Now, assume

$$\limsup_{|x| \to \infty} \frac{g(x)}{x} = \sigma_0 > 0. \tag{10.51}$$

Then there are sequences of positive constants $\{\varepsilon_n\}$ and $\{x_n\}$ satisfying

$$\varepsilon_n \to 0, \quad x_n \to \infty, \quad \text{(as } n \to \infty\text{)},$$

10.4. Duffing Equation of Sub-Linear Type

such that the function $g(x)/x$ attains the local minimuum ε_n at $x = x_n$. Hence, there is an $\alpha_n > 0$ such that

$$|g(x)| < 2\varepsilon_n |x| \qquad (x_n \le x \le x_n + \alpha_n),$$

which together with the condition $(\hat{S}_b^*)$ implies

$$|g'(x)| < 2M\varepsilon_n < \frac{1}{2}, \qquad (x_n \le x \le x_n + \alpha_n),$$

when n is sufficiently large. On the other hand, consider the straight lines L_{ε_n} $(y = \varepsilon_n x)$ and $L_{2\varepsilon_n}$ $(y = 2\varepsilon_n x)$ in the (x, y)-plane. Construct a straight line

$$L_n: \qquad y = \frac{1}{2}(x - x_n) + \varepsilon_n x_n,$$

passing through the point $P_n = (x_n, \varepsilon_n x_n) \in L_{\varepsilon_n}$ with slope $1/2$. Let the line L_n intersects the line $L_{2\varepsilon_n}$ at a point $Q_n = (\tilde{x}_n, 2\varepsilon_n \tilde{x}_n)$. It follows that

$$\tilde{x}_n = x_n + \frac{2\varepsilon_n}{1 - 4\varepsilon_n} x_n = \frac{1 - 2\varepsilon_n}{1 - 4\varepsilon_n} x_n$$

and

$$\varepsilon_n \le \frac{g(x)}{x} \le 2\varepsilon_n \qquad (x_n \le x \le \tilde{x}_n).$$

Now, let us consider the auxiliary equation

$$\dot{x} = y, \quad \dot{y} = -g(x) + \lambda p(t), \tag{10.52}$$

where $\lambda \in [0, 1]$ is a parameter. Note that (10.52) is equivalent to (10.36) when $\lambda = 1$. Let

$$\Gamma: \qquad x = x(t), \quad y = y(t),$$

be the solution of (10.52) satisfying the initial condition $(x(t_0), y(t_0)) = (\tilde{x}_n, 0)$. It can be seen from (10.52) that

$$x(t) < \tilde{x}_n, \quad y(t) > 0, \qquad \text{whenever } 0 < t_0 - t \ll 1.$$

Then, there is a t_1 $(t_1 < t_0)$, such that we have the case (Q_1):

$$x(t_1) = x_n, \quad x_n < x(t) < \tilde{x}_n, \quad 0 < y(t) < 2\varepsilon_n x(t) \qquad (t_1 < t < t_0);$$

or the case (Q_2):

$$y(t_1) = 2\varepsilon_n x_n, \quad x_n < x(t) < \tilde{x}_n, \quad 0 < y(t) < 2\varepsilon_n x_n \qquad (t_1 < t < t_0).$$

Assume first (Q_1) is valid. Then we have

$$\dot{x}(t) = y(t) < 2\varepsilon_n x(t), \qquad t_1 \leq t < t_0,$$

which yields

$$t_0 - t_1 > \frac{1}{2\varepsilon_n} \int_{x_n}^{\tilde{x}_n} \frac{dx}{x} = \frac{1}{2\varepsilon_n} \log \frac{\tilde{x}_n}{x_n}$$

$$= \frac{1}{2\varepsilon_n} \log \frac{1 - 2\varepsilon_n}{1 - 4\varepsilon_n} = 1 + \mathcal{O}(\varepsilon_n) > \frac{1}{2}.$$

It follows that any solution of (10.52) passing through $(\tilde{x}_n, 0)$ cannot be $1/2$-periodic.

Assume next (Q_2) is valid. Then we have

$$0 < g(x(t)) < 2\varepsilon_n x(t), \qquad t \in (t_1, t_0),$$

which implies

$$0 > \dot{y}(t) = -g(x(t)) + \lambda p(t) \geq -2\varepsilon_n x(t) - E \geq -2\varepsilon_n \tilde{x}_n - E.$$

It follows that

$$t_0 - t_1 > \int_0^{2\varepsilon_n x_n} \frac{dy}{2\varepsilon_n \tilde{x}_n + E}$$

$$= \frac{2\varepsilon_n x_n}{2\varepsilon_n \frac{1 - 2\varepsilon_n}{1 - 4\varepsilon_n} x_n + E} \geq \frac{1}{\frac{1 - 2\varepsilon_n}{1 - 4\varepsilon_n} + \frac{E}{2g(x_n)}}$$

$$= 1 + o(1) > \frac{1}{2} \qquad (n = n_0 \gg 1).$$

Therefore, any solution of (10.52) passing through $(\tilde{x}_n, 0)$ cannot be $1/2$-periodic.

It follows from the proposition in the Appendix of [54] that the Duffing equation (10.36) has at least a harmonic solution under the assumptions of Theorem 10.6. We have thus completed the proof. □

10.5 Duffing Equation of Semi-Linear Type

Generally speaking, the existence of periodic solution of Duffing equation depends on the relative positions of the set Λ of resonance points and the set Υ of elasticity.

For example, the super-linear Duffing equation has strong elasticity, such that $\Upsilon = \{\infty\}$ does not overlap the spectrum Λ. Correspondingly, the Poincaré map Φ has strong twist property. We prove the multiplicity of harmonic or subharmonic solutions via the generalized Poincaré-Birkhoff twist theorem.

The sub-linear Duffing equation possesses weak elasticity, such that Υ intersects the spectrum Λ at the boundary point 0. Correspondingly, the Poincaré map Φ has weak twist property. We still prove the multiplicity of subharmonic solutions via the generalized Poincaré-Birkhoff twist theorem.

On the other hand, the semi-linear Duffing equation has the medium (i.e., neither strong nor weak) elasticity, such that Υ overlaps in various ways Λ. The trouble is that there may be no twist property of the Poincaré map Φ. In this case, resonance phenomenon may take place as shown in the following examples.

Example 10.9 Consider the linear Duffing equation

$$\ddot{x} + m^2 x = E \cos \omega t, \tag{10.53}$$

where m is a positive integer, E and ω are positive constants.

It can be seen that the set of elasticity $\Upsilon = m^2$ overlaps the spectrum $\Lambda = \{k^2 : k \in \mathbb{Z}\}$ at $k^2 = m^2$. The general solution of linear differential equation (10.53) can be found by elementary method. It can be seen that the solution is bounded if and only if $\omega \neq m$. Resonance happens to the case $\omega = m$.

Example 10.10 Consider the semi-linear Duffing equation

$$\ddot{x} + m^2 x + h(x) = E \cos mt, \tag{10.54}$$

where m is a positive integer, E is a positive constant and $h(x)$ is a bounded C^1-differentiable function in $x \in \mathbb{R}^1$. Assume

$$\sup_{x \in \mathbb{R}^1} |h(x)| < H,$$

for some constant $H > 0$.

The general solution of this nonlinear Duffing equation is unknown. However, we can still show the unbounded-ness of the solution as follows.

It is apparent that each solution of (10.54) is existent and unique on the interval $(-\infty, \infty)$. Then, if (10.54) has a bounded solution, then the Massera's theorem leads to a 2π-periodic solution $x = \phi(t)$ of (10.54). Hence, we have

$$\ddot{x} + m^2 x = p(t), \tag{10.55}$$

where the function $p(t) = [E \cos mt - h(\phi(t))]$ is 2π-periodic. Of course, $x = \phi(t)$ is a 2π-periodic solution of the linear equation (10.55). Hence, according to the result of Example 10.1, we conclude that

$$a_n = \int_0^{2\pi} [E \cos mt - h(\phi(t))] \cos mt \, dt = 0,$$

which yields

$$E\pi = \int_0^{2\pi} E \cos^2 mt \, dt = \left| \int_0^{2\pi} h(\phi(t)) \cdot \cos mt \, dt \right|$$

$$\leq \int_0^{2\pi} H \, dt = 2\pi H.$$

It leads to $E \leq 2H$. It follows that the equation (10.54) has no bounded solution whenever $E > 2H$, and in this case resonance happens to it.

For general consideration, assume

$$\ddot{x} + g(x) = p(t) \tag{10.56}$$

is a semi-linear Duffing equation without friction. The major subject is to find some non-resonance conditions guaranteeing the existence of periodic solution.

Definition 10.3

(1) Equation (10.56) is said to be *off* the resonance point if the condition

$$\left(\frac{2j\pi}{T}\right)^2 < \alpha \leq \frac{g(x)}{x} \leq \beta < \left(\frac{2j\pi}{T}\right)^2 \tag{10.57}$$

is satisfied for some integer j;

(2) Equation (10.56) is said to be *near* the resonance point if the condition

$$\begin{cases} \dfrac{g(x)}{x} \neq \left(\dfrac{2j\pi}{T}\right)^2, & |x| \geq a_0 \gg 1; \\ \lim\limits_{|x|\to\infty} \dfrac{g(x)}{x} = \left(\dfrac{2j\pi}{T}\right)^2 \end{cases} \quad (10.58)$$

satisfied for some integer j;

(3) Equation (10.56) is said to be *across* the resonance point if the condition

$$\liminf_{|x|\to\infty} \frac{g(x)}{x} < \left(\frac{2j\pi}{T}\right)^2 < \limsup_{|x|\to\infty} \frac{g(x)}{x}, \quad (10.59)$$

is satisfied for some integer j.

For example, consider the Duffing equation

$$\ddot{x} + \mu^2 x + h(x) = p(t),$$

where $p(t)$ is T-periodic and $h(x)$ satisfies the condition

$$xh(x) > 0 \quad \text{and} \quad \lim_{|x|\to\infty} \frac{h(x)}{x} = 0.$$

Then this equation is off the resonance point if $\mu \neq \frac{2m\pi}{T}$ for some integer m, and it is near the resonance point if $\mu = \frac{2m\pi}{T}$ for some integer m.

The Duffing equation

$$\ddot{x} + \tilde{g}(x) = p(t)$$

is across the resonance point, where $\tilde{g}(x)$ is defined in Example 10.8.

10.5.1 *Duffing Equation off Resonance Point*

In 1967, W. Loud gave a non-resonance condition (10.60) when he proved the following existence theorem [92].

Theorem 10.7 *If $g(x)$ is a continuously differentiable odd function and satisfies the condition*

$$(n+\delta)^2 \leq g'(x) \leq (n+1-\delta)^2, \quad x \in \mathbb{R}^1, \quad (10.60)$$

where $n \geq 0$ is an ineger and δ is a small positive number, then the Duffing equation

$$\ddot{x} + g(x) = E\cos t \qquad (E > 0 \text{ is a constant})$$

has a unique 2π-periodic solution. Moreover, the periodic solution is even.

Later, D. Leach generalized the Loud's result as follows [87].

Theorem 10.8 *If $g(x)$ satisfies the Loud condition (10.60), then the Duffing equation (10.56) with $T = 2\pi$) has one and only one 2π-periodic solution.*

Proof. Now, we give a simple proof as follows (see [123]). Let (10.56) be written into the equivalent system

$$\dot{x} = y, \quad \dot{y} = -g(x) + p(t), \qquad (10.61)$$

and let

$$x = x(t) = x(t,\xi,\eta), \quad y = y(t) = y(t,\xi,\eta) \qquad (t \in \mathbb{R}^1)$$

be its solution satisfying the initial condition $(x(0), y(0)) = (\xi, \eta)$.

For the proof of the existence part, it is suffices to prove the following lemma.

Lemma 10.7 *If the condition (10.60) is valid, then there is a number $A > 0$, such that $\xi^2 + \eta^2 \geq A^2$ implies*

$$(x(2\pi,\xi,\eta), y(2\pi,\xi,\eta)) \neq \lambda \cdot (\xi, \eta), \qquad (10.62)$$

where $\lambda \geq 0$ is any constant.

Proof. Consider the inequalities

$$(n+\delta)^2 - \alpha^2 \geq \frac{(n+\alpha)^2}{1-\alpha^2}, \qquad (10.63)$$

$$(n+1-\delta)^2 + \alpha^2 \leq \frac{(n+1-\alpha)^2}{1+\alpha^2}. \qquad (10.64)$$

Since (10.63) and (10.64) are valid in the strict sense when $\alpha = 0$, there must be a sufficiently small positive number $\alpha < 1$ such that (10.63) and

(10.64) hold simultaneously. Also, choose a sufficiently large constant $c > 0$ such that

$$\frac{p_0}{c} \leq \alpha^2, \qquad \text{where } p_0 = \sup_{0 \leq t \leq 2\pi} |p(t)|. \tag{10.65}$$

Using Lemma 10.2, we have a constant $A > 0$ such that $\xi^2 + \eta^2 \geq A^2$ implies

$$x^2(t, \xi, \eta) + y^2(t, \xi, \eta) > c^2, \qquad 0 \leq t \leq 2\pi.$$

Now, we show that this constant A satisfies the property as stated in the Lemma 10.7.

Let

$$x(t, \xi, \eta) = \rho(t) \sin \varphi(t), \quad y(t, \xi, \eta) = \rho(t) \cos \varphi(t).$$

Then $\xi^2 + \eta^2 \geq A^2$ implies that $\rho(t) > c^2$ for $0 \leq t \leq 2\pi$.

On the other hand, we have

$$\dot{\varphi}(t) = \cos^2 \varphi(t) + \frac{[g(\rho(t) \sin \varphi(t)) - p(t)] \sin \varphi(t)}{\rho(t)},$$

which together with (10.60), (10.65) and (10.63) implies

$$\dot{\varphi}(t) \geq \cos^2 \varphi(t) + (n + \delta)^2 \sin^2 \varphi(t) - \alpha^2$$

$$= (1 - \alpha^2) \cos^2 \varphi(t) + [(n + \delta)^2 - \alpha^2] \sin^2 \varphi(t)$$

$$\geq (1 - \alpha^2) \cos^2 \varphi(t) + \frac{(n + \alpha)^2}{1 - \alpha^2} \sin^2 \varphi(t).$$

Similarly, using (10.60), (10.65) and (10.64) implies

$$\dot{\varphi}(t) \leq (1 + \alpha^2) \cos^2 \varphi(t) + \frac{(n + 1 - \alpha)^2}{1 + \alpha^2} \sin^2 \varphi(t).$$

Then, letting k be the integer satisfying

$$2k\pi \leq \varphi(2\pi) - \varphi(0) < 2(k+1)\pi,$$

and using the integral formula

$$\int_0^{2\pi} \frac{dx}{a^2 \cos^2 x + b^2 \sin^2 x} = \frac{2\pi}{ab}$$

where a and b are positive constants, we conclude that

$$\frac{2(k+1)\pi}{n+\alpha} = \int_{\varphi(0)}^{\varphi(0)+2(k+1)\pi} \frac{d\varphi}{(1-\alpha^2)\cos^2\varphi + (n+\alpha)^2\sin^2\varphi/(1-\alpha^2)}$$

$$\geq \int_0^{2(k+1)\pi} \frac{d\varphi}{(1-\alpha^2)\cos^2\varphi + (n+\alpha)^2\sin^2\varphi/(1-\alpha^2)}$$

$$\geq \int_0^{2\pi} \frac{\dot\varphi}{(1-\alpha^2)\cos^2\varphi + (n+\alpha)^2\sin^2\varphi/(1-\alpha^2)} \, dt \geq 2\pi.$$

It follows that $k \geq n - 1 + \alpha$. Since k is an integer, we must have $k \geq n$. Consequently, we have $2n\pi \geq \varphi(2\pi) - \varphi(0)$.

Moreover, we want now to show $2n\pi \neq \varphi(2\pi) - \varphi(0)$. Otherwise, we have $2n\pi = \varphi(2\pi) - \varphi(0)$. It follows that

$$\frac{2n\pi}{n+\alpha} = \int_{\varphi(0)}^{\varphi(2\pi)} \frac{d\varphi}{(1-\alpha^2)\cos^2\varphi + (n+\alpha)^2\sin^2\varphi/(1-\alpha^2)} \geq 2\pi,$$

which yields the contradiction $n \geq n + \alpha$.

Consequently, $2n\pi < \varphi(2\pi) - \varphi(0)$. Similarly, we can prove $\varphi(2\pi) - \varphi(0) < 2(n+1)\pi$.

We have thus proved Lemma 10.7. □

Then consider the Poincaré map Φ of the Duffing equation (10.61) on the disk $\mathfrak{D}_A$ $(x^2+y^2 \leq A^2)$. Lemma 10.7 implies that Φ satisfies the condition of Poincaré-Bohl fixed-point theorem. Hence, there is a fixed-point $(\xi_0, \eta_0) \in \mathfrak{D}_A$, and the Duffing equation (10.61) has a harmonic solution

$$x = x(t, \xi_0, \eta_0), \quad y = y(t, \xi_0, \eta_0).$$

For the proof of the uniqueness part, we need the following lemma, which can be readily proved by using the comparison theorem in the elementary theory of ordinary differential equations.

Lemma 10.8 *If $q(t)$ is a continuous 2π-periodic function satisfying*

$$n^2 < q(t) < (n+1)^2, \quad 0 \leq t \leq 2\pi, \tag{10.66}$$

where n is an integer, then the homogeneous linear equation

$$\ddot u + q(t)u = 0 \tag{10.67}$$

does not have any nontrivial 2π-periodic solution.

Now, suppose that $x = x_1(t)$ and $x = x_2(t)$ are 2π-periodic solutions of (10.56). Then $u = u(t) = x_2(t) - x_1(t)$ is 2π-periodic and satisfies (10.67). Hence, from Lemma 10.8, we obtain $u(t) \equiv 0$; that is, $x_2(t) \equiv x_1(t)$. This proves the uniqueness.

The proof of Theorem 10.7 is thus completed. □

In 1975, Reissig [106] generalized the Loud condition (10.60) to the non-resonance condition (10.57) and proved the following result.

Theorem 10.9 *Duffing equation (10.56) has at least a harmonic solution if the non-resonance condition (10.57) is satisfied.*

Finally, it is remarked that Theorem 10.7, Theorem 10.8 and Theorem 10.9 were originally proved through advanced techniques at the time. Indeed, they can be proved in a simple fashion of Poincaré-Bohl fixed-point theorem as we did in the proof of Theorem 10.8. In what follows, the Poincaré-Bohl fixed-point theorem will be applied further to the Duffing equation.

Associated with Theorem 10.9, W. Li [90] considered the necessary and sufficient condition on the existence and uniqueness of harmonic solution.

10.5.2 Duffing Equation near Resonance Point

Among all the above conditions, the crucial point is to exclude the points of resonance. In other words, the above theorems are valid for the Duffing equation off the resonance point. Therefore, the nonlinear oscillations near or across a resonance point of Duffing equation is really an interesting and complicated problem. Employing the alternative method, Cesari (see [19] and [20]), Lazer and Leach [85] succeeded in proving the existence of harmonic solution of Duffing equation near the resonance point as the following result.

Theorem 10.10 *If*

$$g(x) = \left(\frac{2m\pi}{T}\right)^2 x \pm h(x), \qquad x \in \mathbb{R}^1,$$

where $h(x)$ is a continuous bounded function satisfying the sign condition $(S)_a$ (i.e., $xh(x) > 0$ for $|x| \geq a$),[2] then the Duffing equation (10.56) has

---
[2] It is apparent that in this case the Duffing equation (10.56) is near the resonance point $(2m\pi/T)^2$.

at least a harmonic solution provided

$$\begin{cases} a_m = \int_0^{\frac{2m\pi}{T}} p(t) \cos \frac{2m\pi}{T} dt = 0; \\ b_m = \int_0^{\frac{2m\pi}{T}} p(t) \sin \frac{2m\pi}{T} dt = 0. \end{cases} \quad (10.68)$$

It remains obviously an open question when $h(x)$ is unbounded. In the paper [34], the author proved the existence of harmonic solution of the Duffing equation (10.56) near the resonance point $(2m\pi/T)^2$, where the above function $h(x)$ is unbounded such that $h(x)/x \to 0$ as $|x| \to \infty$. The main result will be reported below in details.

I. Statement of Main Result

Consider the following functional spaces:

$$\mathbb{P} = \{\, p(t) \in C(\mathbb{R}, \mathbb{R}) : \quad p(t + 2\pi) = p(t), \forall\, t \in \mathbb{R}\,\};$$

$$\mathbb{P}_m = \{\, p(t) \in C(\mathbb{R}, \mathbb{R}) : \quad (10.68) \text{ holds}\,\};$$

$$\mathbb{F}_m = \{\, g(x) \in C^1(\mathbb{R}, \mathbb{R}) : \quad m^2 < \alpha \le g'(x) \le \beta < (m+1)^2\,\},$$

$$\mathbb{H}_m = \{\, g(x) \in C^1(\mathbb{R}, \mathbb{R}) : \quad m^2 \le g'(x) \le (m+1)^2\,\},$$

$(m = 0, 1, 2, \cdots)$. Obviously, $\mathbb{F}_m \subset \mathbb{H}_m$. It is clear that if $g(x) \in \mathbb{H}_m$ for $m \ge 1$, then $g(x)$ satisfies the condition $(g_{\pm\infty})$.

If $g(x) \in \mathbb{H}_0$, then assume $g(x)$ satisfies the condition $(g_{\pm\infty})$. The contrary case was studied in an earlier paper of the author [31].

For any $g(x) \in \mathbb{H}_m$, define the characteristic number of $g(x)$ by

$$\chi[g] = \min\left\{\, \sup_{x \in \mathbb{R}} |g(x) - m^2 x|,\ \sup_{x \in \mathbb{R}} |g(x) - (m+1)^2 x|\,\right\}.$$

Note that $\chi[g]$ may be ∞. Define

$$\mathbb{G}_m = \{\, g(x) \in \mathbb{H}_m : \quad \chi[g] = \infty\,\},$$

and

$$\mathbb{K}_m = \{\, g(x) \in \mathbb{H}_m : \quad \chi[g] < \infty\,\},$$

$(m = 0, 1, \cdots)$. Then we have the proper inclusion relations

$$\mathbb{F}_m \subset \mathbb{G}_m \subset \mathbb{H}_m; \quad \mathbb{P}_m \subset \mathbb{P}, \quad (10.69)$$

where

$$\mathbb{H}_m = \mathbb{G}_m \cup \mathbb{K}_m, \quad \mathbb{G}_m \cap \mathbb{K}_m = \emptyset. \tag{10.70}$$

Now, let $X \subset \mathbb{H}_m$ and $Y \subset \mathbb{P}$ be two given subspaces. If for any $g(x) \in X$ and $p(t) \in Y$, such that the Duffing equation (10.56) admits at least one 2π-periodic solution, then the pair $\langle X, Y \rangle$ is said to be *harmonic*; otherwise, it is said to be *nonharmonic*.

In terms of the above mentioned notations, we first restate two known results due to Leach and Lazer.

Theorem A. $\langle \mathbb{F}_m, \mathbb{P} \rangle$ *is harmonic.*

Theorem B. $\langle \mathbb{K}_m, \mathbb{P}_m \rangle$ *is harmonic.*

Next, we state the following three theorems which constitute the main result of the paper [34].

Theorem C. $\langle \mathbb{G}_m, \mathbb{P} \rangle$ *is harmonic.*

Theorem D. $\langle \mathbb{H}_m, \mathbb{P}_m \rangle$ *is harmonic.*

Theorem E. $\langle \mathbb{H}_m, \mathbb{P} \rangle$ *and* $\langle \mathbb{K}_m, \mathbb{P} \rangle$ *are nonharmonic.*

It can be seen from Theorem E and the relations (10.69) and (10.70) that Theorems A and B have been generalized to the utmost extent by Theorems C and D, respectively.

II. Proofs of the Theorems

Let us consider the function $g(x) = m^2 x + h(x)$ defined in Example 10.10 and the periodic function $p(t) = E \cos mt \in \mathbb{P}$, satisfying $E > 2H$. It is noted that $g(x) \in \mathbb{H}_m \cap \mathbb{K}_m$. Then it follows from Example 10.10 that the pair $\langle \mathbb{H}_m, \mathbb{P} \rangle$ and $\langle \mathbb{K}_m, \mathbb{P} \rangle$ are not harmonic. Therefore, Theorem E is proved by the Example 10.10 where $E > 2H$.

On the other hand, it can be seen that Theorem D is an immediate consequence of Theorems B and C. Therefore, we only need to prove Theorem C. For this aim, we need the following preparations.

III. Preliminaries

Let a and α be two positive constants, and let $\xi > 0$ be a large parameter. Assume the definite integral

$$J(\xi) = \int_{-\frac{\pi}{2}+\sigma}^{+\frac{\pi}{2}-\sigma} \cos\theta \cdot \arctan \frac{\xi \cos\theta}{\alpha} \, d\theta,$$

with $\sigma = (a\alpha\pi)/(2\xi)$.

Chapter 10. Duffing Equations of Second Order

Proposition 10.7 *The asymptotic formula*

$$J(\xi) = \pi - \frac{\pi\alpha}{\xi} + \mathcal{O}\left(\frac{1}{\xi^2}\right) \quad (\text{as } \xi \to \infty) \tag{10.71}$$

holds true.

Proof. Integrating by parts, we have

$$J(\xi) = \int_{-\frac{\pi}{2}+\sigma}^{+\frac{\pi}{2}-\sigma} \arctan \frac{\xi \cos\theta}{\alpha} \, d\sin\theta$$

$$= \sin\theta \cdot \arctan \frac{\xi \cos\theta}{\alpha} \bigg|_{-\frac{\pi}{2}+\sigma}^{+\frac{\pi}{2}-\sigma} + \int_{-\frac{\pi}{2}+\sigma}^{+\frac{\pi}{2}-\sigma} \frac{\alpha \xi \sin^2\theta}{\alpha^2 + \xi^2 \cos^2\theta} \, d\theta$$

$$= 2\cos\sigma \cdot \arctan \frac{\xi \sin\sigma}{\alpha} - \frac{\alpha}{\xi}(\pi - 2\sigma) + \frac{\alpha}{\xi}\left(1 + \frac{\alpha^2}{\xi^2}\right) K(\xi),$$

where

$$K(\xi) = \int_{-\frac{\pi}{2}+\sigma}^{+\frac{\pi}{2}-\sigma} \frac{d\theta}{\alpha^2/\xi^2 + \cos^2\theta}.$$

A direct integration yields

$$K(\xi) = \frac{2\xi}{\alpha\sqrt{1+(\alpha^2/\xi^2)}} \arctan\left(\frac{\alpha \cot\sigma}{\xi\sqrt{1+(\alpha^2/\xi^2)}}\right).$$

Hence, for sufficiently large ξ, we obtain

$$K(\xi) = \frac{2\xi}{\alpha}\left[1 + \mathcal{O}\left(\frac{1}{\xi^2}\right)\right] \arctan\left(\frac{\alpha}{\xi}\left[1+\mathcal{O}\left(\frac{1}{\xi^2}\right)\right] \cdot \frac{1+\mathcal{O}\left(\frac{1}{\xi^2}\right)}{\sigma+\mathcal{O}\left(\frac{1}{\xi^2}\right)}\right)$$

$$= \frac{2\xi}{\alpha}\left[1+\mathcal{O}\left(\frac{1}{\xi^2}\right)\right] \arctan\left(\frac{2}{\pi\alpha} + \mathcal{O}\left(\frac{1}{\xi^2}\right)\right)$$

$$= \frac{2\xi}{\alpha} \arctan \frac{2}{\pi\alpha} + \mathcal{O}\left(\frac{1}{\xi}\right).$$

On the other hand, we have

$$\cos\sigma \cdot \arctan \frac{\xi \sin\sigma}{\alpha} = \left[1+\mathcal{O}\left(\frac{1}{\xi^2}\right)\right] \arctan\left(\frac{\xi}{\alpha}\left[\sigma + \mathcal{O}\left(\frac{1}{\xi^2}\right)\right]\right)$$

$$= \left[1 + \mathcal{O}\left(\frac{1}{\xi^2}\right)\right] \arctan\left(\frac{\pi a}{2} + \mathcal{O}\left(\frac{1}{\xi^2}\right)\right)$$

$$= \arctan\frac{\pi a}{2} + \mathcal{O}\left(\frac{1}{\xi^2}\right).$$

Summing up the above results, we arrive at

$$J(\xi) = 2\left(\arctan\frac{\pi a}{2} + \arctan\frac{2}{\pi a}\right) - \frac{\pi \alpha}{\xi} + \mathcal{O}\left(\frac{1}{\xi^2}\right)$$

$$= \pi - \frac{\pi \alpha}{\xi} + \mathcal{O}\left(\frac{1}{\xi^2}\right),$$

which is the desired formula (10.71).

Proposition 10.7 is thus proved. □

Notice that

$$J(\xi) = \int_{-\frac{3\pi}{2}+\sigma}^{-\frac{\pi}{2}-\sigma} \cos\theta \cdot \arctan\frac{\xi\cos\theta}{\alpha}\,d\theta. \tag{10.72}$$

IV. Rotation-Time

The time called the *rotation-time* during which the motion makes a turn around the origin will play an important role in our study.

Consider the solution of (10.61) in polar coordinates. We have

$$\dot{r} = \frac{xy - yg(x) + yp(t)}{r^2}r, \tag{10.73}$$

whenever $r \neq 0$. If $r \geq 1$, we have

$$\left|\frac{xy - yg(x) + yp(t)}{r^2}\right| = \left|\sin\theta\cos\theta\left(1 - \frac{g(x)}{x}\right) + \frac{p(t)}{r}\sin\theta\right|$$

$$\leq 1 + (m+1)^2 + E_0 = q_0,$$

where

$$E_0 = \sup_{0 \leq t \leq 2\pi} |p(t)|.$$

It follows from (10.73) that

$$r_0 e^{-q_0 t} \leq r(t) \leq r_0 e^{q_0 t},$$

whenever $r(t) \geq 1$. Letting $\alpha_1 = e^{4\pi q_0}$ and $r_0 \geq \alpha_1$, we obtain

$$\frac{r_0}{\alpha_1} \leq r(t) \leq \alpha_1 r_0, \qquad 0 \leq t \leq 4\pi. \tag{10.74}$$

Then (10.74) is always assumed to be true below.

On the other hand, as we know before,

$$\dot{\theta}(t) < 0, \qquad 0 \leq t \leq 4\pi, \tag{10.75}$$

when the initial value r_0 is large enough.

Let $\tau_l(r_0, \theta_0)$ be the rotation-time during which the motion passing through the initial point $\langle r_0, \theta_0 \rangle$ will complete l turns around the origin in the time interval $0 \leq t \leq 4\pi$.

One of our main tasks is to estimate the magnitudes of $\tau_m(r_0, \theta_0)$ and $\tau_{m+1}(r_0, \theta_0)$. However, neither the Cartesian coordinates (x, y) nor its polar coordinates $\langle r, \theta \rangle$ are suitable for the purpose.

We will make use of the coordinate transformation

$$u = x, \quad v = \frac{1}{k}y \quad \left(u_0 = x_0, \quad v_0 = \frac{1}{k}y_0\right) \tag{10.76}$$

where $k = m \geq 1$ or $= m+1$, and the corresponding polar coordinates

$$(u, v) = \langle \rho, \varphi \rangle, \qquad \text{with } (u_0, v_0) = \langle \rho_0, \varphi_0 \rangle.$$

According to (10.76), we write the solution $z = (x(t, x_0, y_0), y(t, x_0, y_0))$ in the form

$$u = u(t, u_0, v_0), \quad v = v(t, u_0, v_0) \tag{10.77}$$

or

$$\rho = \rho(t, \rho_0, \varphi_0) > 0, \quad \varphi = \varphi(t, \rho_0, \varphi_0).$$

Since

$$\rho^2 = r^2 \left(\cos^2 \theta + \frac{1}{k^2} \sin^2 \theta\right),$$

we have

$$\frac{1}{k} r \leq \rho \leq r,$$

which together with (10.74) implies

$$\frac{\rho_0}{k\alpha_1} \leq \frac{r_0}{k\alpha_1} \leq \rho(t) \leq \alpha_1 r_0 \leq k\alpha_1 \rho_0, \qquad 0 \leq t \leq 4\pi. \tag{10.78}$$

At the same time, since

$$\varphi(t) = \arctan\left(\frac{1}{k}\tan\theta(t)\right), \qquad \dot\varphi(t) = \frac{k\dot\theta(t)}{k^2\cos^2\theta(t)+\sin^2\theta(t)},$$

the motion $w = w(t) = \langle \rho(t), \varphi(t)\rangle$ moves clockwise around the origin whenever the corresponding motion $z = z(t)$ does. Moreover, since

$$\theta(\tau_l) - \theta(0) = -2l\pi,$$

we have

$$\varphi(\tau_l) - \varphi(0) = \int_0^{\tau_l} \frac{k\dot\theta(t)}{k^2\cos^2\theta(t)+\sin^2\theta(t)}\,dt$$

$$= \arctan\left(\frac{1}{k}\tan\theta(t)\right)\bigg|_{\theta(0)}^{\theta(\tau_l)} = -2l\pi,$$

and vice versa. We have thus proved that the rotation-time $\tau_l(\rho_0, \varphi_0)$ of the motion $w = w(t)$ is equal to the rotation-time $\tau_l(r_0, \theta_0)$ of the motion $z = z(t)$. That is, $\tau_l(\rho_0, \varphi_0) = \tau_l(r_0, \theta_0)$ with the abuse use of notation. However, the estimations of $\tau_m(\rho_0, \varphi_0)$ and $\tau_{m+1}(\rho_0, \varphi_0)$ will be found through the following inequalities.

V. Basic Inequalities

Now, we want to prove the following basic inequalities:

$$\tau_m(\rho_0, \varphi_0) < 2\pi < \tau_{m+1}(\rho_0, \varphi_0), \qquad (10.79)$$

for sufficiently large $\rho_0 > 0$.

We first consider τ_m.

Since $\tau_0 = 0$, the first inequality in (10.79) is naturally valid for $m = 0$. Then assume $m \geq 1$.

Let $g(x) = m^2 x + h(x)$. It follows from $g(x) \in \mathbb{G}_m$ that

(1) $0 \leq h'(x) \leq 2m + 1$;

(2) $h(x)$ is non-decreasing;

(3) $xh(x) \geq 0, \quad \forall x \in \mathbb{R}^1$.

Moreover, we have

$$\lim_{x \to -\infty} h(x) = -\infty \quad \text{or} \quad \lim_{x \to \infty} h(x) = \infty.$$

Without loss of generality, assume that

$$\lim_{x \to -\infty} h(x) \leq -2B \leq 0, \qquad \lim_{x \to \infty} h(x) = \infty, \qquad (10.80)$$

where $B \geq 0$ is a constant. It follows from the monotonicity of $h(x)$ that for any sufficiently large constant $A > 0$ there exists a constant $a > 0$ such that

$$xh(x) > Ax \cdot \arctan x, \qquad x \geq a, \qquad (10.81)$$

and

$$xh(x) \geq Bx \cdot \arctan x, \qquad x \leq -a. \qquad (10.82)$$

Now, we assume that

$$u = x, \quad v = \frac{1}{m}y \qquad \left(u_0 = x_0, \quad v_0 = \frac{1}{m}y_0 \right),$$

with polar coordinates

$$(u, v) = \langle \rho, \varphi \rangle \qquad ((u_0, v_0) = \langle \rho_0, \varphi_0 \rangle).$$

Then we have

$$\dot{u} = mv, \quad \dot{v} = -mu - \frac{1}{m}h(u) + \frac{1}{m}p(t). \qquad (10.83)$$

It follows from (10.78) that

$$\frac{\rho_0}{\alpha} \leq \rho(t) \leq \alpha \rho_0, \qquad o \leq t \leq 4\pi, \qquad (10.84)$$

with $\alpha = m\alpha_1$. And it follows from (10.83) that

$$\dot{\varphi} = -m\Phi, \qquad (10.85)$$

where

$$\Phi = 1 + \frac{1}{m^2 \rho^2} uh(u) - \frac{1}{m^2 \rho} p(t) \cos \varphi.$$

It is obvious that $\Phi > 0$ ($0 \leq t \leq 4\pi$) whenever ρ_0 is large enough. Hence, $\varphi = \varphi(t)$ is monotonically decreasing and its inverse function $t = t(\varphi)$ is univalent and continuously differentiable. We can thus regard Φ as a continuous function of φ. Using the definition of $\tau_m(\rho_0, \varphi_0)$, we have

$$\tau_m(\rho_0, \varphi_0) = \int_{\varphi_0 - 2m\pi}^{\varphi_0} \frac{d\varphi}{m\Phi}. \qquad (10.86)$$

10.5. Duffing Equation of Semi-Linear Type

Let
$$\sigma = \frac{a\alpha\pi}{2\rho_0} \qquad \left(0 < \sigma < \frac{\pi}{10}\right),$$
with sufficiently large $\rho_0 > 0$. Then we have
$$\rho\cos\varphi \geq \frac{\rho_0}{\alpha}\sin\sigma \geq \frac{2\rho_0\sigma}{\alpha\pi} = a, \qquad \varphi \in \left[-\frac{\pi}{2}+\sigma, \frac{\pi}{2}-\sigma\right],$$
and
$$\rho\cos\varphi \leq \frac{-\rho_0}{\alpha}\sin\sigma \geq \frac{-2\rho_0\sigma}{\alpha\pi} = -a, \qquad \varphi \in \left[-\frac{3\pi}{2}+\sigma, -\frac{\pi}{2}-\sigma\right].$$

Without loss of generality, assume $\varphi_0 \in \left[-\frac{\pi}{2}+\sigma, \frac{\pi}{2}-\sigma\right]$ for definiteness since the arguments are similar in other cases.

Now, put (10.86) into the form
$$T_m(\rho_0, \varphi_0) = \frac{1}{m}\sum_{j=0}^{m-1}(I_{1j} + I_{2j} + I_{3j} + I_{4j} + I_{5j}), \qquad (10.87)$$

where
$$I_{1j} = \int_{-2j\pi-\frac{\pi}{2}+\sigma}^{-2j\pi+\varphi_0} \frac{d\varphi}{\Phi},$$

$$I_{2j} = \int_{-2j\pi-\frac{\pi}{2}-\sigma}^{-2j\pi-\frac{\pi}{2}+\sigma} \frac{d\varphi}{\Phi},$$

$$I_{3j} = \int_{-2j\pi-\frac{3\pi}{2}+\sigma}^{-2j\pi-\frac{\pi}{2}-\sigma} \frac{d\varphi}{\Phi},$$

$$I_{4j} = \int_{-2j\pi-\frac{3\pi}{2}-\sigma}^{-2j\pi-\frac{3\pi}{2}+\sigma} \frac{d\varphi}{\Phi},$$

$$I_{5j} = \int_{-2(j+1)\pi+\varphi_0}^{-2j\pi-\frac{3\pi}{2}-\sigma} \frac{d\varphi}{\Phi},$$

($j = 0, 1, \cdots, m-1$). We shall estimate the above integrals respectively.

First, using the above related inequalities, we have
$$I_{1j} = \int_{-2j\pi-\frac{\pi}{2}+\sigma}^{-2j\pi+\varphi_0} \frac{d\varphi}{1 + \frac{1}{m^2\rho^2}[uh(u) - up(t)]}$$

$$< \int_{-2j\pi-\frac{\pi}{2}+\sigma}^{-2j\pi+\varphi_0} \frac{d\varphi}{1+\frac{A}{m^2\rho^2}u\cdot\arctan u - \frac{E_0}{m^2\rho}}$$

$$< \int_{-2j\pi-\frac{\pi}{2}+\sigma}^{-2j\pi+\varphi_0} \frac{d\varphi}{1+\frac{A}{m^2\alpha^2\rho_0^2}\cdot\frac{\rho_0}{\alpha}\cos\varphi\cdot\arctan\frac{\rho_0\cos\varphi}{\alpha} - \frac{\alpha E_0}{m^2\rho_0}}$$

$$< \int_{-\frac{\pi}{2}+\sigma}^{\varphi_0} \frac{d\varphi}{1+\frac{1}{m^2\rho_0}\left[\frac{A}{\alpha^3}\cos\varphi\cdot\arctan\frac{\rho_0\cos\varphi}{\alpha} - \alpha E_0\right]}.$$

In a similar manner, we obtain

$$I_{5j} < \int_{\varphi_0}^{\frac{\pi}{2}-\sigma} \frac{d\varphi}{1+\frac{1}{m^2\rho_0}\left[\frac{A}{\alpha^3}\cos\varphi\cdot\arctan\frac{\rho_0\cos\varphi}{\alpha} - \alpha E_0\right]}.$$

Hence, for sufficiently large $\rho_0 > 0$, we have

$$I_{1j} + I_{5j} < \int_{-\frac{\pi}{2}+\sigma}^{\frac{\pi}{2}-\sigma} \frac{d\varphi}{1+\frac{1}{m^2\rho_0}\left[\frac{A}{\alpha^3}\cos\varphi\cdot\arctan\frac{\rho_0\cos\varphi}{\alpha} - \alpha E_0\right]}$$

$$= \int_{-\frac{\pi}{2}+\sigma}^{\frac{\pi}{2}-\sigma} \Psi\,d\varphi + \mathcal{O}\left(\frac{1}{\rho_0^2}\right),$$

where

$$\Psi = \left\{1 - \frac{1}{m^2\rho_0}\left[\frac{A}{\alpha^3}\cos\varphi\cdot\arctan\frac{\rho_0\cos\varphi}{\alpha} - \alpha E_0\right]\right\}.$$

It follows that

$$I_{1j} + I_{5j} < (\pi - 2\sigma) + \frac{\alpha E_0}{m^2\rho_0}(\pi - 2\sigma) - \frac{AJ(\rho_0)}{m^2\alpha^2\rho_0} + \mathcal{O}\left(\frac{1}{\rho_0^2}\right)$$

$$= (\pi - 2\sigma) + \frac{\alpha\pi E_0}{m^2\rho_0} - \frac{AJ(\rho_0)}{m^2\alpha^2\rho_0} + \mathcal{O}\left(\frac{1}{\rho_0^2}\right),$$

which together with the asymptotic formula (10.71) implies

$$I_{1j} + I_{5j} < (\pi - 2\sigma) - \left(\frac{A}{\alpha^3} - \alpha E_0\right)\frac{\pi}{m^2\rho_0} + \mathcal{O}\left(\frac{1}{\rho_0^2}\right). \quad (10.88)$$

Using a similar technique, we obtain

$$I_{2j} \leq \pi - 2\sigma) - \left(\frac{B}{\alpha^3} - \alpha E_0\right)\frac{\pi}{m^2\rho_0} + \mathcal{O}\left(\frac{1}{\rho_0^2}\right). \quad (10.89)$$

On the other hand, we have

$$I_{2j} = \int_{-2j\pi-\frac{\pi}{2}-\sigma}^{-2j\pi-\frac{\pi}{2}+\sigma} \frac{d\varphi}{1+\mathcal{O}\left(\frac{1}{\rho_0}\right)} = 2\sigma + \mathcal{O}\left(\frac{1}{\rho_0^2}\right), \qquad (10.90)$$

and

$$I_{4j} = \int_{-2j\pi-\frac{3\pi}{2}-\sigma}^{-2j\pi-\frac{3\pi}{2}+\sigma} \frac{d\varphi}{1+\mathcal{O}\left(\frac{1}{\rho_0}\right)} = 2\sigma + \mathcal{O}\left(\frac{1}{\rho_0^2}\right). \qquad (10.91)$$

Then, using (10.87)–(10.91), we obtain

$$\tau_m(\rho_0, \varphi_0) < \sum_{j=0}^{m-1} \left[\frac{2\pi}{m} - \left(\frac{A+B}{\alpha^3} - 2\alpha E_0\right)\frac{\pi}{m^3\rho_0} + \mathcal{O}\left(\frac{1}{\rho_0^2}\right)\right]$$

$$= 2\pi - \left(\frac{A+B}{\alpha^3} - 2\alpha E_0\right)\frac{\pi}{m^2\rho_0} + \mathcal{O}\left(\frac{1}{\rho_0^2}\right).$$

We can take $A > 0$ large enough such that

$$\frac{A+B}{\alpha^3} - 2\alpha E_0 > 0.$$

Therefore, we have

$$\tau_m(\rho_0, \varphi_0) < 2\pi, \qquad (10.92)$$

whenever $\rho_0 > 0$ is large enough.

Next, we need to estimate τ_{m+1}. Since the estimation is similar to the above one, we will be satisfied with a statement of the main steps.

Assume $g(x) = (m+1)^2 x + f(x)$. Since $g(x) \in \mathbb{G}_m$, we have

$$-(2m+1) \le f'(x) \le 0.$$

It follows that $f(x)$ is non-increasing and $xf(x) \le 0$ for $x \in \mathbb{R}^1$. Moreover, $g(x) \in \mathbb{G}_m$ implies

$$\lim_{x \to -\infty} f(x) = \infty \quad \text{or} \quad \lim_{x \to \infty} f(x) = -\infty.$$

Without loss of generality, assume

$$xf(x) < -A_1 x \cdot \arctan x, \qquad x \ge a,$$

and

$$xf(x) \le -B_1 x \cdot \arctan x, \qquad x \le -a.$$

Now, consider the transformation

$$u = x, \quad v = \frac{1}{m+1}y \quad \left(u_0 = x_0, \quad v_0 = \frac{1}{m+1}y_0\right),$$

associated with the polar coordinates

$$(u, v) = \langle \rho, \varphi \rangle \quad ((u_0, v_0) = \langle \rho_0, \varphi_0 \rangle).$$

It follows from (10.56) that

$$\dot{u} = (m+1)v, \quad \dot{v} = -(m+1)u - \frac{1}{m+1}[f(u) - p(t)], \quad (10.93)$$

with the motion

$$u = u(t, u_0, v_0), \quad v = v(t, u_0, v_0),$$

or

$$\rho = \rho(t, \rho_0, \varphi_0) > 0, \quad \varphi = \varphi(t, \rho_0, \varphi_0).$$

Similarly, we obtain

$$\tau_{m+1}[\rho_0, \varphi_0] = \int_{\varphi_0 - 2(m+1)\pi}^{\varphi_0} \frac{d\varphi}{(m+1)\Phi},$$

where

$$\Phi = 1 + \frac{1}{(m+1)^2 \rho^2} u f(u) - \frac{1}{(m+1)^2 \rho} p(t) \cos\varphi > 0,$$

for $0 \le t \le 4\pi$ and sufficiently large ρ_0. Then, in a similar manner, we prove that

$$\tau_{m+1}(\rho_0, \varphi_0) > 2\pi,$$

which together with (10.92) implies (10.79). Therefore, we have proved

$$\tau_m(r_0, \theta_0) < 2\pi < \tau_{m+1}(r_0, \theta_0). \quad (10.94)$$

VI. Final Step

It is clear that (10.94) implies

$$-2(m+1)\pi < \theta(2\pi, r_0, \theta_0) - \theta_0 < -2m\pi, \quad (10.95)$$

whenever r_0 is large enough. Let

$$x_1 = x(2\pi, x_0, y_0), \quad y_1 = y(2\pi, x_0, y_0).$$

It follows from (10.95) that

$$(x_1, y_1) \neq \lambda(x_0, y_0), \qquad \forall \lambda \geq 0,$$

whenever $r_0 = \sqrt{x_0^2 + y_0^2} > 0$ is large enough. Consider the closed disk

$$\mathcal{D}_\kappa: \qquad \sqrt{x^2 + y^2} = \kappa,$$

where $\kappa > 0$ is a sufficiently large constant. Hence, the Poincaré map of the Duffing equation (10.61) on $\mathcal{D}_\kappa$ satisfies the condition of Poincaré-Bohl fixed-point theorem. There is at least a fixed-point $(\xi_0, \eta_0) \in \mathcal{D}_\kappa$, such that

$$x = x(t, \xi_0, \eta_0), \qquad y = y(t, \xi_0, \eta_0)$$

is a harmonic solution of the Duffing equation (10.61).

The proof of Theorem C is thus completed. □

10.5.3 Duffing Equation across Resonance Point

It is known above that resonance phenomenon may happen to a semi-linear Duffing equation near resonance point. However, we will demonstrate below that a semi-linear Duffing equation (10.61) across resonance points may be not resonant in the sense that it still has infinitely many harmonic solutions. This shows once more the complexity of the Duffing equation at resonance.

1. Hypotheses

Let us assume first the following conditions.

(**H$_1$**): Let $g(x) \in C^1(\mathbb{R}^1, \mathbb{R}^1)$, and let $K > 0$ be a constant, such that

$$|g'(x)| \leq K, \qquad x \in \mathbb{R}^1;$$

(**H$_2$**): There exist two constants $A_0 > 0$ and $M_0 > 0$, such that

$$x^{-1} g(x) \leq A_0, \qquad |x| \geq M_0.$$

Then consider an auxiliary equation

$$\ddot{x} + g(x) = 0,$$

or its equivalent system

$$\dot{x} = y, \qquad \dot{y} = -g(x). \qquad (10.96)$$

This is a planar autonomous system whose orbits are the curves determined by the following first integral

$$V(x,y) := \frac{1}{2}y^2 + G(x) = c, \quad \text{where } G(x) = \int_0^x g(x)\,dx, \quad (10.97)$$

where c is an integral constant. The hypothesis (H_2) obviously implies

$$\lim_{|x|\to\infty} G(x) = \infty \quad \text{and} \quad \lim_{|x|+|y|\to\infty} V(x,y) = \infty.$$

It follows that $V^{-1}(c)$ is compact. On the other hand, (H_2) yields that there exist constant $c_0 > 0$ and $A_1 > 0$ such that if $c \geq c_0$, then

$$y^2 + xg(x) \geq A_1(x^3 + y^2), \quad (x,y) \in V^{-1}(c). \quad (10.98)$$

Note that the left-hand member of the last inequality is just the directional derivative of $V(x,y)$ along the vector (x,y). Therefore, there is no critical point in $V^{-1}(c)$ for $c \geq c_0$, and $V^{-1}(c)$ is a compact one-dimensional manifold. Furthermore, $V^{-1}(c)$ is starlike about the origin. We have thus proved

Lemma 10.9 *If (H_2) holds, then $V^{-1}(c)$ is a closed curve for $c \geq c_0$ which is starlike about the origin o.*

In the sequel, we denote the curve $V^{-1}(c)$ by Γ_c. It follows from Lemma 10.9 that each Γ_c intersects the x-axis at two points: $(h(c),0)$ and $(-h_1(c),0)$, where $h(c) > 0$ and $h_1(c) > 0$ are uniquely determined by the formula

$$G(h(c)) = G(-h_1(c)) = c.$$

Let $(x(t), y(t))$ be any solution of (10.96) whose orbit is Γ_c $(c \geq c_0)$. Clearly, this solution is periodic. Let $\tau(c)$ denote the least positive period of this solution. It follows from (10.97) that

$$\tau(c) = \sqrt{2}\int_{-h_1(c)}^{h(c)} \frac{du}{\sqrt{c - G(u)}}. \quad (10.99)$$

To obtain the desired result, we need another hypotheses.

$(\mathbf{H_3})$: There exist a constant $\alpha > 0$, an integer $m > 0$, and two sequences $\{a_k\}$ and $\{b_k\}$ such that $a_k \to \infty$ and $b_k \to \infty$ as $k \to \infty$, satisfying

$$\tau(a_k) < \frac{2\pi}{m} - \alpha, \quad \tau(b_k) > \frac{2\pi}{m} + \alpha. \quad (10.100)$$

2. Main Result

Then we are in a position to state the following result.

Theorem 10.11 *Assume the above hypotheses* (H_1), (H_2) *and* (H_3) *hold. Then the Duffing equation (10.61) has infinitely many harmonic solutions.*

Proof. The proof will be proceeded in the following steps.

Step 1. Let Γ_{a_k} and Γ_{b_k} be the closed starlike curves defined as above, where the specified parameters a_k, $b_k \geq c_0$ are given in (H_3), for $k \geq n_0$. We may rearrange $\{a_k\}$ and $\{b_k\}$, if necessary, so that

$$a_k < b_k < a_{k+1}, \qquad \text{for } k \geq n_0.$$

Assume Γ_{a_k} and Γ_{b_k} bound the (closed) annulus $\mathcal{A}_k$, for $k \geq n_0$.

Let $\mathcal{T} : \mathbb{R}^2 \to \mathbb{R}^2$ be the Poincaré map induced by the Duffing equation (10.61). It has been already known that $\mathcal{T}$ is area-preserving. For the proof of Theorem 10.11, it suffices to prove that $\mathcal{T}$ is twisted on the annulus $\mathcal{A}_k$. Then we only need to apply the generalized Poincaré-Birkhoff twist theorem on $\mathcal{A}_k$ for $k \geq n_0$.

Step 2. Consider the equivalent system of the semi-linear Duffing equation (10.61); that is,

$$\dot{x} = y, \qquad \dot{y} = -g(x) + p(t). \tag{10.101}$$

Let $x = x(t, x_0, y_0)$, $y = y(t, x_0, y_0)$ be the solution of the system (10.101) satisfying the initial condition $(x(0), y(0)) = (x_0, y_0)$. It can be easily shown that the solution exists on the whole t-axis for any given initial point $(x_0, y_0) \in \mathbb{R}^2$. Therefore, the Poincaré map $\mathcal{T}$ is well defined by the formula

$$(x_0, y_0) \mapsto (x(2\pi, x_0, y_0), y(2\pi, x_0, y_0)).$$

As we know, $\mathcal{T}$ is an area-preserving homeomorphism of $\mathbb{R}^2$.

Let $\langle r(t, r_0, \theta_0), \theta(t, r_0, \theta_0) \rangle$ be the polar coordinates of the solution $(x(t, x_0, y_0), y(t, x_0, y_0))$ through the initial condition $(x_0, y_0) = \langle r_0, \theta_0 \rangle$. Then we obtain

$$\begin{cases} \dot{r} = r\cos\theta(t)\sin\theta(t) - g(r\cos\theta)\sin\theta + p(t)\sin\theta, \\ \dot{\theta} = -\sin^2\theta - \dfrac{1}{r}[g(r\cos\theta)\cos\theta - p(t)\cos\theta], \end{cases} \tag{10.102}$$

whenever $r > 0$. For sufficiently large $r_0 > 0$, we have

$$\mathscr{T}: \quad r_1 = r(2\pi, r_0, \theta_0), \qquad \theta_1 = \theta(2\pi, r_0, \theta_0) + 2j\pi,$$

where j is an arbitrarily integer. It can be seen that

$$\Theta(r_0, \theta_0) := \theta(2\pi, r_0, \theta_0) - \theta_0$$

is well-defined whenever $r(t, r_0, \theta_0) > 0$ for $t \in [0, 2\pi]$, and satisfies

$$\Theta(r_0, \theta_0 + 2\pi) = \Theta(r_0, \theta_0) + 2\pi.$$

Then the Poincaré map $\mathscr{T}$ can be written in the form

$$r_1 = r(2\pi, r_0, \theta_0), \quad \theta_1 = \theta_0 + \Theta(r_0, \theta_0) + 2j\pi, \qquad (10.103)$$

where $\Theta(r_0, \theta_0)$ are continuous in $\langle r_0, \theta_0 \rangle$ and 2π-periodic about θ_0.

On the other hand, when (10.101) is autonomous (i.e., $p(t) \equiv 0$), let $\langle \rho, \varphi \rangle = \langle r, \theta \rangle$. Then (10.102) becomes

$$\begin{cases} \dot{\rho} = \rho \cos \varphi(t) \sin \varphi(t) - g(\rho \cos \varphi) \sin \varphi, \\ \dot{\varphi} = -\sin^2 \varphi - \dfrac{1}{\rho} g(\rho \cos \varphi) \cos \varphi. \end{cases} \qquad (10.104)$$

Then, for sufficiently large $\rho_0 > 0$, the Poincaré map of (10.104) is given by

$$\mathscr{T}_0: \qquad \langle \rho_0, \varphi_0 \rangle \;\mapsto\; \langle \rho_1, \varphi_1 \rangle,$$

where

$$\rho_1 = \rho(2\pi, \rho_0, \varphi_0), \qquad \varphi_1 = \varphi(2\pi, \rho_0, \varphi_0) + 2j\pi,$$

where j is an arbitrarily integer. Letting

$$\Phi(\rho_0, \varphi_0) := \varphi(2\pi, \rho_0, \varphi_0) - \varphi_0 \qquad (\rho_0 \gg 1),$$

then we have

$$\mathscr{T}_0: \quad \rho_1 = \rho(2\pi, \rho_0, \varphi_0), \quad \varphi_1 = \varphi_0 + \Phi(\rho_0, \varphi_0) + 2j\pi,$$

where $\Phi(\rho_0, \varphi_0)$ are continuous in $\langle \rho_0, \varphi_0 \rangle$ and 2π-periodic about φ_0.

Step 3. Let

$$\beta = \min\{2\pi, m\alpha A_1\},$$

where m, α and A_1 are given in (H_3) and (10.98) respectively.

Lemma 10.10 $\Phi(\rho_0, \varphi_0)$ satisfies the following inequalities

$$\begin{cases} \Phi(\rho_0, \varphi_0) \leq -2m\pi - \beta, & \langle \rho_0, \varphi_0 \rangle \in \Gamma_{a_k}; \\ \Phi(\rho_0, \varphi_0) \geq -2m\pi + \beta, & \langle \rho_0, \varphi_0 \rangle \in \Gamma_{b_k}, \end{cases} \quad (10.105)$$

for sufficiently large $k > 0$.

Proof. Let $\langle \rho(t, \rho_0, \varphi_0), \varphi(t, \rho_0, \varphi_0) \rangle$ be the solution of (10.104) through the initial point $\langle \rho_0, \varphi_0 \rangle \in \Gamma_{a_k}$. It follows from (10.98) that

$$\dot{\varphi}(t, \rho_0, \varphi_0) \leq A_1, \quad (10.106)$$

whenever ρ_0 is large enough.

Note that $\langle \rho(t, \rho_0, \varphi_0), \varphi(t, \rho_0, \varphi_0) \rangle$ is a periodic solution of period $\tau(a_k)$. Then we have

$$\Phi(\rho_0, \varphi_0) = \varphi(2\pi, \rho_0, \varphi_0) - \varphi_0 = -2l\pi - \sigma,$$

where $l \geq 0$ is some integer and σ is a constant ($0 \leq \sigma < 2\pi$). Equivalently, there is a constant t_σ ($0 \leq t_\sigma < \tau(a_k)$), such that

$$l\tau(a_k) + t_\sigma = 2\pi,$$

which means that the motion rotates clockwise an angle σ during the time interval $[0, t_\sigma]$. It follows that

$$2\pi = l\tau(a_k) + t_\sigma < (l+1)\tau(a_k) \leq (l+1)\left(\frac{2\pi}{m} - \alpha\right),$$

which implies that $l \geq m$. If $l \geq m + 1$, we have

$$\Phi(\rho_0, \varphi_0) = -2l\pi - \sigma \leq -2l\pi \leq -2(m+1)\pi. \quad (10.107)$$

Now, assume $l = m$. Then we have

$$t_\sigma = 2\pi - m \cdot \tau(a_k) \geq 2\pi - m\left(\frac{2\pi}{m} - \alpha\right) = m\alpha. \quad (10.108)$$

On the other hand, using

$$-\sigma = \int_{l \cdot \tau(a_k)}^{l \cdot \tau(a_k) + t_\sigma} \dot{\varphi}(t, \rho_0, \varphi_0)\, dt \leq -A_1 t_\sigma \leq -m\alpha A_1,$$

we obtain

$$\Phi(\rho_0, \varphi_0) = -2l\pi - \sigma \leq -2m\pi - m\alpha A_1, \quad (10.109)$$

which together with (10.107) yields the first inequality of (10.105).

Chapter 10. Duffing Equations of Second Order

The second inequality of (10.105) can be proved in a similar manner. We have thus proved Lemma 10.10. □

Lemma 10.11 *There is a constant $R_0 > 0$, such that*

$$|\Theta(r_0, \theta_0) - \Phi(r_0, \theta_0)| < \beta, \qquad (10.110)$$

whenever $r_0 \geq R_0$.

Proof. Remember that

$$x = x(t, x_0, y_0), \qquad y = y(t, x_0, y_0)$$

is the solution of (10.101) through the initial point (x_0, y_0). Now, assume

$$x = \tilde{x}(t, x_0, y_0), \qquad y = \tilde{y}(t, x_0, y_0)$$

is the solution of (10.96) through the initial point (x_0, y_0). Letting

$$u = x(t, x_0, y_0) - \tilde{x}(t, x_0, y_0), \qquad v = y(t, x_0, y_0) - \tilde{y}(t, x_0, y_0),$$

then we obtain

$$\dot{u} = v, \qquad \dot{v} = p(t) - g'(\sigma(t))u,$$

where $\sigma(t) = \tilde{x}(t, x_0, y_0) + \lambda(t)[x(t, x_0, y_0) - \tilde{x}(t, x_0, y_0)]$ ($0 \leq \lambda(t) \leq 1$). Assume $\eta(t) = \sqrt{u^2(t) + v^2(t)}$. It follows that

$$\eta\dot{\eta} = uv + p(t)v - g'(\sigma(t))uv,$$

which implies

$$|\dot{\eta}| \leq \frac{1}{2}(1 + K)\eta + B,$$

where B is a bound $|p(t)|$ for $t \in [0, 2\pi]$ and K is the constant given in (H_1). It follows from $\eta(0) = 0$ that

$$\eta(t) \leq K_0 := \frac{2B}{K+1}[e^{(K+1)\pi} - 1],$$

for $t \in [0, 2\pi]$.

Then let $\psi(t) = \theta(t, r_0, \theta_0) - \varphi(t, r_0, \theta_0)$. It is clear that if $|\psi(t)| < \pi$, then $\psi(t)$ is just the angle between the vectors $(x(t), y(t))$ and $(\tilde{x}(t), \tilde{y}(t))$. By the law of cosines, we have

$$\cos\psi(t) = \frac{r^2(t) + \rho^2(t) - \eta^2(t)}{2r(t)\rho(t)} \geq 1 - \frac{K_0^2}{2r(t)\rho(t)}.$$

On the other hand, we have $r(t) \geq \rho(t) - \eta(t) \geq \rho(t) - K_0$. Therefore, under the assumption $|\psi(t)| < \pi$ and $\rho(t) - K_0 > 0$, we have

$$\cos \psi(t) \geq 1 - \frac{K_0^2}{2(\rho(t) - K_0)\rho(t)}.$$

Note that $\rho(t) > 0$ ($0 \leq t \leq 2\pi$) is sufficiently large if $\rho_0 > 0$ is large enough. Therefore, for arbitrarily small constant $\delta > 0$ ($\delta < \beta$), we have

$$|\psi(t)| < \delta, \quad \text{if } |\psi(t)| < \pi, \tag{10.111}$$

whenever $\rho_0 > 0$ is large enough. Since (11.43) holds for $t = 0$ and $\psi(t)$ varies continuously in $t \in [0, 2\pi]$, we conclude that (11.43) holds for $t \in [0, 2\pi]$. Then, letting $t = 2\pi$, we obtain the desired inequality (10.110).

Lemma 10.11 is thus proved. $\square$

Step 4. It follows from Lemma 10.11 and Lemma 10.10 that

$$\begin{cases} \Theta(r_0, \theta_0) < -2m\pi, & \langle r_0, \theta_0 \rangle \in \Gamma_{a_k}; \\ \Theta(r_0, \theta_0) > -2m\pi, & \langle r_0, \theta_0 \rangle \in \Gamma_{b_k}. \end{cases}$$

This proves that the area-preserving Poincaré map $\mathscr{T}$ is twisted on $\mathcal{A}_k$. It follows from the generalized Poincaré-Birkhoff twist theorem that there are at least two different fixed points $(\breve{u}_{ik}, \breve{v}_{ik}) \in \mathcal{A}_k$ ($i = 1, 2$) of $\mathscr{T}$ in $\mathcal{A}_k$ for sufficiently large $k \geq n_0$. Then, we obtain two harmonic solutions

$$x = x(t, \breve{u}_{ik}, \breve{v}_{ik}), \quad y = y(t, \breve{u}_{ik}, \breve{v}_{ik}) \quad (i = 1, 2)$$

passing through the initial point $(\breve{u}_{ik}, \breve{v}_{ik}) \in \mathcal{A}_k$ ($i = 1, 2$), for sufficiently large $k \geq n_0$.

The proof of Theorem 10.11 is thus completed. $\square$

The improved results of this section can be found in the paper [55].

3. A Concrete Example

As an application, consider the following semi-linear Duffing equation

$$\ddot{x} + \tilde{g}(x) = p(t), \tag{10.112}$$

where $\tilde{g}(x)$ is defined in Example 10.8, and $p(t)$ is a periodic function with least period 2π. It can be seen that the conditions (H_1), (H_2) and (H_3) of Theorem 10.11 are satisfied. Hence, the Duffing equation (10.112) has infinitely many harmonic solutions.

Chapter 11

Some Special Problems

In the theory of dynamical systems, there are many interesting problems which call our attention.

11.1 Reeb's Problem

11.1.1 Statement of the Problem

Consider the autonomous system of differential equations of n-th order

$$\frac{du}{dt} = F(u), \qquad u \in \mathbb{R}^n, \tag{11.1}$$

where $F(u)$ is a smooth vector field with the origin $\mathfrak{o} \in \mathbb{R}^n$ as an isolated singular point. It is obvious that if $n = 1$, each nontrivial solution in some neighborhood of $\mathfrak{o}$ tends positively or negatively to the singular point. In what follows, assume $n \geq 2$.

The system (11.1) can be put in the following form

$$\frac{du}{dt} = Au + H(u), \qquad u \in \mathbb{R}^n, \tag{11.2}$$

where $A = F'_u(\mathfrak{o})$ is the Jacobian matrix, and $H(u)$ is the nonlinear term, satisfying the condition $H(u) = o(|u|)$; that is,

$$\lim_{|u| \to 0} \frac{|H(u)|}{|u|} = 0.$$

We have the linearized system

$$\frac{du}{dt} = Au, \qquad u \in \mathbb{R}^n, \tag{11.3}$$

which has the characteristic equation

$$\det[\lambda E - A] \equiv \lambda^n - \text{tr}[A]\lambda^{n-1} + \cdots + (-1)^n \det[A] = 0. \quad (11.4)$$

Consider respectively the following cases.
- Assume A is a non-degenerate matrix (i.e., $\det[A] \neq 0$) and

(1) $n = 2k+1$ is an odd number. In this case, the characteristic equation (11.4) has at least a real root $\lambda = \lambda_1 \neq 0$. It follows that the linearized system (11.3) has the non-trivial solution

$$u = u_0 e^{\lambda_1 t},$$

which positively (or negatively) tends to $\mathfrak{o}$ as $t \to +\infty$ if $\lambda_1 < 0$ (or $\lambda_1 > 0$). Correspondingly, we can show without much difficulty that the nonlinear system (11.2) has a similar nontrivial solution.

(2) $n = 2k$ is an even number. In this case, the characteristic equation (11.4) may have no real root. Correspondingly, the linearized system (11.3) may have no nontrivial solution tending to the singular point $\mathfrak{o}$ neither as $t \to +\infty$ nor as $-\infty$. For example, each nontrivial solution of the linear system

$$\frac{dx_j}{dt} = y_j, \qquad \frac{dy_j}{dt} = -x_j \qquad (j = 1, \cdots, k)$$

is kept a positive distance from the singular point $\mathfrak{o}$ since it satisfies

$$\sum_{j=1}^{k}(x_j^2 + y_j^2) = c^2 > 0.$$

- Assume A is a degenerate matrix (i.e., $\det[A] = 0$) and

(3) $n = 2k$ is an even number. In this case, consider the example

$$\frac{dx_j}{dt} = y_j(x_j^2 + y_j^2), \qquad \frac{dy_j}{dt} = -x_j(x_j^2 + y_j^2) \qquad (j = 1, \cdots, k), (11.5)$$

which has the first integral

$$\sum_{j=1}^{k}(x_j^2 + y_j^2) = c^2 > 0.$$

Hence, each nontrivial solution of (11.5) is kept a positive distance from the singular point $\mathfrak{o}$;

(4) $n = 2k + 1$ is an odd number. For $n \geq 3$, it is an open question:

"Does the spatial system (11.2) admit at least a nontrivial solution which tends positively or negatively to the isolated singular point o ?"

In 1959, V. Nemytskii reported especially the following problem (see, Uspehi Mat. Nauk., (New Series), Vol. 86, no. 2, 1959), proposed by Reeb in the $ICM/1958$ at Edinburgh for a desired solution.

"Construct an autonomous system of differential equations in the space $\mathbb{R}^3$, such that $o \in \mathbb{R}^3$ is a unique singular point and each other orbit is kept a positive distant from the singular point o."

In 1963, the present author gave a positive response to the problem by constructing an example [25], which will be restated as follows.

11.1.2 Construction of the Example

We are going to construct some figure blocks, which will constitute the desired example.

Step 1: Block F_0^*.

In the closed region

$$F_0: \quad x^2 + y^2 + z^2 \leq 1, \quad z \geq 1/\sqrt{2}, \quad (x, y, z) \in \mathbb{R}^3,$$

define a unit-vector field

$$V(x, y, z) = (0, 0, -1), \quad (x, y, z) \in F_0.$$

Note that the direction of V is downward. Then, the region F_0 together with the vector field V is defined as the block F_0^* (for simplicity, all the figures in this construction are referred to the book [51]).

Consider the leaning cylindric regions

$$F_s: \quad a_{s+1} \leq z \leq a_s, \quad z \geq \sqrt{x^2 + y^2},$$

with the constants

$$a_s = \sqrt{2}\left(1 - \sum_{k=1}^{s} 2^{-k}\right), \quad s = 1, 2, \cdots.$$

Let A_s and A_{s+1} be the top-face and the bottom-face of the boundary ∂F_s, respectively; namely,

$$A_j: \quad z = a_j, \quad \sqrt{x^2 + y^2} \leq a_j, \quad j = s, s+1.$$

Consider the horizontal sections of F_s

$$B_s = \left\{(x,y,z) \in F_s : \quad z = a_{s+1} + \frac{\sqrt{2}}{3} 2^{-s}\right\}$$

and

$$C_s = \left\{(x,y,z) \in F_s : \quad z = a_{s+1} + \frac{\sqrt{2}}{3} 2^{-s-1}\right\}.$$

Now we modify the method of Füller [66] a little to construct a vector field in F_s as follows:

(1) In the boundary ∂F_s and the section C_s, define the vector field satisfying $V = (0, 0, -1)$. Note that the direction of V is downward.

(2) Denote by $\mathbf{o}_s$ the center of B_s. In the disk B_s, construct four circles at the center $\mathbf{o}_s$ with radiuses

$$r_i = \frac{i}{4}\left(a_{s+1} + \frac{\sqrt{2}}{3} 2^{-s-1}\right), \quad i = 1, 2, 3, 4,$$

respectively. It follows that B_s is divided into a circular disk $B_s^{(0)}$ centered at $\mathbf{o}_s$ with radius r_1 and four annuli $B_s^{(i)}$ ($i = 1, 2, 3, 4$). Then define a downward unit-vector field $V = (0, 0, -1)$ in $B_s^{(0)}$ and $B_s^{(4)}$, and a vector field in $B_s^{(2)}$ such that all the orbits in $B_s^{(2)}$ are centric circles centered at $\mathbf{o}_s$. Finally, using the principle of linear extension along each radial line of B_s, we extend the vector field in $B_s^{(0)}$, $B_s^{(2)}$ and $B_s^{(4)}$ into $B_s^{(1)}$ and $B_s^{(3)}$.

(3) In the region between A_{s+1} and C_s, define the vector field such that all orbits on A_{s+1} come from the hole $Q_s \subset C_s$, where Q_s is seated in the vertical projection of $B_s^{(2)}$ to C_s.

(4) Let F_s' be the region F_s between the sections C_s and B_s. It is noted that we have defined a vector field on the boundary of $\partial F_s'$. Then, using the principle of linear extension along the z-direction, we obtain the vector field in F_s'.

Let F_s'' be the region F_s between the sections B_s and A_s. Similarly, using the principle of linear extension along the z-direction, we obtain the vector field in F_s''.

In this manner, we can define a piecewise smooth unit-vector field V in the region F_s. The region F_s together with the constructed vector field V is denoted by F_s^*, called the figure block. It is emphasized that each orbit in F_s which starts from the top-face A_s can not arrive at bottom-face A_{s+1}.

Step 3: Block G^*.

Consider the region

$$G: \quad 0 < \sqrt{x^2 + y^2} \leq \frac{\sqrt{2}}{2}, \quad 0 \leq z \leq \sqrt{x^2 + y^2},$$

which does not contain the origin o of $\mathbb{R}^3$. Define a vector field

$$V(x,y,z) = \frac{1}{\sigma}(-y[\sqrt{x^2+y^2}-z], x[\sqrt{x^2+y^2}-z], -z^2)$$

for $(x,y,z) \in G$, where

$$\sigma = \sqrt{(x^2+y^2)(\sqrt{x^2+y^2}-z^2)^2 + z^4} > 0,$$

so that V is a unit-vector field. The region G with the vector field V is called the figure block G^*. It is obvious that for any point $p = (x_0, y_0, z_0)$ of the surface $z = \sqrt{x^2 + y^2}$ (in the region G), the orbit Γ_p positively enters the region G on the cylinder $x_0^2 + y_0^2 = r_0^2 (= x_0^2 + y_0^2)$, and spirally tends to the closed orbit: $x_0^2 + y_0^2 = r_0^2$ and $z = 0$.

Step 4: Block H^*.

In the region

$$H: \quad \sqrt{x^2+y^2} \geq \frac{\sqrt{2}}{2}, \quad x^2 + y^2 + z^2 \leq 1, \quad z \geq 0,$$

construct a vector field $V = (u, v, w)$, where

$$u = \sigma_1[y - x(\sqrt{x^2+y^2} - 1/\sqrt{2})][\sqrt{x^2+y^2}-z],$$
$$v = \sigma_1[-x - y(\sqrt{x^2+y^2} - 1/\sqrt{2})][\sqrt{x^2+y^2}-z],$$
$$w = -\sigma_1 z^2,$$

where the function $\sigma_1 = \sigma_1(x,y,z)$ is so chosen that V is a unit-vector field. The region H with the vector field V is called the figure block H^*. It follows from $(u,v,w) \cdot (x,y,z) < 0$ that each orbit of H^* starting from the sphere $x^2 + y^2 + z^2 = 1$ enters the inside of the sphere and tends to the closed orbit $\Gamma = \{\sqrt{x^2+y^2} = 1/\sqrt{2}, \quad z = 0\}$.

Step 5: Block J^*.

It can be easily verified that the figure blocks

$$G^*, \ H^*, \ F_s^* \quad (s = 0, 1, 2, \cdots)$$

coalesce in the region

$$I: \quad 0 < x^2 + y^2 + z^2 \leq 1, \quad z \geq 0,$$

such that they constitute a figure block I^*, defined by a continuous unit-vector field V. Using V and its symmetric vector field with respect to the plane $z = 0$, we obtain a continuous unit-vector field, denoted again by V, in the unit-ball

$$J: \quad 0 < x^2 + y^2 + z^2 \leq 1,$$

with deleted center $\mathbf{o}$. The region J together with the vector field V is called the figure block J^*. It can be seen that the orbit of J^* starting from the sphere $x^2 + y^2 + z^2 = 1$ enters the unit-ball and each orbit in J tends to a closed orbit (or it is itself a closed orbit).

Let us define the vector field $V = (X, Y, Z)$ in the solid unit-ball

$$K: \quad 0 \leq x^2 + y^2 + z^2 \leq 1,$$

by

$$X(0,0,0) = Y(0,0,0) = Z(0,0,0) = 0,$$

and

$$\begin{aligned} X(x,y,z) &= (x^2 + y^2 + z^2)u(x,y,z), \\ Y(x,y,z) &= (x^2 + y^2 + z^2)v(x,y,z), \quad (x,y,z) \in J, \\ Z(x,y,z) &= (x^2 + y^2 + z^2)w(x,y,z), \end{aligned}$$

where (u, v, w) is the vector field of J^*. It is easy to verify that this vector field V is continuous in K and has a unique singular point $\mathbf{o}$, and the orbit passing through each point in K is unique. Moreover, each nonsingular orbit is kept a positive distance from the singular point $\mathbf{o}$. The ball K together with the vector field V is called the figure block K^*.

Step 7: Finally, we obtain the desired example

$$\frac{dx}{dt} = X(x,y,z), \quad \frac{dy}{dt} = Y(x,y,z), \quad \frac{dz}{dt} = Z(x,y,z), \quad (11.6)$$

which resolves the Reeb's problem.

In the end of this section, we give a remark that the origin $\mathbf{o}$ of $\mathbb{R}^3$ is an isolated singular point of the system (11.6) and there are infinitely many small closed orbits in the neighborhood of $\mathbf{o}$. If we apply the method

Chapter 11. Some Special Problems

of Schweitzer's example to the above construction instead of the method of Füller, then the closed orbits of (11.6) will be replaced by the Denjoy's minimal sets.

11.2 Birkhoff's Conjecture

11.2.1 B-Recurrent but Non-Almost Periodic Motion

In the theory of dynamical systems, Birkhoff discovered a new class of recurrent motions (which will be called the B-recurrent motions) and proved that an almost periodic motion is B-recurrent. To show the difference between them, Birkhoff gave an example of differential equations which possesses the B-recurrent but not almost periodic motion. However, that example is not continuous. For the naturalness of motion, Birkhoff wanted to construct an analytic example of B-recurrent motion and conjectured the existence of such motions [8].

In 1981, this author constructed the following example which gives an answer to the above-mentioned Birkhoff's conjecture.

1) First let

$$a_1 = 1, \qquad p_1 = 1.$$

Then, define the series

$$a_{n+1} = 2^{2(p_n+q_n)}, \qquad n = 1, 2, \cdots,$$

by induction, where the positive integers p_n and q_n are defined by the continued irreducible fractions

$$\frac{p_n}{q_n} = \frac{1}{a_1} + \frac{1}{a_2} + \cdots + \frac{1}{a_n}, \qquad n = 1, 2, \cdots.$$

It follows that the infinite continued fraction

$$\lambda = \frac{1}{a_1} + \frac{1}{a_2} + \cdots + \frac{1}{a_n} + \cdots \qquad (11.7)$$

is a positive irrational number. Moreover, it satisfies the inequality

$$|p_n - \lambda q_n| \leq 2^{-2(p_n+q_n)}, \qquad n = 1, 2, \cdots. \qquad (11.8)$$

2) Next, using the trigonometric series yields the real function

$$F(u,v) = \sum_{k,j=-\infty}^{\infty} C_{kj} e^{iku} e^{ijv}, \quad (u,v) \in \mathbb{R}^2, \tag{11.9}$$

where the coefficients are defined by a constant

$$C_{00} > 0$$

and

$$C_{kj} = 2^{-2(|k|+|j|)}, \quad |k|+|j| > 0. \tag{11.10}$$

Note that

$$F(u,v) = C_{00} + F_0(u,v),$$

where

$$F_0(u,v) = \sum_{|k|+|j|>0} C_{kj} e^{iku} e^{ijv}, \quad (u,v) \in \mathbb{R}^2.$$

Using the inequality (11.10), we conclude that $F_0(u,v)$ is a 2π-periodic and continuous function with respect to u and v. Since $F_0(u,v)$ is independent of C_{00}, we can choose the constant $C_{00} > 0$ sufficient large such that the function $F(u,v)$ is positive for $(u,v) \in \mathbb{R}^2$.

3) Then, we claim that $F(u,v)$ is analytic in $(u,v) \in \mathbb{R}^2$.
For this aim, consider the complex domain

$$\mathbb{D}: \quad u = u_1 + iu_2, \quad v = v_1 + iv_2,$$

where $u_1, v_1 \in \mathbb{R}^1$ are the real parts, and $u_2, v_2 \in \mathbb{R}^1$ are the imaginary parts satisfying the condition:

$$|u_2| < \log 2, \quad |v_2| < \log 2. \tag{11.11}$$

Now, formally differentiating the expression (11.9), we obtain

$$\frac{\partial^{m+n} F}{\partial u^m \partial v^n} = \sum_{k,j=-\infty}^{\infty} i^{m+n} k^m j^n C_{kj} e^{iku} e^{ijv}. \tag{11.12}$$

It follows from (11.11) that

$$\sum_{k,j=-\infty}^{\infty} |i^{m+n} k^m j^n C_{kj} e^{iku} e^{ijv}| \le \sum_{|k|+|j|>0} |k|^m |j|^n 2^{-(|k|+|j|)} e^{|ku_2|} e^{|jv_2|}$$

Chapter 11. Some Special Problems 321

$$\le \sum_{|k|+|j|>0}(|k|+|j|)^{m+n}2^{-(|k|+|j|)} = \sum_{r=1}^{\infty}8r^{m+n+1}2^{-r} < \infty,$$

which implies that the series (11.12) is absolutely and uniformly convergent. Hence, the partial derivative $\frac{\partial^{m+n}F}{\partial u^m \partial v^n}$ exists continuously in the domain $\mathbb{D}$. It follows from the Hartogs theorem [13] in the theory of several complex variables that $F(u,v)$ is an analytic function in the complex domain $\mathbb{D}$. In particular, $F(u,v)$ is analytic in the real variables $(u,v) \in \mathbb{R}^2$.

4) Finally, we have the system of differential equations

$$\frac{du}{dt} = \frac{1}{F(u,v)}, \qquad \frac{dv}{dt} = \frac{\lambda}{F(u,v)}, \qquad (11.13)$$

on the torus $\mathbb{T}^2$, where $\lambda > 0$ is an irrational number defined in (11.7). Then an analytic dynamical system f^t on the torus $\mathbb{T}^2$ is defined by (11.13).

From (11.13), we obtain an integral

$$v - \lambda u = v_0,$$

which represents the orbit of f^t passing through the initial point $(0, v_0)$ when $t = 0$. Therefore, when the motion $f^t(0, v_0)$ comes from the initial point $(0, v_0)$ to the point (u, v) of the orbit, it spends the time

$$\tau(u, v_0) = \int_0^u F(\xi, \lambda\xi + v_0)\, d\xi.$$

Consider the function

$$T(u, v_0) = \tau(u, v_0) - \tau(u, 0) = \int_0^u \Phi(\xi, v_0)\, d\xi, \qquad (11.14)$$

where the integrand $\Phi(\xi, v_0) = F(\xi, \lambda\xi + v_0) - F(\xi, \lambda\xi)$ is almost periodic in $\xi \in \mathbb{R}^1$, with the Fourier series

$$\Phi(\xi, v_0) = \sum_{|k|+|j|>0} C_{kj} e^{i(k+j\lambda)\xi}(e^{ijv_0} - 1). \qquad (11.15)$$

In the Bohr theory of almost periodic functions, we have the following Propositions (P_1) and (P_2) (see [14]).

(P_1): If the indefinite integral $T(u, v_0)$ is almost periodic in $u \in \mathbb{R}^1$, then its Fourier series is obtained by integrating the Fourier series (11.15) term by term; that is,

$$T(u, v_0) = \sum_{|k|+|l|>0} \frac{C_{kl}(e^{ilv_0} - 1)e^{i(k+\lambda l)u}}{i(k + \lambda l)}.$$

Moreover, the series of the square modules of its coefficients is convergent; that is,

$$\sum_{|k|+|l|>0} \left| \frac{C_{kl}(e^{ilv_0} - 1)}{i(k + \lambda l)} \right|^2 = \sum_{|k|+|l|>0} \frac{C_{kl}^2}{(k + \lambda l)^2} \sin^2\left(\frac{lv_0}{2}\right) < \infty. \quad (11.16)$$

(P_2): If the indefinite integral $T(u, v_0)$ is not almost periodic in $u \in \mathbb{R}^1$, then it is unbounded with respect to $u \in \mathbb{R}^1$.

Now, we claim that for almost all $v_0 \in \mathbb{R}^1$, the series (11.16) is divergent.

Assume the contrary. Then there is a sufficiently large constant $N > 0$, such that the set

$$E = \{v_0 \in \mathbb{R}^1 : \ |v_0| < N \text{ and the series (11.16) is convergent}\}$$

has a finite positive measure $\nu(E)$ ($0 < \nu(E) < 2N$).

Then, for $v_0 \in E$, (11.16) is a convergent series with positive terms. It follows that integrating (11.16) term by term yields a convergent series

$$\sum_{|k|+|l|>0} \int_E \frac{C_{kl}^2}{(k + \lambda l)^2} \sin^2\left(\frac{lv_0}{2}\right) dv_0$$

$$= \sum_{|k|+|l|>0} \frac{C_{kl}^2}{(k + \lambda l)^2} \int_E \frac{1 - \cos lv_0}{2} dv_0$$

$$= \frac{\nu(E)}{2} \sum_{|k|+|l|>0} (1 - \alpha_l) \frac{C_{kl}^2}{(k + \lambda l)^2}, \quad (11.17)$$

with $0 < \nu(E) < \infty$ and

$$\alpha_l = \frac{1}{\nu(E)} \int_0^{2\pi} \varphi_E(v_0) \cos lv_0 \, dv_0,$$

where $\varphi_E(v_0)$ is the characteristic functions of the set E. It follows from the properties of Fourier coefficients that

$$\alpha_l \to 0, \quad \text{as } l \to \pm\infty.$$

Then, we get

$$1 - \alpha_l > \frac{1}{2}, \quad \text{for sufficiently large } |l|.$$

On the other hand, since (11.17) is convergent, we have

$$(1 - \alpha_l)\frac{C_{kl}^2}{(k + \lambda l)^2} < \frac{1}{2}$$

whenever $|k| + |l|$ is sufficiently large. It follows that

$$\frac{C_{kl}^2}{(k + \lambda l)^2} < 1 \quad \text{for sufficiently large } |l|. \tag{11.18}$$

However, from (11.8) and the definition of C_{kl} it follows that

$$\frac{C_{p_n\{-q_n\}}^2}{(p_n - \lambda q_n)^2} > \left(\frac{2^{-2(p_n+q_n)}}{2^{-2(p_n+q_n)}}\right) = 1, \quad n = 1, 2, \cdots,$$

which contradicts (11.18). This contradiction proves the desired conclusion: $\nu(E) = 0$.

Therefore, the series (11.16) is divergent for almost all $v_0 \in \mathbb{R}^1$. It follows that $T(u, v_0) = \tau(u, v_0) - \tau(u, 0)$ is an unbounded function of u for almost all $v_0 \in \mathbb{R}^1$. Therefore, for any given positive constant K, there are a sufficient small $v_0 > 0$ and a sufficiently large $\bar{u}$, such that

$$|\tau(\bar{u}, v_0) - \tau(\bar{u}, 0)| > K.$$

It follows that the motion $f^t(0,0)$ is Liapunov unstable, and thus it is not almost periodic.

On the other hand, it is clear that every orbit of f^t is dense on the torus $\mathbb{T}^2$ and, therefore, $\mathbb{T}^2$ is a minimal set of f^t. It follows that every motion of f^t is B-recurrent. On the other hand, (11.13) is a system of analytic differential equations. Therefore, this example gives a positive response to the Birkhoff's conjecture on the existence of B-recurrent but not almost periodic motions of the analytic dynamical system.

Moreover, we will prove in the final section of this chapter that the above analytic flow f^t is weak mixing (more complicated than chaotic).

In addition, as a supplement to the above example, we would like to mention the following interesting result due to Prof. W. Huang:

"*The set*

$$\Lambda = \{\lambda \in \mathbb{R}^1 : \text{ the motion of (11.13) is not almost periodic}\}$$

has the Lebesgue measure $\nu(\Lambda) = 0$ (see [123])."

11.2.2 Nemytskii Problem

In the theory of dynamical systems, sometimes, it is difficult to construct an example in finite dimensional spaces with desired property, but it may be not hard to realize it in the abstract Bebytov space. For example, analogous to the Birkhoff's conjecture, Nemytskii proposed a problem [99] to find an analytic recurrent but not almost periodic motion in the Bebytov dynamical system.

Nevertheless, the above answer to the Birkhoff's conjecture implies a natural reply to the Nemytskii problem as follows.

Consider the Bebytov space of continuous functions

$$\mathbb{B} = \{\varphi: \quad \mathbb{R}^1 \;\to\; \mathbb{R}^1; \quad x \;\mapsto\; \varphi(x)\,\}$$

endowed with the metric

$$\varrho(\varphi_1, \varphi_2) = \sup_{X>0} \min\left\{\max_{|x|\leq X} |\varphi_1(x) - \varphi_2(x)|, \frac{1}{X}\right\}$$

for any $\varphi_1, \varphi_2 \in \mathbb{B}$. It follows that $\mathbb{B}$ is a complete metric space. Then the Bebytov dynamical system $g(\cdot, t)$ is defined by

$$g(\varphi(x), t) = \varphi(x+t) \quad (t \in \mathbb{R}^1), \qquad \text{for } \varphi \in \mathbb{B}.$$

It is not hard to prove the following Propositions:

(A) For any $\varepsilon > 0$, $\varrho(\varphi_1, \varphi_2) < \varepsilon$ if and only if

$$\max_{|x|\leq 1/\varepsilon} |\varphi_1(x) - \varphi_2(x)| < \varepsilon;$$

(B) It is clear that $g(\varphi(x), t)$ is an almost periodic motion if and only if $\varphi(x)$ is an almost periodic function [100].

Let $f^t(p) = (\xi(t,p), \eta(t,p))$ be the motion of (11.13) passing through the initial point $p \in \mathbb{T}^2$. It follows from the criterion of Bochner (see [77]) that $z = f^t(p)$ is an almost periodic motion if and only if $\xi(t,p)$ and $\eta(t,p)$ are almost functions. Since $z = f^t(p)$ is not an almost periodic motion, then there is at least one component, say $\xi(t,p)$, which is not an almost periodic function. Then, by defining

$$\varphi_p(x) = \xi(x,p), \qquad \text{for } p \in \mathbb{T}^2 \quad (t \in \mathbb{R}^1),$$

$$\Phi(p) = p_1, \qquad \text{for } p = (p_1, p_2) \in \mathbb{T}^2,$$

$$\Psi(f^t(p)) = \Phi(f^t(p), x), \qquad \text{for } p \in \mathbb{T}^2,$$

we obtain

$$\Psi(f^t(p)) = \Phi(f^{x+t}(p)) = \xi(x+t, p) = \varphi_p(x+t) = g(\varphi_p(x), t).$$

This means that Ψ maps the motion $f^t(p)$ of (11.13) into the motion $g(\varphi_p(x), t)$ of the Bebytov dynamical system. It follows from the above Proposition (B) that the analytic motion $g(\varphi_{\bar{p}}(x), t)$ is not an almost periodic motion.

Now, we claim that $g(\varphi_{\bar{p}}(x), t)$ is a B-recurrent motion in the Bebytov dynamical system.

In fact, for any given $\varepsilon > 0$, and any $|x| < 1/\varepsilon$, the continuity of $f^x(p)$ with respect to the initial point p implies the existence of a constant $\delta = \delta(\varepsilon) > 0$ such that

$$|f^x(p) - f^x(\bar{p})| < \varepsilon, \qquad \text{for } |x| < 1/\varepsilon,$$

provided that

$$|p - \bar{p}| < \delta, \qquad \text{for } p \in \mathbb{T}^2.$$

Then we have

$$|\xi(x, p) - \xi(x, \bar{p})| < \varepsilon, \qquad \text{for } |x| < 1/\varepsilon,$$

which yields

$$|\varphi_p(x) - \varphi_{\bar{p}}(x)| < \varepsilon, \qquad \text{for } |x| < 1/\varepsilon.$$

Then, from Proposition (A), we obtain

$$\varrho(\varphi_p(\cdot), \varphi_{\bar{p}}(\cdot)) < \varepsilon, \qquad \text{whenever } |p - \bar{p}| < \delta, \ p \in \mathbb{T}^2. \qquad (11.19)$$

On the other hand, since $f^t(\bar{p})$ is a B-recurrent motion of (11.13), then for $\delta(\varepsilon) > 0$ there is a relatively dense set $\{\tau_n\}$ with respect to a constant $L(\delta(\varepsilon)) > 0$, such that

$$|f^{\tau_n}(\bar{p}) - \bar{p}| < \delta(\varepsilon).$$

It follows from (11.19) that

$$\varrho(\varphi_{f^{\tau_n}(\bar{p})}(\cdot), \varphi_{\bar{p}}(\cdot)) < \varepsilon. \qquad (11.20)$$

Note that

$$\varphi_{f^{\tau_n}(\bar{p})}(x) = \xi(x, f^{\tau_n}(\bar{p})) = \Phi(f^x(f^{\tau_n}(\bar{p}))) = \Phi(f^{x+\tau_n}(\bar{p}))$$

$$= \xi(x + \tau_n, \bar{p}) = \varphi_{\bar{p}}(x + \tau_n) = g(\varphi_{\bar{p}}(x), \tau_n).$$

It follows from (11.20) that

$$\varrho(g(\varphi_{\bar{p}}(\cdot), \tau_n), \varphi_{\bar{p}}(\cdot)) < \varepsilon,$$

where $\{\tau_n\}$ is a relatively dense set with respect to $L(\delta(\varepsilon)) > 0$.

We have thus proved that $g(\varphi_{\bar{p}}(x), t)$ is an analytic B-recurrent but not almost periodic motion in the Bebytov dynamical system. This is a desired solution to the above mentioned Nemytskii problem.

11.3 Morse's Conjecture

11.3.1 C^∞ - Flows

1) A flow on the n-dimensional compact manifold $\mathcal{M}$ is termed to be *topological transitivity* if it has a dense orbit in $\mathcal{M}$. G. Birkhoff gave the first example of such transitivity on the torus $\mathbb{T}^2$.

Under the metric transitivity, as first defined by G. Birkhoff and P. Smith [12], the only invariant subsets of $\mathcal{M}$ are sets with measure which is 0 or mes($\mathcal{M}$). It is known that the metric transitivity implies the topological transitivity. However, it is an open question whether or not the converse is true. M. Morse pointed out in 1946 [95] and in 1973 [96] that a proof or a disproof of this converse is much needed, and the relative importance of the ergodic theorem of Birkhoff could hardly be assessed until this converse should be found out. Then, Morse conjectured that the converse theorem was probably true for analytic systems or systems with some degree of analytic regularity.

We understand the meaning of Morse conjecture in the following sense. The hypothesis of metric transitivity is rather severe than the condition of topological transitivity.

In 1991, the present author constructed an almost analytic C^∞-flow on the n-dimensional torus $\mathbb{T}^n$ ($n \geq 2$), such that it is topologically transitive but not metrically transitive [43]. Therefore, the above Morse conjecture (i.e., the metric transitivity implies the topological transitivity) is incorrect for C^∞-flows on the manifold of dimension (≥ 2).

2) Let $\mathbb{T}^n$ be the n-dimensional torus ($n \geq 2$).[1] Our purpose is to construct an almost everywhere analytic C^∞ flow on $\mathbb{T}^n$, such that it is topologically transitive but not metrically transitive.

---
[1] The Morse conjecture obviously holds true for the one-dimensional manifold.

Chapter 11. Some Special Problems

Take a constant vector
$$v_0 = (1, \pi, \cdots, \pi^{n-1}),$$
and define a vector field V on $\mathbb{T}^n$ as follows:
$$V(x) = v_0, \quad \text{for } x \in \mathbb{T}^n.$$
Then we have a system of differential equations
$$\frac{dx}{dt} = V(x), \quad x \in \mathbb{T}^n, \tag{11.21}$$
which defines the analytic flow
$$f^t(x) = x + tv_0 \tag{11.22}$$
on $\mathbb{T}^n$. It can be easily verified that each orbit of (11.22) is dense on $\mathbb{T}^n$. For definiteness, consider the motion $z = f^t(\mathbf{o})$ with the dense orbit $\Gamma_{\mathbf{o}}$ in $\mathbb{T}^n$, where $\mathbf{o}$ is the original point of the torus.

Now, take two sections of the torus $\mathbb{T}^n$:
$$S_1 = \{x \in \mathbb{T}^n : x_1 = 0\} \quad \text{and} \quad S_2 = \{x \in \mathbb{T}^n : x_1 = \pi\}.$$
Then the orbit $\Gamma_{\mathbf{o}}$ intersects S_1 and S_2 at infinitely many points
$$p_j = f^{2j\pi}(\mathbf{o}) \quad \text{and} \quad q_j = f^{2(j+1/2)\pi}(\mathbf{o}), \quad j \in \mathbb{Z}^1,$$
respectively. Notice that $q_j = p_j + \pi v_0$, and the interval
$$[p_j, q_j] = \{x \in \mathbb{T}^n : x = p_j + sv_0 \ (0 \le s \le \pi)\}$$
is an arc of $\Gamma_{\mathbf{o}}$ between S_1 and S_2 ($0 \le x_1 \le \pi$) with end-points p_j and q_j.

Denote by $\nu(\cdot)$ the n-dimensional Lebesgue measure and by ν^* the $(n-1)$-dimensional Lebesgue measure. Then we have
$$\tau = \nu(\mathbb{T}^n) = (2\pi)^n, \quad \nu(S_1) = \nu(S_2) = 0,$$
and
$$\nu^*(S_1) = \nu^*(S_2) = (2\pi)^{n-1} = \frac{\tau}{2\pi}.$$
Now, take a positive constant σ such that
$$\sum_{k=0}^{\infty} \frac{2\sigma}{1+k^2} = \frac{\tau}{4\pi}.$$

For a given integer k, let B_k be an open ball in S_1 centered at p_k having the measure

$$\nu^*(B_k) = \frac{\sigma}{1+k^2}.$$

It follows that

$$B = \bigcup_{k \in \mathbb{Z}^1} B_k$$

is an open set in S_1 having the measure

$$0 < \nu^*(B) \leq \frac{\tau}{4\pi}.$$

Hence,

$$F_1 = S_1 \setminus B$$

is a closed set in S_1 having the measure

$$\frac{\tau}{4\pi} < \nu^*(F_1) < \frac{\tau}{2\pi}.$$

Then, associated to the set F_1 in S_1, define the set

$$F_2 = \{x \in \mathbb{T}^n : \quad x = p + \pi v_0, \quad \forall \, p \in F_1 \},$$

which is a closed set in S_2 with measure

$$\nu^*(F_2) = \nu^*(F_1).$$

It is clear that the flow f^t joins the set F_1 and F_2 in the set

$$E = \{x \in \mathbb{T}^n : \quad x = p + sv_0, \quad 0 \leq s \leq \pi, \; \forall \, p \in F_1 \},$$

which is a closed set between S_1 and S_2 ($0 \leq x_1 \leq \pi$) with measure

$$\nu(E) = \pi \nu^*(F_1).$$

Hence, we get

$$\frac{\tau}{4} < \nu(E) < \nu(\mathbb{T}^n). \tag{11.23}$$

It is noticed that the set $F = F_1 \cup F_2$ is a closed set in $\mathbb{T}^n$ with measure

$$\nu(F) = 0.$$

Since the set F is compact in $\mathbb{T}^n$, there is a C^∞-function

$$\theta: \quad \mathbb{T}^n \quad \to \quad [0, \infty), \quad \text{with } \theta^{-1}(0) = F, \tag{11.24}$$

(see [74], for example); that is,

$$\theta(x) = 0 \quad \text{for } x \in F \quad \text{and} \quad \theta(x) > 0 \quad \text{for } x \in \mathbb{T}^n \setminus F.$$

Furthermore, using the Whiteney theorem [74], we can make the above function $\theta(x)$ to be almost everywhere analytic in $\mathbb{T}^n$ without destroying (11.24) and the C^∞-continuity property.

Therefore,

$$W(x) = \theta(x) V(x), \quad x \in \mathbb{T}^n,$$

is an almost everywhere analytic C^∞-vector field, which is singular at every point in F. Note that the point $y \in \mathbb{T}^n \setminus F$ is nonsingular and the vector $W(y)$ has the same direction of $V(y)$.

Now, consider the system of differential equations

$$\frac{dx}{dt} = W(x), \quad x \in \mathbb{T}^n, \tag{11.25}$$

which defines an almost everywhere analytic C^∞-flow

$$g^t: \quad \mathbb{T}^n \quad \to \quad \mathbb{T}^n \quad (t \in \mathbb{R}^1).$$

The definition of F implies that the orbit $\Gamma_\circ$ does not meet any singular point of (11.25) (i.e., $\Gamma_\circ \cap F = \emptyset$). Hence, the dense orbit $\Gamma_\circ$ agrees with the orbit of the motion $z = g^t(\circ)$. It follows that g^t is topologically transitive on $\mathbb{T}^n$.

On the other hand, since every point of F is a singular point of g^t, we have

$$g^t(F) = F, \quad t \in \mathbb{R}^1. \tag{11.26}$$

For given $x_0 \in (E \setminus F)$, there is a unique open interval

$$(p, q) = \{ x \in E : \quad x = p + s v_0, \quad 0 < s < \pi \}$$

containing x_0, where p is some point of F_1 and $q = p + \pi v_0$ is a point of F_2. Since this interval (p, q) is a sub-arc of the orbit of the motion $f^t(x_0)$ and does not contain the singular point of g^t, it is the orbit of the motion $g^t(x_0)$. Thus, we get

$$g^t((p,q)) = (p, q), \quad \forall\, t \in \mathbb{R}^1,$$

which together with (11.26) implies that E is an invariant closed set of g^t. It follows from (11.23) that g^t is not metrically transitive on $\mathbb{T}^n$. Therefore, this counterexample disproves the above Morse conjecture in case of C^∞-flow with infinitely many singular points on $\mathbb{T}^n$ ($n \geq 2$).

Finally, it is remarked that the above method of construction for the counterexample on $\mathbb{T}^n$ is naturally available on any n-dimensional manifold ($n \geq 2$).

11.3.2 Analytic Flows

We will consider the flows on two dimensional surfaces with finitely many singular points. The main result can be mentioned as follows.

Theorem 11.1 *If f^t is a C^1-flow on the closed surface $\mathfrak{F}$ having a finite number of singular points, then the Morse conjecture is valid for f^t (i.e., the topological transitivity implies the metric transitivity on closed surfaces).*

Proof. In fact, using the condition of topological transitivity of f^t implies that Lemmas 8.3 - 8.6 (see the corresponding proofs). Therefore, f^t has the following properties:

(1) All the equilibrium points of f^t are saddles;
(2) f^t has no closed orbit;
(3) f^t has no one-sided contour;
(4) f^t has no nontrivial quasi-minimal set which is a proper subset of $\mathfrak{F}$.

Assume there is a compact invariant set Q of f^t having the measure $\nu(Q)$, satisfying $0 < \nu(Q) < \nu(\mathfrak{F})$.

It follows that Q is a proper subset of $\mathfrak{F}$.

The above Property (4) implies that there is no nontrivial $P^\pm$-recurrent orbit in Q.

Since Q is a compact invariant set, there is at leat a minimal set M in Q. It follows from the Properties (1), (2) and (4) that M is a saddle point q_1 of f^t. Moreover, there is an orbit Γ_1 positively tending to q_1 and negatively tending to some saddle, say q_2. It follows from the Property (3) that $q_2 \neq q_1$. Similarly, there is an orbit Γ_2 positively tending to q_2 and negatively tending to some saddle, say q_3. It follows from the Property (3) that $q_2 \neq q_2$ and q_1, and so on. Since f^t has a finite number of saddle points, we obtain m saddles

$$q_1, q_2, \cdots, q_m \in Q,$$

and $m-1$ separatrices

$$\Gamma_1, \ \Gamma_2, \ \cdots, \ \Gamma_{m-1} \subset Q,$$

such that Γ_i joins from q_{i+1} to q_i ($i = 1, 2, \cdots, m-1$) and there is an orbit Γ_m which positively tends to q_m and, however, it can not tend negatively to any saddle point. Therefore, Γ_m is a nontrivial P^--recurrent orbit which is contained in a nontrivial quasi-minimal set in Q. However, it is in conflict with the Property (4). This contradiction proves that f^t is metrical transitivity.

Theorem 11.1 is thus proved. □

As a direct consequence of Theorem 11.1, we prove the Morse conjecture for analytic flows on 2-dimensional closed surfaces.

Theorem 11.2 *The Morse conjecture is valid for analytic flows on closed surfaces.*

However, it is still an open question whether or not the Morse conjecture is valid for analytic flows on higher dimensional manifolds (i.e., manifolds of dimension ≥ 3).

11.4 Kolmogorov's Problem

In a memorial paper on Kolmogorov [2], Arnold mentioned an intuitive description for the turbulence by Landau in 1946. We sketch it as follows:

In low speed the motion is in equilibrium state near the boundary layer. When the speed increases indefinitely, the motion will turn consequently to be oscillating with single frequency, then double frequencies, multiple frequencies, and even infinitely many frequencies (that is the turbulence).

For the phenomenon of turbulence, Kolmogorov considered an analogy between the above Landau theory and the theory of dynamical systems: the equilibrium state corresponds to the singular point, the oscillation of single frequency corresponds to the closed orbit, the oscillation of double (or multiple) frequencies corresponds to the two (or higher)-dimensional invariant torus, and the oscillation of infinitely many frequencies (or the turbulence) corresponds the infinite dimensional invariant torus.

Then, Arnold said in the paper [2], "Kolmogorov devised the *method of accelerated convergence* not for the sake of those remarkable applications to problems of classical mechanics to which it leads, but for the sake of investigating the possibility of realizing the special set-theoretic pathology in systems on a two-dimensional torus (mixing). Kolmogorov did not solve the problem he posed to himself on realizing of mixing on weakly perturbed invariant tori since, on the tori that he found, his method automatically constructs angular coordinates that vary uniformly under the motion of a phase point.[2] The question of the existence of invariant tori carrying the mixing flows in generic systems close to integrable ones remains unsolved even today."

In 2003, the present author published a paper [50] to show the existence of nearly integrable analytic Hamiltonian system admitting minimal torus of weak mixing.

According to the definition of mixing in literature (for example, [115]), the term 'mixing' in Kolmogorov's problem may be understood as 'strong mixing' or 'weak mixing'. In this sense, the result in [50] is merely a partial answer to the above Kolmogorov problem.

However, this author proves recently [52] that the strong mixing flow cannot exist on a minimal torus.

The last result is a negative answer to the Kolmogorov's problem if the term 'mixing' is understood as strong mixing, while the result of [50] is a positive answer if the term 'mixing' is considerd as weak mixing. Therefore, the above Kolmogorov's problem has been solved in either case.

11.4.1 Auxiliary Lemmas

To report the result of [50], let us consider the analytic system (11.13) again with more details.

Assume the projection

$$P: \quad \mathbb{R}^2 \quad \to \quad \mathbb{T}^2 = \mathbb{R}^2/\mathrm{mod}\,(2\pi).$$

When (11.13) defines an analytic dynamical system f^t on the torus $\mathbb{T}^2$, it defines also an analytic dynamical system $\tilde{f}^t$ on the plane $\mathbb{R}^2$, satisfying

$$P \circ \tilde{f}^t = f^t, \qquad t \in \mathbb{R}^1.$$

Let $z \in \mathbb{R}^2$ and $\zeta = P(z) \in \mathbb{T}^2$. Then the orbit $\tilde{\Gamma}_z$ of $\tilde{f}^t$ and the orbit Γ_ζ of f^t satisfy the relation

$$P(\tilde{\Gamma}_z) = \Gamma_\zeta.$$

---

[2] It means that the motion is Liapunv stable and is thus almost periodic.

Consider the planar vector field

$$V(z) = (G(u,v), \lambda G(u,v)), \quad \text{for } z = (u,v) \in \mathbb{R}^2,$$

for $G(u,v) = 1/F(u,v)$, where $F(u,v) > 0$ is the analytic function defined by (11.9) as above. Then $G(u,v)$ is analytic in $(u,v) \in \mathbb{R}^2$. Using the Lagrange formula

$$G(u_1,v_1) - G(u_2,v_2) = G'_u(u_2+\theta\Delta u, v_2+\Delta v)\Delta u + G'_v(u_2+\theta\Delta u, v_2+\Delta v)\Delta v,$$

where

$$\Delta u = u_1 - u_2, \quad \Delta v = v_1 - v_2, \quad \text{and} \quad \theta \in (0,1),$$

we conclude that $V(z)$ is globally Lipschitzian:

$$|V(z_1) - V(z_2)| \leq L|z_1 - z_2|, \quad \forall\, z_1, z_2 \in \mathbb{R}^2, \tag{11.27}$$

with some Lipschitzian constant $L > 0$.

Lemma 11.1 *For any given constant τ, there is a constant $B = B(\tau) > 0$, such that*

$$|\tilde{f}^\tau(z_1) - \tilde{f}^\tau(z_2)| \geq B|z_1 - z_2|, \quad \forall\, z_1, z_2 \in \mathbb{R}^2.$$

Proof. For given points z_1 and z_2 in $\mathbb{R}^2$, consider

$$p_1 = \tilde{f}^\tau(z_1) \quad \text{and} \quad p_2 = \tilde{f}^\tau(z_2) \tag{11.28}$$

as initial points, such that

$$\tilde{f}^t(p_1) = p_1 + \int_0^t V(\tilde{f}^s(p_1))\, ds, \quad t \in \mathbb{R}^1$$

and

$$\tilde{f}^t(p_2) = p_2 + \int_0^t V(\tilde{f}^s(p_2))\, ds, \quad t \in \mathbb{R}^1.$$

It follows that

$$|\tilde{f}^t(p_1) - \tilde{f}^t(p_2)| \leq |p_1 - p_2| + L\left|\int_0^t |\tilde{f}^s(p_1) - \tilde{f}^s(p_2)|\, ds\right|. \tag{11.29}$$

Let

$$\xi(t) = \left|\int_0^t |\tilde{f}^s(p_1) - \tilde{f}^s(p_2)|\, ds\right|.$$

Then we have

$$\begin{cases} \xi'(t) - L\xi(t) \leq |p_1 - p_2|, & \text{if } t \geq 0, \\ -\xi'(t) - L\xi(t) \leq |p_1 - p_2|, & \text{if } t < 0, \end{cases}$$

with $\xi(0) = 0$. It follows that

$$\xi(t) \leq \begin{cases} \dfrac{|p_1 - p_2|}{L}(e^{Lt} - 1), & \text{if } t \geq 0; \\ \dfrac{|p_1 - p_2|}{L}(e^{-Lt} - 1), & \text{if } t < 0, \end{cases}$$

which together with (11.29) implies

$$|\tilde{f}^t(p_1) - \tilde{f}^t(p_2)| \leq \begin{cases} |p_1 - p_2|e^{Lt}, & \text{if } t \geq 0, \\ |p_1 - p_2|e^{-Lt}, & \text{if } t < 0. \end{cases}$$

It is equivalent to

$$|p_1 - p_2| \geq \begin{cases} e^{-Lt}|\tilde{f}^t(p_1) - \tilde{f}^t(p_2)|, & \text{if } t \geq 0, \\ e^{Lt}|\tilde{f}^t(p_1) - \tilde{f}^t(p_2)|, & \text{if } t < 0. \end{cases}$$

It follows from (11.28) that

$$|\tilde{f}^\tau(z_1) - \tilde{f}^\tau(z_2)| \geq \begin{cases} e^{-Lt}|\tilde{f}^{t+\tau}(z_1) - \tilde{f}^{t+\tau}(z_2)|, & \text{if } t \geq 0, \\ e^{Lt}|\tilde{f}^{t+\tau}(z_1) - \tilde{f}^{t+\tau}(z_2)|, & \text{if } t < 0. \end{cases}$$

Then, letting $t = -\tau$, we get

$$|\tilde{f}^\tau(z_1) - \tilde{f}^\tau(z_2)| \geq \begin{cases} e^{L\tau}|z_1 - z_2|, & \text{if } \tau \leq 0, \\ e^{-L\tau}|z_1 - z_2|, & \text{if } \tau > 0. \end{cases}$$

Taking the constant

$$B = \begin{cases} e^{-L\tau}, & \text{if } \tau > 0, \\ e^{L\tau}, & \text{if } \tau \leq 0, \end{cases}$$

we obtain the desired inequality.

The proof of Lemma 11.1 is thus complete. □

Since $G(u,v) = 1/F(u,v) > 0$ is a continuous function in $(u,v) \in \mathbb{T}^2$, there are constants $A_2 > A_1 > 0$, such that

$$A_1 < G(u,v) < A_2, \quad (u,v) \in \mathbb{R}^2.$$

Chapter 11. Some Special Problems

In what follows, let $\{\xi_j\}$ and $\{t_j\}$ be any sequences in $\mathbb{R}^1$ with $t_j \to \infty$ as $j \to \infty$. Assume σ is a given constant with $|\sigma| < 1$.

Lemma 11.2 *Given a constant $N > 0$, there is a constant $\tau_0 = \tau_0(N) > 0$, such that*

$$|\tilde{f}^{t_j}(\xi_j, \lambda\xi_j) - (\xi_j, \lambda\xi_j + \sigma)| > N, \qquad \text{when } t_j > \tau_0.$$

In other words, we have

$$|\tilde{f}^{t_j}(\xi_j, \lambda\xi_j) - (\xi_j, \lambda\xi_j + \sigma)| \;\to\; \infty \qquad \text{as } t_j \to \infty.$$

Proof.
Let $(u(t), v(t)) = \tilde{f}^t(\xi_j, \lambda\xi_j)$. Then we have

$$u(t) = \xi_j + \int_0^t G(u(s), v(s))\,ds, \quad v(t) = \lambda\left(\xi_j + \int_0^t G(u(s), v(s))\,ds\right),$$

which yields

$$|\tilde{f}^{t_j}(\xi_j, \lambda\xi_j) - (\xi_j, \lambda\xi_j + \sigma)|$$

$$= \sqrt{\left[\int_0^{t_j} G(u(s),v(s))\,ds\right]^2 + \left[\int_0^{t_j} \lambda G(u(s),v(s))\,ds - \sigma\right]^2}$$

$$> \sqrt{(A_1 t_j)^2 + (\lambda A_1 t_j - |\sigma|)^2} > \sqrt{(A_1 t_j)^2 + (\lambda A_1 t_j - 1)^2},$$

where $t_j > 1/(\lambda A_1)$ and $|\sigma| < 1$. It follows that

$$|\tilde{f}^{t_j}(\xi_j, \lambda\xi_j) - (\xi_j, \lambda\xi_j + \sigma)| > N, \qquad \text{for } t_j > \tau_0,$$

where

$$\tau_0 = \max\left\{\frac{1}{\lambda A_1}, \frac{\lambda + \sqrt{\lambda^2 + (1+\lambda^2)N^2}}{(1+\lambda^2)A_1}\right\}.$$

The proof of Lemma 11.2 is thus completed. $\square$

Now, denote by $\mathfrak{S}_\eta$ the strip in $\mathbb{R}^2$ bounded by the straight lines

$$L_0 = \{(u,v) \in \mathbb{R}^2 : \quad v = \lambda u\}$$

and

$$L_\eta = \{(u,v) \in \mathbb{R}^2 : \quad v = \lambda u + \eta\}.$$

Let $\mathcal{C}$ be a continuous curve having finite length in the torus $\mathbb{T}^2$. We call $\mathcal{C}$ a curve of type $\mathfrak{S}_\eta$ if it has a continuous lifting-branch $J \subset \mathfrak{S}_\eta$ with finite length, such that $P(J) = \mathcal{C}$.

Assume $\mathcal{C} \subset \mathbb{T}^2$ is a curve of type $\mathfrak{S}_\eta$ and let J be the corresponding lifting-branch (with finite length) in $\mathfrak{S}_\eta$. Then define the *norm* of $\mathcal{C}$ by

$$\|\mathcal{C}\| = \sup_{z, w \in J} |z - w|.$$

Let $\mathfrak{D}_\varepsilon(p)$ be the ε-disk in $\mathbb{T}^2$ centered at the point p.

Lemma 11.3 *Assume $\mathcal{C} \subset \mathbb{T}^2$ is a curve of type $\mathfrak{S}_\sigma$. Then, for any given constant $\varepsilon > 0$, we have*

$$\mathcal{C} \cap \mathfrak{D}_\varepsilon(p) \neq \emptyset, \qquad \text{for all } p \in \mathbb{T}^2,$$

provided that $|\sigma|$ is small enough and $\|\mathcal{C}\|$ is large enough.

Proof. Let J be the lifting branch of $\mathcal{C}$ in $\mathfrak{S}_\sigma$ with finite branch. It follows that J is bounded between the lines L_0 and L_σ. Since J has finite length, it is also bounded between two vertical lines the straight lines

$$K_a = \{(u, v) \in \mathbb{R}^2 : \quad u = a, \quad v \in \mathbb{R}^1\}$$

and

$$K_b = \{(u, v) \in \mathbb{R}^2 : \quad u = b, \quad v \in \mathbb{R}^1\},$$

with some constants a and b ($b > a$), such that J connects K_a and K_b.

Take the points

$$z_a = (a, \lambda a), \qquad z_b = (b, \lambda b) \in L_0$$

and

$$w_a = (a, \lambda a + \sigma), \qquad w_b = (b, \lambda b + \sigma) \in L_\sigma.$$

Then we have a quadrangle $\mathfrak{Q}$ with vertices z_a, z_b, w_b and w_a, such that $\tilde{J} \subset \mathfrak{Q}$. It can be seen that the line-segment from z_a to w_b is the longest diagonal of $\mathfrak{Q}$. Then, by the definition of $\|\mathcal{C}\|$, we get

$$\|\mathcal{C}\| \leq |z_a - w_b| \leq |z_a - z_b| + |z_b - w_b| \leq |z_a - z_b| + |\sigma|. \quad (11.30)$$

Denote by $L_0[z_a, z_b]$ the line-segment of L_0 from z_a to z_b. Hence, the length of $L_0[z_a, z_b]$ is $|z_a - z_b|$. When $\|\mathcal{C}\|$ is large and $|\sigma|$ is small, we have

$|\sigma| < b-a$. It follows from (11.30) together with $(b-a) < \sqrt{1+\lambda^2}(b-a) = |z_a - z_b|$ that

$$\|\mathcal{C}\| \leq |z_a - z_b| + (b-a) < 2|z_a - z_b|.$$

Therefore, $|z_a - z_b|$ is sufficiently large whenever $\|\mathcal{C}\|$ is large enough. On the other hand, since $P(L_0)$ is dense in $\mathbb{T}^2$, we have

$$P(L_0[z_a, z_b]) \cap \mathfrak{D}_{\varepsilon/2}(p) \neq \emptyset, \qquad \text{for all } p \in \mathbb{T}^2, \tag{11.31}$$

provided that $\|L_0[z_a, z_b]\|$ is sufficiently large (or $|z_a - z_b|$ is large enough). It follows from $J \subset \mathfrak{Q}$ that

$$J \subset \mathcal{N}_{\varepsilon/2}(L_0[z_a, z_b]),$$

where $\mathcal{N}_{\varepsilon/2}(L_0[z_a, z_b])$ is the $\varepsilon/2$-neighborhood of $L_0[z_a, z_b]$, whenever $|\sigma|$ is sufficiently small. Then we get

$$\mathcal{C} \subset P(\mathcal{N}_{\varepsilon/2}(L_0[z_a, z_b])) = \mathcal{N}_{\varepsilon/2}(P(L_0[z_a, z_b])),$$

which together with (11.31) implies

$$\mathcal{C} \cap \mathfrak{D}_\varepsilon(p) \neq \emptyset, \qquad \forall\, p \in \mathbb{T}^2,$$

whenever $|\sigma|$ is small enough and $\|\mathcal{C}\|$ is sufficiently large.

Lemma 11.3 is thus proved. $\square$

Let $z_s = (s, \lambda s) \in L_0$ and $w_s = (s, \lambda s + \eta) \in L_\eta$. It follows that when the motion $\tilde{f}^t(w_s)$ comes from the initial point w_s at $t = 0$ to the point $(u, v) = (u, \lambda u + \eta) \in L_{w_s}$, it spends the time

$$\tau(u, s, \eta) = \int_s^u F(x, \lambda x + \eta)\, dx. \tag{11.32}$$

Since (11.13) is an autonomous system and $w_s \in L_\eta$, we have

$$\begin{cases} \tau(u, s, \eta) = \tau(u, 0, \eta) - \tau(s, 0, \eta), \\ \tau(u, s, 0) = \tau(u, 0, 0) - \tau(s, 0, 0). \end{cases} \tag{11.33}$$

Now, consider the function

$$T(u, s, \eta) = \tau(u, s, \eta) - \tau(u, s, 0) \tag{11.34}$$

of the variable $u \in \mathbb{R}^1$, where s and η are given as parameters. It follows from (11.33) and (11.34) that

$$T(u, s, \eta) = [\tau(u, 0, \eta) - \tau(u, 0, 0)] - [\tau(s, 0, \eta) - \tau(s, 0, 0)],$$

which yields
$$T(u,s,\eta) = T(u,0,\eta) - T(s,0,\eta), \tag{11.35}$$

where the term $T(s,0,\eta)$ is independent of the variable $u \in \mathbb{R}^1$.

Now, using (11.32) and (11.34), we have
$$T(u,0,\eta) = \int_0^u \Phi(x,\eta)\,dx,$$

where the integrand $\Phi(x,\eta) = F(x, \lambda x + \eta) - F(x, \lambda x)$ has the Fourier series
$$\Phi(x,\eta) = \sum_{k,j=-\infty}^{\infty} C_{kj} e^{i(k+\lambda j)x}(e^{ij\eta} - 1),$$

which agrees with the Fourier series (11.15). Therefore, we have proved the following conclusion.

Lemma 11.4 *For almost all $\eta \in \mathbb{R}^1$, $T(u,0,\eta)$ is an unbounded function of $u \in \mathbb{R}^1$.*

Indeed, it follows that f^t is Liapunov unstable in the minimal torus $\mathbb{T}^2$.

11.4.2 Weak Mixing

Let us state the definition of (strong or weak) mixing flows in literature (see [115], for example), which possess more complicated dynamical behavior than that of chaotic flows.

Let ψ^t be a flow on the n-dimensional manifold $\mathfrak{M}$, and let $\mathcal{K} \subset \mathfrak{M}$ be a compact invariant (nonempty) set of ψ^t. Denote by $\mathcal{O}_\mathcal{K}$ the set of all (nonempty) relatively open sets in $\mathcal{K}$. Then, for any $D_1, D_2 \in \mathcal{O}_\mathcal{K}$, define the 'meeting time set'
$$\mathfrak{m}(D_1, D_2) = \{ t \in \mathbb{R}^1 : \psi^t(D_1) \cap D_2 \neq \emptyset \}.$$

It is apparent that $\mathfrak{m}(D_1, D_2) = \mathfrak{m}(D_2, D_1)$.

Definition 11.1 *For any given $D_1, D_2 \in \mathcal{O}_\mathcal{K}$, if there is a constant $T_0 > 0$, such that*
$$\mathfrak{m}(D_1, D_2) \supset \{ t \in \mathbb{R}^1 : |t| \geq T_0 \},$$

then ψ^t is called a strong mixing flow in $\mathcal{K}$.

Definition 11.2 For any given D_1, G_1, D_2, $G_2 \in \mathcal{O}_\mathcal{K}$, if the inequality

$$\mathfrak{m}(D_1, G_1) \cap \mathfrak{m}(D_2, G_2) \neq \emptyset$$

is valid, then ψ^t is called a *weak mixing* flow in $\mathcal{K}$.

It is apparent that if ψ^t is strong mixing in $\mathcal{K}$ then it is weak mixing in $\mathcal{K}$. Furthermore, we can prove easily that if ψ^t is weak mixing in $\mathcal{K}$ then it is chaotic in $\mathcal{K}$.

Now, we are in position to prove the following result.

Theorem 11.3 *The flow f^t defined by the analytic system (11.13) is weak mixing on the minimal torus $\mathbb{T}^2$.*

Proof. It has been proved that $\mathbb{T}^2$ is a minimal torus of f^t. We only need to show that f^t is weak mixing on $\mathbb{T}^2$.

Let D_1, G_1, D_2 and G_2 be nonempty open sets in $\mathbb{T}^2$. Then consider the straight line

$$L_0: \quad v = \lambda u,$$

on the plane $\mathbb{R}^2$. Since the projection $P(L_0)$ is a dense curve on the torus $\mathbb{T}^2$, the straight line L_0 intersects the liftings $P^{-1}(D_1)$ and $P^{-1}(G_1)$, which are open sets in $\mathbb{R}^2$. Hence, $P^{-1}(D_1)$ and $P^{-1}(G_1)$ contains components $\check{D}_1$ and $\check{G}_1$, such that

$$L_0 \cap \check{D}_1 \neq \emptyset \quad \text{and} \quad L_0 \cap \check{G}_1 \neq \emptyset,$$

respectively. It follows that for sufficiently small $\sigma > 0$, there are points

$$x_a = (a, \lambda a) \in L_0 \cap \check{D}_1 \quad \text{and} \quad y_a = (a, \lambda a + \sigma) \in L_\sigma \cap \check{D}_1, \quad (11.36)$$

and

$$x_b = (b, \lambda b) \in L_0 \cap \check{G}_1 \quad \text{and} \quad y_b = (b, \lambda b + \sigma) \in L_\sigma \cap \check{G}_1, \quad (11.37)$$

such that

$$I_a \subset \check{D}_1 \quad \text{and} \quad I_b \subset \check{G}_1,$$

where $I_a = [x_a, y_a]$ is the closed interval with endpoints x_a and y_a and $I_b = [x_b, y_b]$ is the closed interval with endpoints x_b and y_b. Note that $x_a, x_b \in L_0$ and $y_a, y_b \in L_\sigma$.

Using Lemma 11.4, we conclude that there is a sufficiently small constant $\eta > 0$, such that $T(u, 0, \eta)$ is unbounded with respect to $u \in \mathbb{R}^1$. It follows that there is a sequence $\{u_j\}$ tending to infinity with the property that $T(u_j, 0, \eta)$ approaches infinity. Without loss of generality, assume

$$u_j \to +\infty \quad \text{and} \quad T(u_j, 0, \eta) \to +\infty, \quad \text{as } j \to +\infty,$$

for definiteness. From (11.35), we have

$$T(u_j, s, \eta) = T(u_j, 0, \eta) - T(s, 0, \eta) \to \infty, \quad \text{as } j \to \infty,$$

where the sequence $\{u_j\}$ is independent of s and it depends indeed merely on η. Hence, we have

$$\lim_{j \to \infty} T(u_j, a, \eta) = +\infty, \qquad \lim_{j \to \infty} T(u_j, b, \eta) = +\infty.$$

It is noted that

$$\tilde{f}^{\tau(u_j, a, 0)}(x_a) = (u_j, \lambda u_j) \in L_0, \qquad \tilde{f}^{\tau(u_j, a, \eta)}(y_a) = (u_j, \lambda u_j + \eta) \in L_\eta.$$

It can be seen from (11.32) that $\tau(u, s, \eta)$ is monotonically increasing with respect to u. Hence, when $u_j \to \infty$ as $j \to \infty$, we have

$$\tau_j = \tau(u_j, 0, \eta) \to \infty, \qquad \text{as } j \to \infty.$$

Consider the continuous curves

$$J_a = \tilde{f}^{\tau_j}(I_a) \quad \text{and} \quad J_b = \tilde{f}^{\tau_j}(I_b), \tag{11.38}$$

which are the images of the intervals I_a and I_b under the map $\tilde{f}^{\tau_j}$. It is noted that J_a and J_b are continuous curves of finite length in the strip $\mathfrak{S}_\sigma$, and (11.38) implies

$$\tilde{f}^{\tau_j}(x_a) \in J_a, \qquad \tilde{f}^{\tau_j}(x_b) \in J_b. \tag{11.39}$$

Note that $x_a = (a, \lambda a) \in I_a$ and $y_a = (a, \lambda a + \eta) \in I_a$. With (11.33) and (11.34) in mind, we get the following formula

$$\tau_j = \tau(u_j, 0, \eta) = \tau(u_j, a, \eta) + \tau(a, 0, \eta)$$
$$= T(u_j, a, \eta) + \tau(u_j, a, 0) + \tau(a, 0, \eta),$$

which together with the definition of τ_j implies

$$\tilde{f}^{\tau_j}(x_a) = \tilde{f}^{\tau(a,0,\eta)} \circ \tilde{f}^{T(u_j,a,\eta)} \circ \tilde{f}^{\tau(u_j,a,0)}(a, \lambda a)$$

$$= \tilde{f}^{\tau(a,0,\eta)} \circ \tilde{f}^{T(u_j,a,\eta)}(u_j, \lambda u_j).$$

It follows from (11.39) that
$$\tilde{f}^{\tau_j}(x_a) = \tilde{f}^{\tau(a,0,\eta)} \circ \tilde{f}^{T(u_j,a,\eta)}(u_j, \lambda u_j) \in J_a. \tag{11.40}$$

Similarly, using
$$\tau_j = \tau(u_j, 0, \eta) = \tau(u_j, a, \eta) + \tau(a, 0, \eta)$$
implies
$$\tilde{f}^{\tau_j}(y_a) = \tilde{f}^{\tau(a,0,\eta)} \circ \tilde{f}^{\tau(u_j,a,\eta)}(a, \lambda a + \eta)$$
$$= \tilde{f}^{\tau(a,0,\eta)}(u_j, \lambda u_j + \eta) \in J_a. \tag{11.41}$$

Hence, we have
$$\|P(J_a)\| \geq |\tilde{f}^{\tau_j}(x_a) - \tilde{f}^{\tau_j}(y_a)|$$
$$= |\tilde{f}^{\tau(a,0,\eta)} \circ \tilde{f}^{T(u_j,a,\eta)}(u_j, \lambda u_j) - \tilde{f}^{\tau(a,0,\eta)}(u_j, \lambda u_j + \eta)|,$$

which together with Lemma 11.1 implies
$$\|P(J_a)\| \geq B|\tilde{f}^{T(u_j,a,\eta)}(u_j, \lambda u_j) - (u_j, \lambda u_j + \eta)|, \tag{11.42}$$

where $B = B(\tau(a, 0, \eta))$ is independent of u_j. Since
$$T(u_j, a, \eta) \to \infty, \quad \text{as } j \to \infty,$$

Lemma 11.2 implies that
$$|\tilde{f}^{T(u_j,a,\eta)}(u_j, \lambda u_j) - (u_j, \lambda u_j + \eta)| \to \infty, \quad \text{as } j \to \infty.$$

Hence, using (11.42), we conclude that $\|P(J_a)\|$ is sufficiently large if τ_j is large enough. It follows that the projection
$$C_a = P(J_a) \subset \mathbb{T}^2$$

is a continuous curve of type $\mathfrak{S}_\eta$ having sufficiently large norm $\|C_a\|$ when j is large enough. Therefore, using Lemma 11.3, we conclude that
$$C_a \cap D_2 \neq \emptyset,$$

that is,
$$P(J_a) \cap D_2 \neq \emptyset.$$

Then we have
$$P(\tilde{f}^{\tau(u_j,0,\eta)}(I_a)) \cap D_2 \neq \emptyset,$$
which yields
$$\tilde{f}^{\tau(u_j,0,\eta)}(P(I_a)) \cap D_2 \neq \emptyset.$$
It follows from $I_a \subset \check{D}_1$ and $P(\check{D}_1) \subset D_1$ that
$$\tilde{f}^{\tau(u_j,0,\eta)}(D_1) \cap D_2 \neq \emptyset, \quad \text{for sufficiently large } j > 0. \quad (11.43)$$
This proves that $\tau(u_j,0,\eta) \in \mathfrak{m}(D_1,D_2)$ for sufficiently large $j > 0$.

With a similar process from (11.36) to (11.43), we arrive from (11.37) at
$$\tilde{f}^{\tau(u_j,0,\eta)}(G_1) \cap G_2 \neq \emptyset, \quad \text{for sufficiently large } j > 0,$$
which means that $\tau(u_j,0,\eta) \in \mathfrak{m}(G_1,G_2)$ for sufficiently large $j > 0$.

Therefore, we have
$$\tau(u_j,0,\eta) \in \mathfrak{m}(D_1,D_2) \cap \mathfrak{m}(G_1,G_2) \neq \emptyset,$$
which proves Theorem 11.3. $\square$

11.4.3 Nearly Integrable Hamiltonian Systems

Now, consider the system
$$\frac{du}{dt} = \frac{C_{00}}{C_{00} + F_0(u,v)}, \quad \frac{dv}{dt} = \frac{\lambda C_{00}}{C_{00} + F_0(u,v)}, \quad (11.44)$$
which is equivalent to the system (11.13) with $F(u,v) = C_{00} + F_0(u,v) > 0$ and $C_{00} > 0$ as defined above. Then $\varepsilon = 1/C_{00}$ is a small parameter if C_{00} is sufficiently large. Using (11.45), we have the system
$$\frac{du}{dt} = 1 + \varepsilon E(u,v,\varepsilon), \quad \frac{dv}{dt} = \lambda(1 + \varepsilon E(u,v,\varepsilon)), \quad (11.45)$$
where
$$E(u,v,\varepsilon) = \frac{-F_0(u,v)}{1 + \varepsilon F_0(u,v)}$$
is analytic in $(u,v,\varepsilon) \in \mathbb{T}^2 \times [-\alpha,\alpha]$. Since (11.45) is equivalent to (11.13), $\mathbb{T}^2$ is a minimal torus of weak mixing for (11.45) as $\varepsilon \in [-\alpha,\alpha]$.

Finally, consider the Hamiltonian system

$$\frac{dz}{dt} = -\frac{\partial H}{\partial \theta}, \quad \frac{d\theta}{dt} = \frac{\partial H}{\partial z}, \tag{11.46}$$

with the Hamiltonian

$$H(z, \theta, \varepsilon) = \frac{1}{2}(I^2 + J^2) + (I + \lambda J)[1 + \varepsilon E(u, v, \varepsilon)],$$

where $z = (I, J) \in \mathbb{R}^2$ and $\theta = (u, v) \in \mathbb{T}^2$ are, respectively, the action and angular variables, and $\varepsilon \geq 0$ is a small parameter, and the irrational number λ and the analytic function $E(u, v, \varepsilon)$ are defined above. It is apparent that the Hamiltonian system (11.46) is nearly integrable. Moreover, for the small parameter $\varepsilon > 0$, it has an invariant minimal torus $z = 0$, $\theta \in \mathbb{T}^2$ of weak mixing.

Therefore, we have proved the result.

Theorem 11.4 *The nearly integrable Hamiltonian system (11.46) has a minimal torus $\mathbb{T}^2$ of weak mixing.*

11.4.4 Strong Mixing

Now, we prove the following result.

Theorem 11.5 *A flow cannot be strong mixing in the minimal torus.*

Proof. Let $\mathbb{T}^2$ be a minimal torus of f^t. Then f^t is ergodic on $\mathbb{T}^2$ and the Poincaré rotation number $\lambda = \mathrm{rot}(f^t)$ is irrational. Consider the flow g^t defined by the system of differential equations

$$\frac{du}{dt} = 1, \quad \frac{dv}{dt} = \lambda, \quad (u, v) \in \mathbb{T}^2, \tag{11.47}$$

which implies that the Poincaré rotation number $\mathrm{rot}(g^t)$ is equal to $\lambda = \mathrm{rot}(f^t)$. It follows from the Theorem 1.3 of Chapter 6 in the book [3] that the flow f^t is topologically equivalent to the flow g^t. In other words, we have the following commutative diagram

$$\begin{array}{ccc} \mathbb{T}^2 & \xrightarrow{h} & \mathbb{T}^2 \\ f^t \downarrow & & \downarrow g^t \\ \mathbb{T}^2 & \xrightarrow{h} & \mathbb{T}^2, \end{array}$$

where h is a homeomorphism of the torus $\mathbb{T}^2$. Then, we have

$$f^t(S) = h^{-1} \circ g^t \circ h(S) \tag{11.48}$$

for any subset S in $\mathbb{T}^2$.

Now, assume the contrary of Theorem 11.5.

Hence, there is a strong mixing flow f^t in $\mathbb{T}^2$. It follows from the definition of strong mixing that for any given open set S in $\mathbb{T}^2$, there is a constant $\tau_0 = \tau_0(S) > 0$, such that

$$f^t(S) \cap S \neq \emptyset, \qquad \forall\, t > \tau_0. \tag{11.49}$$

Now, let us consider the vertical lines

$$M_n: \quad u = 2n\pi,$$
$$N_n: \quad u = 2(n + \tfrac{1}{2})\pi, \qquad (n \in \mathbb{Z}^1)$$

in the covering space $\mathbb{R}^2$. Then the projections

$$\mathcal{C}_1 = P(M_n) \quad \text{and} \quad \mathcal{C}_2 = P(N_n) \qquad (\forall\, n \in \mathbb{Z}^1)$$

are two meridian circles, which cut the torus $\mathbb{T}^2$ into two disjoint open regions

$$U_1 = \{(u,v) \in \mathbb{T}^2 : \quad 0 < u' < \pi \quad \text{where } u' = u \mod 2\pi\},$$

and

$$U_2 = \{(u,v) \in \mathbb{T}^2 : \quad \pi < u'' < 2\pi \quad \text{where } u'' = u \mod 2\pi\}.$$

Note that $U_1 \cap U_2 = \emptyset$.

With the positive direction of u-axis in mind, we conclude that $\mathcal{C}_1$ is the left boundary of U_1 and $\mathcal{C}_2$ is the right boundary, and $\mathcal{C}_1$ is the right boundary of U_2 and $\mathcal{C}_2$ is the left boundary.

On the other hand, using the property of g^t, we obtain

$$g^{(2k+\frac{1}{2})\pi}(U_1) = U_2, \qquad g^{(2k+\frac{1}{2})\pi}(U_2) = U_1, \qquad \forall\, k \in \mathbb{R}^1. \tag{11.50}$$

Using (11.48) with $S = h^{-1}(U_1)$, we obtain

$$f^t(S) \cap S = h^{-1} \circ g^t \circ h(S) \cap S$$

$$= h^{-1} \circ g^t(U_1) \cap h^{-1}(U_1) = h^{-1}(g^t(U_1) \cap U_1),$$

for any $t \in \mathbb{R}^1$. However, when $t = t_k = 2(k + \frac{1}{2})\pi$, it follows from (11.50) and $U_1 \cap U_2 = \emptyset$ that

$$h^{-1}(g^t(U_1) \cap U_1) = h^{-1}(U_2 \cap U_1) = h^{-1}(\emptyset) = \emptyset,$$

which implies

$$f^t(S) \cap S = \emptyset, \quad \text{when } t = t_k = (2k + \frac{1}{2})\pi.$$

It is in conflict with (11.49). This contradiction proves that f^t cannot be strong mixing in the minimal torus $\mathbb{T}^2$.

Theorem 11.5 is thus proved. □

Therefore, the considered Kolmogorov's problem is solved by Theorems 11.4 and 11.5 despite of the weak mixing case or the strong mixing case.

Furthermore, it follows from Theorem 11.5 that

" *If f^t is a flow of strong mixing on the torus $\mathbb{T}^2$, then it possesses not only a dense orbit but also admits at least a singular point in $\mathbb{T}^2$.*"

Conversely, it is easy to prove that if f^t possesses a dense orbit and admits at least a singular point in $\mathbb{T}^2$, then it is strong mixing on $\mathbb{T}^2$.

As shown above, the minimal torus cannot carry the strong mixing flow. However, it is an open question whether or not a general (non-torus) minimal set can carry the strong mixing flow.

11.5 Brillouin Focusing System

11.5.1 *A Boundary-Value Problem*

The so-called Brillouin focusing system is governed by a boundary-value problem of nonlinear periodic differential equation

$$\frac{d^2 y}{dx^2} + a(1 + \cos 2x)y = \frac{1}{y}, \qquad (11.51)$$

with the boundary condition

$$y(0) = y(\pi) > 0, \quad y'(0) = y'(\pi) = 0. \qquad (11.52)$$

where a is a parameter in the interval: $0 < a \leq 1$.

In the application, one has first to choose the parameter a, such that the boundary-value problem (11.51)+(11.52) admits a positive solution. However, the mathematical treatment of this boundary-value problem so far has been limited to approximate calculations. The right side of the equation

(11.51) is usually replaced by 0, so that the Mathieu's equation is encountered. In this case, the spectrum of the corresponding boundary-value problem is known to be discrete. However, we will see that the spectrum of the boundary-value problem (11.51)+(11.52) is continuous and contains some interval $0 < a \leq \chi_*$. In view of this essential difference, it seems that the above formal approximation in application is not justified (see the references of the paper [26]).

By the qualitative method of differential equations, the existence of positive solution of (11.51)+(11.52) is proved in [26] when $0 < a \leq 1/4$.

In 1978, Ye and Wang [120] improved the above result as follows:

$$0 < a < 8/(\pi^2 + 4) \approx 0.5768.$$

Recently, M. Zhang [122] studied the boundary-value problem for general differential equation

$$\frac{d^2y}{dx^2} + f(y)\frac{dy}{dx} + g(x,y) = 0,$$

by means of a new method. As a particular case, he proved the existence of positive solution of the boundary-value problem (11.51)+(11.52) whenever

$$0 < a \leq \chi_0 \approx 0.6128.$$

However, the approximate calculations (see the references of [26]) shows that the spectrum of (11.51)+(11.52) contains the interval $0 < a \leq \chi_* \approx 1$.

In what follows, we introduce the result of [26] for brevity.

11.5.2 Preliminary Works

Before proving some lemmas, we first remark that the conditions for local existence and uniqueness of solution are evidently satisfied by differential equation (11.51). However, it is not clear that the solution should globally exist on the interval $0 \leq x \leq \pi$ since (11.51) is singular at $y = 0$.

Lemma 11.5 *If differential equation (11.51) has a positive solution $y = y(x)$ on the interval $0 \leq x \leq \pi/2$, satisfying the boundary condition*

$$y'(0) = 0, \qquad y'(\frac{\pi}{2}) = 0, \qquad (11.53)$$

then it has a positive solution $y = \varphi(x)$ on the interval $0 \leq x \leq \pi$, satisfying the boundary condition (11.52).

Chapter 11. Some Special Problems

Proof. Writing $\pi - x$ for x in the equation (11.51); that is,

$$\frac{d^2y}{dx^2}(x) + 2a\cos^2 x \cdot y(x) = \frac{1}{y(x)}, \qquad 0 \le x \le \frac{\pi}{2}, \tag{11.54}$$

we obtain

$$\frac{d^2y}{dx^2}(\pi - x) + 2a\cos^2 x \cdot y(\pi - x) = \frac{1}{y(\pi - x)}, \tag{11.55}$$

for $\frac{\pi}{2} \le x \le \pi$.

Let

$$\varphi(x) = \begin{cases} y(x), & \text{if } 0 \le x \le \frac{\pi}{2}, \\ y(\pi - x), & \text{if } \frac{\pi}{2} < x \le \pi. \end{cases}$$

Since $y'(\frac{\pi}{2}) = 0$, it is easy to see that $y = \varphi(x) > 0$ is continuously differentiable on the interval $0 \le x \le \pi$. It follows from (11.54) and (11.55) that $y = \varphi(x)$ is a positive solution of (11.51). Moreover, it satisfies obviously the boundary condition

$$\varphi(0) = \varphi(\pi)(= y(0)), \qquad \varphi'(0) = \varphi'(\pi)(= y'(0)) = 0;$$

that is, $y = \varphi(x)$ satisfies the boundary condition (11.52). We have thus proved that $y = \varphi(x)$ is a positive solution of (11.51)+(11.52).

The proof of Lemma 11.5 is complete. □

Lemma 11.6 *Suppose that $a \le \frac{1}{2}$ and that $y = u(x) > 0$ is a solution of (1) satisfying the initial condition*

$$u(0) = \frac{1}{\sqrt{2a}}, \qquad y'(0) = 0,$$

then $y = u(x)$ exists on the interval $0 \le x \le \frac{\pi}{2}$ and satisfies the following inequalities:

i) $\quad 0 < u(x) < \dfrac{1}{\sqrt{2a}} \sec x, \qquad 0 < x < \dfrac{\pi}{2};$

ii) $\quad u'(x) > 0, \qquad\qquad\qquad 0 < x \le \dfrac{\pi}{2}.$

Proof. Letting

$$G(x) = \frac{1}{\sqrt{2a}}\sec x, \qquad 0 \le x < \frac{\pi}{2},$$

then we have
$$G'(x) = \frac{1}{\sqrt{2a}} \tan x \sec x,$$
$$G''(x) = \frac{1}{\sqrt{2a}}(1 + 2\tan^2 x)\sec x,$$

Using $u(0) = \frac{1}{\sqrt{2a}}$ yields
$$u''(0) = \frac{1}{u(0)} - 2au(0) = 0.$$

Then we have
$$u''(0) < G''(0) = \frac{1}{\sqrt{2a}},$$

which together with $u(0) = G(0)$ and $u'(0) = G'(0)$ implies that there is a point x_1 $(0 < x_1 \leq \frac{\pi}{2})$ such that
$$0 < u(x) < G(x), \qquad 0 < x < x_1. \tag{11.56}$$

It follows that when $0 < x < x_1$, we have
$$u''(x) = \frac{1}{u(x)} - 2a\cos^2 x \cdot u(x) = \frac{2a\cos^2 x}{u(x)}[G^2(x) - u^2(x)] > 0.$$

Hence, $u'(x)$ is monotonically increasing. Since $u'(0) = 0$, we have
$$u'(x) > 0, \qquad 0 < x < x_1. \tag{11.57}$$

Hence, $u(x)$ is monotonically increasing and satisfies
$$u(x) > u(0) = \frac{1}{\sqrt{2a}}, \qquad 0 < x < x_1. \tag{11.58}$$

It follows that
$$u''(x) = \frac{1}{u(x)} - 2a\cos^2 x \cdot u(x) < \frac{1}{u(x)} < \sqrt{2a}, \tag{11.59}$$

which implies
$$\begin{cases} u'(x) < \sqrt{2a}x_1, \\ u(x) < u(0) + \sqrt{2a}x_1^2, \end{cases} \quad \text{when } 0 < x < x_1. \tag{11.60}$$

Chapter 11. Some Special Problems 349

On the other hand, the monotonic properties of $u(x)$ and $u'(x)$ together with (11.60) assure that the left limit-values $u(x_1)$ and $u'(x_1)$ exist. Therefore, the solution $y = u(x)$ can be prolonged on the interval $0 \leq x \leq x_1$.

To prove the lemma, in view of (11.56) and (11.57), it is sufficient to show that one can take $x_1 = \dfrac{\pi}{2}$.

In fact, assume the contrary. Then there is a constant x_1 ($0 < x_1 < \dfrac{\pi}{2}$) such that (11.56) is valid, but

$$u(x_1) = G(x_1). \tag{11.61}$$

Then it follows from (11.56) and (11.61) immediately that

$$u'(x_1) \geq G'(x_1). \tag{11.62}$$

On the other hand, (11.57) yields (11.59). Therefore, when $a \leq \dfrac{1}{2}$, we have

$$u''(x) < \sqrt{2a} \leq \frac{1}{\sqrt{2a}} < G''(x), \qquad 0 < x < x_1.$$

Integrating this inequality on the interval $[0, x_1]$, we obtain

$$u'(x_1) < G'(x_1),$$

which is in conflict with (11.62).

Lemma 11.6 is thus proved by the contradiction. □

Lemma 11.7 *Suppose that* $a \leq \dfrac{1}{4}$ *and* $y = v(x) > 0$ *is a solution of the differential equation (11.51) satisfying the initial condition*

$$v(0) = v_0 > \frac{1}{\sqrt{2a}}, \qquad v'(0) = 0, \tag{11.63}$$

then $y = v(x)$ *exists on the interval* $0 \leq x \leq \dfrac{\pi}{2}$ *and satisfies the inequality:*

$$v(x) > u(x), \qquad 0 \leq x \leq \frac{\pi}{2}. \tag{11.64}$$

Proof. Similar to the derivation of (11.60) from (11.58), we have

$$v'(x) < \sqrt{2a} \cdot \pi, \qquad v(x) < v_0 + \sqrt{2a} \cdot \pi^2, \tag{11.65}$$

provided that $v(x) > \dfrac{1}{\sqrt{2a}}$. Moreover, it follows from (11.65) that

$$v''(x) = \dfrac{1}{v(x)} - 2a\cos^2 x \cdot v(x) > -2a\cos^2 x \cdot v(x) > -M,$$

where $M = 2a(v_0 + \sqrt{2a}\pi^2)$. Hence

$$v'(x) > -M\pi.$$

Then we have

$$v(x) < v_0 + \sqrt{2a}\cdot \pi^2, \qquad -M\pi < v'(x) < \sqrt{2a}\cdot \pi, \qquad (11.66)$$

provided that $v(x) > \dfrac{1}{\sqrt{2a}}$.

On the other hand, the existence theorem of solution together with (11.63) implies that there is an $\alpha > 0$ such that

$$v(x) > u(x), \qquad 0 \le x < \alpha. \qquad (11.67)$$

Therefore, Lemma 11.6 yields

$$v(x) > u(x) \ge \dfrac{1}{\sqrt{2a}}, \qquad 0 \le x < \alpha. \qquad (11.68)$$

Let

$$\alpha_1 = \sup\{\alpha\}, \qquad \text{where } \alpha \text{ satisfies (11.67)}.$$

It suffices to show that $\alpha_1 > \dfrac{\pi}{2}$.

Assume the contrary. Then, there exists an α $\left(0 < \alpha \le \dfrac{\pi}{2}\right)$ such that (11.67) and consequently (11.68) hold. Moreover, when $x \to \alpha - 0$, we have the possibilities:

(i) $v(\alpha) = u(\alpha)$;
(ii) $v(x)$ or $v'(x)$ has no limit.

Since (18) implies that (16) holds on $0 \le x < \alpha$, then the continuation theorem of solution excludes the possibility (ii).

Now, we exclude the possibility (i) as follows.

It is noted that the differential equation (11.51) is unchanged when x is replaced by $-x$. Since $u'(0) = v'(0) = 0$, we can prolong the solutions $y = u(x)$ and $y = v(x)$ from $0 \le x \le \alpha$ to the interval $-\alpha \le x \le 0$ such that

$$u(-x) = u(x), \quad v(-x) = v(x), \qquad -\alpha \le x \le \alpha. \qquad (11.69)$$

Letting
$$z = v(x) - u(x), \qquad -\alpha \le x \le \alpha,$$
then we have
$$\frac{d^2 z}{dt^2} + Q(x)z = 0, \qquad (11.70)$$
where
$$Q(x) = \frac{1}{u(x)v(x)} + 2a\cos^2 x > 0, \qquad -\alpha \le x \le \alpha.$$
It follows from (11.68) and (11.69) that
$$0 < Q(x) < 2a + 2a\cos^2 x \le 4a \le 1. \qquad (11.71)$$

On the other hand, using (11.68), (11.69) and the hypothesis of contradiction (i.e., the possibility (i)), we conclude that $x = -\alpha$ and $x = \alpha$ are two consecutive zeros of the solution $z = z(t)$. Hence, using (11.71) and the comparison theorem, we have
$$2\alpha > \pi.$$
This contradicts $\alpha \le \frac{\pi}{2}$. Hence, we have proved that $\alpha_1 > \frac{\pi}{2}$.
The proof of Lemma 11.7 is thus complete. □

Lemma 11.8 *Suppose that $a \le \frac{1}{4}$ and $y = w(x) > 0$ is a solution of the differential equation (11.51) satisfying the initial condition*
$$w(0) = w_0 > v_0, \qquad w'(0) = 0,$$
then $y = w(x)$ exists on the interval $0 \le x \le \frac{\pi}{2}$ and satisfies the inequality:
$$w(x) > v(x), \qquad 0 \le x \le \frac{\pi}{2}. \qquad (11.72)$$

Proof. The proof is similar to that of Lemma 11.7. □

Lemma 11.9 *Suppose that $a \le \frac{1}{4}$. Then there exists at least a positive solution $y = w(x)$ of differential equation (1) on $0 \le x \le \frac{\pi}{2}$ satisfying the following condition*
$$w'(0) = 0, \qquad w'\left(\frac{\pi}{2}\right) < 0.$$

Proof. Let us take the solution $y = v(x)$ of (11.51) defined in Lemma 11.7. Since $v(0) > G(0)$ and $v'(0) = 0$, there exists a positive constant $\sigma < \dfrac{\pi}{2}$, such that

$$v(x) > G(x), \qquad 0 \leq x < \sigma. \tag{11.73}$$

Hence, when $0 \leq x < \sigma$, we have

$$v''(x) = \frac{1}{v(x)} - 2a\cos^2 x \cdot v(x) < \frac{1}{G(x)} - 2a\cos^2 x \cdot G(x) = 0.$$

It follows that

$$v'(x) < 0, \quad v(x) < v_0 \qquad (0 < x < \sigma). \tag{11.74}$$

Using (11.73), (11.74) together with the property

$$G'(x) > 0, \qquad \lim_{x \to \frac{\pi}{2}} G(x) = \infty,$$

we can take, without loss of generality, the above constant σ such that the relations (11.73) and

$$v(\sigma) = G(\sigma) \tag{11.75}$$

hold true.

Similarly, for the solution $y = w(x)$ of differential equation (11.51) defined in Lemma 11.8, there exists a constant $\delta < \dfrac{\pi}{2}$ such that the relations

$$w(x) > G(x), \qquad 0 \leq x < \delta, \tag{11.76}$$

and

$$w(\delta) = G(\delta)$$

hold true.

Similar to the relation (11.74), we have

$$w'(x) < 0, \qquad w(x) < w_0 \qquad (0 < x < \delta). \tag{11.77}$$

It follows from (11.72) that

$$\delta > \sigma. \tag{11.78}$$

Then let $v(x)$ and δ be fixed. The inequality (11.77) implies

$$w''(x) = \frac{1}{w(x)} - 2a\cos^2 x \cdot w(x) > -2aw_0.$$

Hence, we have
$$w'(x) > -2aw_0 x, \qquad 0 < x < \delta,$$
which yields
$$w(x) > w_0 - aw_0 x^2 > \left(1 - \frac{a\pi^2}{4}\right) w_0 > 0, \qquad 0 < x < \delta.$$
It follows that
$$w''(x) = \frac{1}{w(x)} - 2a\cos^2 x \cdot w(x)$$
$$< \frac{1}{\left(1 - \frac{a\pi^2}{4}\right) w_0} - a\left(1 - \frac{a\pi^2}{4}\right) w_0 (1 + \cos 2x).$$
Integrating this inequality from 0 to $\delta > 0$, we have
$$w'(\delta) < \frac{\delta}{\left(1 - \frac{a\pi^2}{4}\right) w_0} - a\left(1 - \frac{a\pi^2}{4}\right) \left(\delta + \frac{1}{2}\sin 2\delta\right) w_0. \qquad (11.79)$$
It follows from $\sin 2\delta > 0$ and (11.78) that
$$w'(\delta) < \frac{\pi}{\left(1 - \frac{a\pi^2}{4}\right) w_0} - a\left(1 - \frac{a\pi^2}{4}\right) a\sigma w_0,$$
which implies
$$w'(\delta) < -\sqrt{\frac{a}{2}}\pi, \qquad (11.80)$$
provided that
$$w_0 > \kappa = \max\left\{v_0, \; \frac{\pi + \sqrt{\pi^2 + 8\sigma\pi}}{\sqrt{8a} \cdot \sigma} \cdot \frac{4}{4 - a\pi^2}\right\}.$$
It is noted that the above-defined constant $\kappa > 0$ is independent of the solution $y = w(x)$. It follows from (11.72) that
$$w(x) > \frac{1}{\sqrt{2a}}, \qquad 0 \le x \le \frac{\pi}{2},$$
which yields
$$w''(x) = \frac{1}{w(x)} - 2a\cos^2 x \cdot w(x) \le \frac{1}{w(x)} < \sqrt{2a}, \qquad 0 \le x \le \frac{\pi}{2}.$$

Integrating this inequality from δ to $\frac{\pi}{2}$, we obtain

$$w'\left(\frac{\pi}{2}\right) < w'(\delta) + \sqrt{2a}\left(\frac{\pi}{2} - \delta\right) < w'(\delta) + \sqrt{\frac{a}{2}} \cdot \pi,$$

which together with (30) implies $w'\left(\frac{\pi}{2}\right) < 0$.

The proof of Lemma 11.9 is thus completed. □

11.5.3 Main Theorem

Finally, we are in position to prove the following result.

Theorem 11.6 *When $0 < a \leq \frac{1}{4}$, the boundary-value problem (11.51)+(11.52) possesses a positive solution $y = y(x)$ on the interval $0 \leq x \leq \pi$.*

Proof. Let $y = \varphi(x; c)$ be the solution of the differential equation (11.51) satisfying the initial condition

$$y(0) = c, \qquad y'(0) = 0.$$

It follows from Lemmas 11.6 and 11.7 that $y = \varphi(x; c)$ exists on the interval $0 \leq x \leq \frac{\pi}{2}$ for $c \geq \frac{1}{\sqrt{2a}}$. Then, we have

$$H(c) = \varphi'\left(\frac{\pi}{2}; c\right), \qquad c \geq \frac{1}{\sqrt{2a}}.$$

Using the continuous dependence of solution on the initial values implies that $H(c)$ is a continuous function. Moreover, Lemmas 11.6 and 11.9 imply that there are constants

$$c_1 = u_0, \qquad c_2 = w_0 > u_0$$

such that

$$H(c_1) > 0 \quad \text{and} \quad H(c_2) < 0.$$

It follows that there is a constant $c_0 \in (c_1, c_2)$ satisfying

$$H(c_0) = 0.$$

Therefore, the solution $y = \varphi(x; c_0)$ satisfies the boundary condition (11.53). It follows from Lemma 11.5 that $y = \varphi(x; c_0)$ is a positive solution of (11.51)+(11.52).

Theorem 11.6 is thus proved. □

Remark 11.1 *The method used above is valid for the general equation*

$$\frac{d^2y}{dx^2} + a[F(x)]y = \frac{1}{y} + \frac{\varepsilon}{y^3},$$

where $\varepsilon > 0$ is a small parameter, and $F(x) \geq 0$ is π-periodic with the property that $F(x) = F(\pi - x) \neq 0$ for $x \neq \frac{\pi}{2}$.

11.6 A Retarded Equation

11.6.1 The Bernfeld-Haddock Conjecture

In order to explain the problem considered, we quote the following passage from a paper written by S. Bernfeld and J. Haddock [7]:

> "······ Thus, this leads to an interesting question: What happens if a functional differential equation has an ordinary and functional part, but the ordinary part does not dominate the functional part? As a starting point, what can be said about the equation
>
> $$x'(t) = -x^\theta(t) + x^\theta(t-r) \qquad (11.81)$$
>
> where $r > 0$ is a constant retard and $\theta > 0$ is the quotient of odd integers? It is easy to see that for any such θ, any constant function is a solution. For $\theta \geq 1$, one can often construct a Liapunov functional and employ an invariant principle to show that each solution tends to a constant as $t \to \infty$. Also a clever result of Cooke and Yorke [22] can be employed to obtain this result. However, for $\theta < 1$ neither technique seems to work. After having made several attempts at examining the case $\theta < 1$, we are led to the following conjecture.
>
> **Conjecture.** *Each solution for the scalar equation*
>
> $$x'(t) = -x^{\frac{1}{3}}(t) + x^{\frac{1}{3}}(t-r) \qquad (11.82)$$
>
> *tends to a constant as $t \to \infty$.*"

In the paper [33], the author gave a positive response to this conjecture. In fact, more general equation is considered there as mentioned below.

11.6.2 Properties of the Retarded Equation

Before proving the conjecture, we analyze the properties of some retarded differential equations. In what follows, assume the functions $F(u)$ and $G(u)$ are continuous in $u \in \mathbb{R}^1$, and moreover $F(u)$ is monotone increasing with respect to $u \in \mathbb{R}^1$.

Proposition 11.1 *Consider the ordinary differential equation*

$$u' = -F(u) + K \qquad \left(' = \frac{d}{dt}\right), \qquad (11.83)$$

where K is a constant, and the initial condition

$$u(t_0) = u_0. \qquad (11.84)$$

Then the solution of Cauchy problem (11.83)+(11.84) exists uniquely [3] *on the infinite interval $[t_0, \infty)$.*

Proof. It follows from the Peano's existence theorem that the Cauchy problem (11.83)+(11.84) has a local solution $u = u(t)$. Assume its maximal existence right-interval is $[t_0, h)$.

We want to prove $[t_0, h) = [t_0, \infty)$ (i.e., $h = \infty$).

In fact, assume first $-F(u) + K > 0$ for $u \in \mathbb{R}^1$. Using

$$u'(t) = -F(u(t)) + K > 0, \qquad t \in [t_0, t_0 + h),$$

we conclude that $u(t) \geq u(t_0)$ for $t \in [t_0, t_0 + h)$. Then, the inequality

$$u'(t) - u'(t_0) = F(u(t_0)) - F(u(t)) \leq 0$$

on the interval $[t_0, t]$ ($t_0 < t < t_0 + h$) implies

$$u(t_0) \leq u(t) \leq u(t_0) + u'(t_0)(t - t_0), \qquad t \in [t_0, t_0 + h).$$

It follows that the solution $u = u(t)$ is bounded on $[t_0, t_0 + h)$. Employing the continuation theorem of solution, we conclude that $h = \infty$.

Then, assume $-F(u) + K < 0$ for $u \in \mathbb{R}^1$. In a similar manner, we can prove the conclusion $h = \infty$.

---
[3] A correction is published in the recent paper [121].

Finally, assume that $-F(u)+K$ changes its sign. Then there is a unique constant α, such that

$$\begin{cases} -F(u) + K < 0, & \text{for } u > \alpha; \\ -F(u) + K = 0, & \text{for } u = \alpha; \\ -F(u) + K > 0, & \text{for } u < \alpha. \end{cases}$$

If $u_0 = \alpha$, then $u = u(t) \equiv \alpha$ exists on $[t_0, \infty)$. It follows that $h = \infty$.

If $u_0 < \alpha$, then the right uniqueness of solution guarantees that $u(t) \leq \alpha$ for $t \in [t_0, t_0 + h)$. Hence, we have

$$u'(t) = -F(u(t)) + F(\alpha) \geq 0, \qquad t \in [t_0, t_0 + h).$$

It follows that $u = u(t)$ is a monotone increasing function with an upper bound α on the interval $[t_0, h)$. The continuation theorem of solution yields the conclusion $h = \infty$.

If $u_0 > \alpha$, we reach the same conclusion in a similar manner.

The proof of Proposition 11.1 is thus complete. □

Proposition 11.2 *Consider the ordinary differential equation*

$$u' = -F(u) + g(t), \tag{11.85}$$

where $g(t)$ is a continuous function on $t_0 \leq t \leq t_0 + r$, and the initial condition

$$u(t_0) = u_0. \tag{11.86}$$

Then solution of the Cauchy problem (11.85)+(11.86) exists uniquely on the interval $t_0 \leq t \leq t_0 + r$.

Proof. Let K_1 and K_2 be two constants such that

$$K_1 \leq g(t) \leq K_2, \qquad t \in [t_0, t_0 + r].$$

Then Proposition 11.2 can be easily proved by using Proposition 11.1 together with the usual technique of differential inequalities. □

Proposition 11.3 *Let $u = u(t; t_0, K)$ be the solution of Cauchy problem (11.83)+(11.84), where $t_0 \in \mathbb{R}^1$ and $K \in \mathbb{R}^1$. Then, for any given constant $\alpha > 0$, the function $\varphi(K) := u(t_0 + \alpha; t_0, K)$ is independent of t_0, and continuous in K.*

Proof. Since $u(t; t_0, K) = u(t - t_0; 0, K)$, we have
$$\varphi(K) = u(t_0 + \alpha; t_0, K) = u(\alpha; 0, K),$$
which is independent of t_0.

Now, we are going to prove that $\varphi(K)$ is continuous in K.

Assume the contrary. Then $\varphi(K)$ is discontinuous at some point $K = K_0$. It follows that there is a constant $\varepsilon_0 > 0$ and a sequence $\{K_n\}$ tending to K_0 as $n \to \infty$ such that
$$|\varphi(K_n) - \varphi(K_0)| \geq \varepsilon_0, \qquad \text{as } n \to \infty;$$
that is,
$$|u(t_0 + \alpha; t_0, K_n) - u(t_0 + \alpha; t_0, K_0)| \geq \varepsilon_0 \qquad \text{as } n \to \infty. \quad (11.87)$$

Without loss of generality, assume that $K_0 - 1 \leq K_n \leq K_0 + 1$. Using the comparison theorem yields
$$u(t; t_0, K_0 - 1) \leq u(t; t_0, K_n) \leq u(t; t_0, K_0 + 1), \qquad t \in [t_0, \infty).$$

It follows that the sequence of functions $\{u(t; t_0, K_n)\}$ is uniformly bounded on the compact interval $[t_0, t_0 + \alpha]$ with respect to K_n. Then, using (11.83), we conclude that $\{u'(t; t_0, K_n)\}$ is also uniformly bounded on $[t_0, t_0 + \alpha]$. Therefore, $\{u(t; t_0, K_n)\}$ is an equi-continuous and bounded sequence in $t \in [t_0, t_0 + \alpha]$. By employing Ascoli's lemma, the existence of a uniformly convergent subsequence is established, which we might well still denote as $\{u(t; t_0, K_n)\}$. Let
$$v(t) = \lim_{n \to \infty} u(t; t_0, K_n), \qquad t \in [t_0, t_0 + \alpha].$$

Then, the uniform convergence implies that $u = v(t)$ is a solution of the initial problem (11.83)+(11.84) with $K = K_0$. Letting $n \to \infty$ in (11.87), we obtain
$$|v(t_0 + \alpha) - u(t_0 + \alpha; t_0, K_0)| \geq \varepsilon_0 > 0. \quad (11.88)$$

On the other hand, by definition, $u = u(t; t_0, K_0)$ is a solution of of (11.83)+(11.84) with $K = K_0$. Then, from the uniqueness of solution, we have
$$v(t) = u(t; t_0, K_0), \qquad t \in [t_0, t_0 + \alpha],$$
which contradicts the inequality (11.88).

The proof of Proposition 11.3 is thus completed. $\square$

Proposition 11.4 *Consider the ordinary differential equation*

$$u' = -F(u) + F(C + \varepsilon), \qquad (11.89)$$

where C is a constant and ε is a parameter ($0 \le \varepsilon \le 1$), and the initial condition

$$u(t_0) = u_0, \qquad \text{with } u_0 < C. \qquad (11.90)$$

Let $u = u(t; t_0, \varepsilon)$ be the solution of the Cauchy problem (11.89)+(11.90), and $\alpha > 0$ be a given constant. Then there is a positive constant μ independent of t_0 and ε such that

$$(C + \alpha) - u(t; t_0, \varepsilon) \ge \mu > 0, \qquad t \in [t_0, t_0 + \alpha].$$

Proof. eqnarray Using Proposition 11.3 with $K = F(C+\varepsilon)$, we conclude that

$$\psi(\varepsilon) = (C + \varepsilon) - u(t_0 + \alpha; t_0, \varepsilon)$$

is a continuous function of ε, which is independent of t_0. Note that $u(t_0; t_0, \varepsilon) = u_0 < C + \varepsilon$, and that both $u = u(t; t_0, \varepsilon)$ and $u = C + \varepsilon$ are solutions of (11.89). It follows from the uniqueness theorem of solution that

$$u(t; t_0, \varepsilon) < C + \varepsilon, \qquad t \in [t_0, \infty), \qquad (11.91)$$

which implies $\psi(\varepsilon) > 0$. Hence, the minimal value

$$\mu = \min_{0 \le \varepsilon \le 1} \psi(\varepsilon)$$

is positive and independent of t_0 and ε.

On the other hand, from (11.89) and (11.91) we get $u'(t; t_0, \varepsilon) > 0$, so that

$$u(t; t_0, \varepsilon) \le u(t_0 + \alpha; t_0, \varepsilon), \qquad t \in [t_0, t_0 + \alpha].$$

It follows that

$$(C + \varepsilon) - u(t; t_0, \varepsilon) \ge \psi(\varepsilon) \ge \mu, \qquad t \in [t_0, t_0 + \alpha].$$

This completes the proof of Proposition 11.4. □

Proposition 11.5 *Consider the ordinary differential equation*

$$u' = -F(u) + F(C - \varepsilon), \qquad (11.92)$$

where C is a constant and ε is a parameter $(0 \leq \varepsilon \leq 1)$, and the initial condition

$$u(t_0) = u_0, \qquad \text{with } u_0 > C. \tag{11.93}$$

Let $u = u(t; t_0, \varepsilon)$ be the solution of the Cauchy problem (11.92)+(11.93), and $\alpha > 0$ be a given constant. Then there is a positive constant ν independent of t_0 and ε such that

$$u(t; t_0, \varepsilon) - (C - \varepsilon) \geq \nu > 0, \qquad t \in [t_0, t_0 + \alpha].$$

Proof. The proof is much the same as that of Proposition 11.4, and is thus omitted. □

Now, we are ready to analyze the retarded differential equation

$$x'(t) = -F(x(t)) + G(x(t-r)), \tag{11.94}$$

where $r > 0$ is a constant lag. Let us consider the initial condition

$$x(t) = f(t), \qquad t \in [-r, 0], \tag{11.95}$$

where $f(t)$ is a given continuous function in the interval $[-r, 0]$.

Proposition 11.6 *The initial-value problem (11.94)+(11.95) has a unique solution on the interval $[-r, \infty)$.*

Proof. This proposition is an immediate consequence of Proposition 11.2, where $g(t) = f(t)$ on $[-r, 0]$, and then, by induction, let $g(t) = G(x(t-r))$ on the interval $[nr, (n+1)r]$ for $n = 0, 1, 2, \cdots$. □

In the subsequent, the properties of (11.94) will be analyzed in more details.

Lemma 11.10 *If $F(u) \geq G(u)$, then none of the solutions of (11.94) can be monotone increasing on the interval $[a, a + 2r]$, where a is any constant.*

Proof. Assume it were not the case. Then there is at least a solution $x = x(t)$ of (11.94), which is monotone increasing on $[a, a + 2r]$ for some constant a. It follows that

$$x(t - r) < x(t), \qquad t \in [a + r, a + 2r],$$

which together with the hypothesis $F(u) \geq G(u)$ implies

$$G(x(t-r)) \leq F(x(t-r)) < F(x(t)), \qquad t \in [a+r, a+2r].$$

Hence, using (11.94), we have

$$x'(t) = -F(x(t)) + G(x(t-r)) < 0, \qquad t \in [a+r, a+2r],$$

which implies that $x = x(t)$ is monotone decreasing on $[a+r, a+2r]$. This is a contradiction.

Lemma 11.10 is thus proved by contradiction. □

Lemma 11.11 *If $F(u) \leq G(u)$, then none of the solutions of (11.94) can be monotone decreasing on the interval $[a, a+2r]$, where a is any constant.*

Proof. The proof is similar to that of Lemma 11.10. □

In the following, let $x = x(t)$ be any solution of the retarded differential equation (11.94), and let $I_n = [nr, (n+1)r]$ for $n = 0, 1, 2, \cdots$. Since $x = x(t)$ is a continuous function on $[-r, \infty)$, we have constants α_n and β_n in I_n, such that

$$\begin{cases} A_n = \max_{t \in I_n} x(t) = x(\alpha_n) \\ B_n = \min_{t \in I_n} x(t) = x(\beta_n) \end{cases} \quad (n = 0, 1, 2, \cdots).$$

Lemma 11.12 *If $F(u) \geq G(u)$ and there is a point $\xi \in I_{n+1}$ such that $x'(\xi) = 0$, then $x(\xi) \leq A_n$.*

Proof. Since $x'(\xi) = 0$, we obtain from (11.94) that $F(x(\xi)) = G(x(\xi - r))$, which implies $F(x(\xi)) = G(x(\xi - r)) \leq F(x(\xi - r))$. Since $F(u)$ is monotone increasing, $x(\xi) \leq x(\xi - r)$. It follows from $\xi - r \in I_n$ and $x(\xi - r) \leq A_n$ that $x(\xi) \leq A_n$.

Lemma 11.12 is thus proved. □

Lemma 11.13 *If $F(u) \leq G(u)$ and there is a point $\xi \in I_{n+1}$ such that $x'(\xi) = 0$, then $x(\xi) \geq B_n$.*

Proof. The proof is omitted since it is similar to that of Lemma 11.12. □

Lemma 11.14 *If $F(u) \geq G(u)$ and there is an integer $m \geq 0$ such that $A_m < A_{m+1}$, then $A_{m+1} < A_{m+2}$ and $x = x(t)$ is monotone increasing on the interval I_{m+2}.*

Proof. Since $x(\alpha_{m+1}) = A_{m+1}$ is the maximal value of $x(t)$ on I_{m+1}, $x(\alpha_{m+1})$ is also the maximal value of $x(t)$ on $I_m \cup I_{m+1}$.

If $(m+1)r \leq \alpha_{m+1} < (m+2)r$, then α_{m+1} is an interior point of $I_m \cup I_{m+1}$. It follows that $x'(\alpha_{m+1}) = 0$. Then Lemma 11.12 yields $x(\alpha_{m+1}) \leq A_{m+1}$, namely, $A_{m+1} \leq A_m$. This inequality is in conflict with $A_m < A_{m+1}$.

We have thus proved that $\alpha_{m+1} = (m+2)r$. In other words, $x = x(t)$ attains its maximal value on I_{m+1} at the right-hand point of I_{m+1}. Hence, we have $x'(\alpha_{m+1}) = x'((m+2)r) \geq 0$. If $x'(\alpha_{m+1}) = 0$, then Lemma 11.12 implies that $x(\alpha_{m+1}) \leq A_m$ (i.e., $A_{m+1} \leq A_m$). This is in conflict with the assumption $A_m < A_{m+1}$. It follows that $x'(\alpha_{m+1}) = x'((m+2)r) > 0$. Then it is obviuous that a point ξ can be found in I_{m+2} such that $x(\xi) > x((m+2)r) = x(\alpha_{m+1}) = A_{m+1}$. Hence we have proved that $A_{m+2} \geq x(\xi) > A_{m+1}$ (i.e., $A_{m+1} < A_{m+2}$).

Finally, we have to prove that $x = x(t)$ is monotone increasing on I_{m+2}. In fact, since $x'((m+2)r) > 0$, there exists $\sigma > (m+2)r$ such that

$$x'(t) > 0, \qquad t \in [(m+2)r, \sigma), \tag{11.96}$$

and $x'(\sigma) = 0$ for $\sigma < \infty$. If $\sigma < (m+3)r$, then $\sigma \in I_{m+2}$ and $x(\sigma) > x((m+2)r) = A_{m+1}$. On the other hand, by Lemma 11.12, $x'(\sigma) = 0$ implies $x(\sigma) \leq A_{m+1}$. This leads to a contradiction. Hence we have $\sigma > (m+3)r$, and the inequality (11.96) ensure that $x = x(t)$ is monotone increasing on I_{m+2}.

The proof of Lemma 11.14 is now completed. $\square$

Lemma 11.15 *If $F(u) \leq G(u)$ and there is an integer $m \geq 0$ such that $B_m > B_{m+1}$, then $B_{m+1} > B_{m+2}$ and $x = x(t)$ is monotone decreasing on the interval I_{m+2}.*

Proof. The proof is similar to that of Lemma 11.14. $\square$

11.6.3 Answer to the Conjecture

With the above preliminaries, we are ready to prove the following theorems, which give directly the answer of the Bernfeld-Haddock's conjecture.

Theorem 11.7 *If $F(u) \geq G(u)$, then the solution $x = x(t)$ of the retarded differential equation (11.94) tends to either $-\infty$ or a constant as $t \to \infty$.*

Proof. We shall first prove

$$A_0 \geq A_1 \geq \cdots \geq A_n \geq A_{n+1} \geq \cdots\cdots . \tag{11.97}$$

Assume the contrary. Then there is at least an integer $m \geq 0$ such that $A_m < A_{m+1}$. Hence, by using Lemma 11.14, we conclude that $A_{m+1} <$

A_{m+2} and $x = x(t)$ is monotone increasing on I_{m+2}. Furthermore, the inequality $A_{m+1} < A_{m+2}$ together with Lemma 11.14 implies in turn that $A_{m+2} < A_{m+3}$ and $x = x(t)$ is monotone increasing on I_{m+3}. We have thus deduced that $x = x(t)$ is monotone increasing on $I_{m+2} \cup I_{m+3}$. This conclusion is incompactible with Lemma 11.10. The inequality (11.97) is thus proved by contradiction. It is apparent that (11.97) implies

$$A = \limsup_{t \to \infty} x(t) = \lim_{n \to \infty} A_n < \infty. \tag{11.98}$$

If $A = -\infty$, then we naturally obtain

$$\lim_{t \to \infty} x(t) = -\infty.$$

Suppose now that $A > -\infty$ (i.e., A is a constant). Let

$$B = \liminf_{t \to \infty} x(t).$$

It is obvious that $B \leq A$.

Then the proof of the theorem is reduced to that of the equality $B = A$. Otherwise, we have $B < A$. Then there is a constant H satisfying $B < H < A$. According to the definitions of A and B, for any given positive integer N, there are arbitrarily large integers $n > N$ and numbers $\tau_n \in I_n$ such that $x(\tau_n) = H$. Since

$$[\tau_n, \tau_n + 2r] \subset I_n \cup I_{n+1} \cup I_{n+2},$$

we have

$$[\tau_n - r, \tau_n + r] \subset I_{n-1} \cup I_n \cup I_{n+1},$$

which together with $A_{n-1} \geq A_n \geq A_{n+1}$ yields

$$x(t - r) \leq A_{n-1}, \quad t \in [\tau_n, \tau_n + 2r]. \tag{11.99}$$

Let $A_{n-1} = A + \varepsilon_n$. We can assume $0 \leq \varepsilon_n \leq 1$ in view of (11.98). It follows from (11.99) and the hypotheses of the theorem that

$$x'(t) \leq -F(x(t)) + F(A + \varepsilon_n), \quad t \in [\tau_n, \tau_n + 2r]. \tag{11.100}$$

Denote by $u = u(t; \tau_n, \varepsilon_n)$ the solution of the initial-value problem

$$x'(t) = -F(x(t)) + F(A + \varepsilon_n), \quad u(\tau_n) = H. \tag{11.101}$$

Note that $H < A$. Then, from Proposition 11.4, we have

$$(A + \varepsilon_n) - u(t; \tau_n, \varepsilon_n) \geq \mu, \quad t \in [\tau_n, \tau_n + 2r], \tag{11.102}$$

where μ is a positive constant independent of τ_n and ε_n. By employing the usual technique of differential inequality, we obtain from (11.100) and (11.101) that

$$x(t) \leq u(t; \tau_n, \varepsilon_n), \qquad t \in [\tau_n, \tau_n + 2r].$$

It follows from (11.102) that

$$A_{n-1} - x(t) \geq \mu, \qquad t \in [\tau_n, \tau_n + 2r]. \tag{11.103}$$

Note that $\alpha_{n+1} \in [\tau_n, \tau_n + 2r]$. A substitution of $t = \alpha_{n+1}$ into (11.103) yields

$$A_{n-1} - A_{n+1} \geq \mu > 0, \tag{11.104}$$

for some arbitrarily large integers n. The inequality (11.104) contradicts (11.98).

The proof of Theorem 11.7 is thus completed. □

Remark 11.2 *We cannot exclude, in general, the possibility*

$$\lim_{t \to \infty} x(t) = -\infty.$$

For example, consider the following retarded differential equation

$$x'(t) = -(1 + e^{x(t)}) + \frac{1}{1 + x^2(t-r)}. \tag{11.105}$$

Here we have

$$F(u) = 1 + e^u, \qquad G(u) = \frac{1}{1+u^2},$$

and $F(u) \geq G(u)$. It follows from (11.105) that

$$x'(t) \leq -(1 + e^{x(t)}) + 1 = -e^{x(t)}.$$

Then

$$x(t) \leq -\log[(t - t_0) + e^{-x(t_0)}]. \qquad t \geq t_0,$$

which implies that each solution of (11.105) tends to $-\infty$ as $t \to \infty$.

Theorem 11.8 *If $F(u) \leq G(u)$, then the solution $x = x(t)$ of the retarded differential equation (11.94) tends to either $+\infty$ or a constant as $t \to \infty$.*

Proof. This theorem can be proved by the similar method as in the proof of Theorem 11.7. □

Remark 11.3 We cannot exclude, in general, the possibility

$$\lim_{t\to\infty} x(t) = \infty.$$

For example, consider the following retarded differential equation

$$x'(t) = \frac{-x(t)}{1+x^2(t)} + 2e^{x^2(t-r)}$$

is a case in point.

Remark 11.4 If either $F(u) \le G(u)$ or $F(u) \ge G(u)$, then the retarded differential equation (11.94) has no nontrivial periodic solution.

Theorem 11.9 Consider the retarded differential equaion

$$x'(t) = -F(x(t)) + F(x(t-r)). \tag{11.106}$$

Then each solution $x = x(t)$ of (11.106) tends to a constant as $t \to \infty$.

Proof. The conclusion of the theorem is an immediate consequence of Theorem 11.7 and Theorem 11.8. □

Corollary 11.1 Each solution of the retarded differential equation (11.81) tends to a constant as $t \to \infty$. In particular, the above Bernfeld-Haddock's conjecture holds true.

11.7 Periodic Lotka-Volterra System

11.7.1 Introduction

In 1989, A. Hausrath and R. Manasevich [73] considered the Lotka-Volterrra system with periodical perturbations

$$\begin{cases} \dfrac{dp}{dt} = (1+h_1(t))p - (1+h_3(t))pq, \\ \dfrac{dq}{dt} = -\mu(1+h_2(t))q + \mu(1+h_4(t))pq, \end{cases} \quad (p>0,\ q>0), \tag{11.107}$$

where $\mu > 0$ is a parameter and $h_i(t)$ ($i = 1, \cdots, 4$) are continuous periodic functions with minimal period $T > 0$. Using the method of generalized Poincaré-Birkhoff twist theorem, they proved that the system 11.107) has at least $2n$ periodic solutions of minimal period T under the conditions:

i) $\quad \|h_i\| = \sup_{t \in \mathbb{R}^1} |h_i(t)| \leq h_0 \ll 1, \quad (i = 1, \cdots, 4);$

ii) $\quad 2n\pi < \sqrt{uT} < 2(n+1)\pi,$

together with some other conditions.

In the paper [61], the authors studied the general periodic Lotka-Volterra system

$$\begin{cases} \dfrac{dp}{dt} = H_1(t)p - H_3(t)pq, \\ \dfrac{dq}{dt} = -H_2(t)q + H_4(t)pq, \end{cases} \quad (p > 0, \ q > 0), \quad (11.108)$$

where the coefficient functions $H_1(t), \cdots, H_4(t)$ are positive, continuous and periodic with minimal period $T > 0$. As defined before, the T-periodic solution of (11.108) is termed to be *harmonic*, and the nT-periodic solution of (11.108) is termed to be *subharmonic of order n* if nT is the minimal period of the solution, where $n > 1$ is an integer. The following two theorems are the main result of [61].

Theorem 11.10 *The general periodic Lotka-Volterra system (11.108) has at least one harmonic solution.*

Theorem 11.11 *There is an integer $N > 1$ such that for any given integer $n \geq N$, the periodic Lotka-Volterra system (11.108) possesses at least $2S_n$ subharmonic solutions of order n, with the property that*

$$1 \leq S_N \leq \cdots \leq S_n \leq S_{n+1} \leq \cdots \quad (\to \infty).$$

Before proving the above results, we give the following example, which shows that Theorem 11.10 can not be improved in general case; and the integer N in Theorem 11.11 may be larger than 2, such that there exists no subharmonic solution of order m ($2 \leq m \leq N - 1$).

Example 11.1 Letting

$$H(t) = H_i(t) := 1 + \frac{1}{2}\cos\frac{2\pi}{T}t \quad (i = 1, \cdots, 4),$$

and $p = e^x$ and $q = e^y$ in (11.108), we have

$$\begin{cases} \dfrac{dx}{dt} = H(t)(1 - e^y), \\ \dfrac{dy}{dt} = -H(t)(1 - e^x). \end{cases} \quad (11.109)$$

Then, this system is transformed into

$$\begin{cases} \dfrac{dx}{ds} = (1 - e^y), \\ \dfrac{dy}{ds} = -(1 - e^x). \end{cases} \qquad (11.110)$$

by using the substitution

$$s = s(t) = t + \frac{T}{4\pi} \sin \frac{2\pi}{T} t.$$

It can easily be shown that the planar autonomous system (11.110) has a singular point at the origin o and all other orbits are cycles surrounding o. Hence, o is a center. Let Γ_c be the closed orbit of (11.110) passing through the initial point $(c, 0)$, and let $\tau(c) > 0$ be the minimal period of Γ_c. Using the method of [116], we can easily prove that $\tau(c)$ is continuous and monotonically increasing in c, such that

$$\lim_{c \to 0^+} \tau(c) = 2\pi, \qquad \lim_{c \to \infty} \tau(c) = \infty. \qquad (11.111)$$

On the other hand, let

$$x = \phi(s, x_0, y_0), \qquad y = \psi(s, x_0, y_0) \qquad (11.112)$$

be the solution of (11.110) satisfying the initial condition

$$x_0 = \phi(0, x_0, y_0), \qquad y_0 = \psi(0, x_0, y_0).$$

Then it is a periodic solution of minimal period $\tau(c)$ if $(x_0, y_0) \in \Gamma_c$. It is noticed that (11.112) is the trivial solution (i.e., the singular point o) when $(x_0, y_0) = o$.

Hence,

$$x = \phi(s(t), x_0, y_0), \qquad y = \psi(s(t), x_0, y_0) \qquad (11.113)$$

is the general solution of (11.109). It is obvious that if (11.113) is nT-periodic, then we have

$$s(nT) = k\tau(c), \qquad \text{whenever } (x_0, y_0) \in \Gamma_c,$$

with some positive integer k. It follows that

$$nT = k\tau(c),$$

which implies
$$T \geq \frac{1}{n}\tau(c) \geq \frac{2\pi}{n}.$$

It follows that when $n = 1$, we have $T \geq 2\pi$. In this case, if $T < 2\pi$, the system (11.109) has no nontrivial T-periodic solution and the only harmonic solution is the trivial solution $(x, y) = (0, 0)$.

In a similar manner, we conclude that if $T < \frac{2\pi}{N}$ with $N > 1$, the system (11.109) has no subharmonic solution of order m $(2 \leq m \leq N - 1)$.

11.7.2 Existence of Harmonic Solution

Proof of Theorem 11.10. Through the substitution
$$p = e^x, \qquad q = e^y,$$
the system (11.108) is transformed to the following system
$$\begin{cases} \dfrac{dx}{dt} = H_1(t) - H_3(t)e^y, \\ \dfrac{dy}{dt} = -H_2(t) + H_4(t)e^x, \end{cases} \quad (x, y) \in \mathbb{R}^2, \qquad (11.114)$$

which is evidently conservative. Denote by
$$x = x(t, x_0, y_0), \qquad y = y(t, x_0, y_0) \qquad (11.115)$$
the solution of (11.114) satisfying the initial condition
$$x_0 = x(0, x_0, y_0), \qquad y_0 = y(0, x_0, y_0).$$

The solution (11.115) exists on a local interval J depending on the initial point (x_0, y_0). However, it is not obvious that the existence interval J is the whole t-axis.

Lemma 11.16 *For any initial point (x_0, y_0), the solution (11.115) exists on the whole t-axis.*

Proof. It suffices to prove that the solution (11.115) exists on the interval $[0, \infty)$ since the existence proof on the interval $(-\infty, 0]$ is similar.

Assume the contrary. Then there is a solution $(x, y) = (x(t), y(t))$ of (11.114) exists on a finite interval $[0, a)$, such that
$$\lim_{t \to a} (x^2(t) + y^2(t)) = \infty. \qquad (11.116)$$

On the other hand, using (11.114) yields

$$\frac{dx}{dt} \le H_1(t), \quad \frac{dy}{dt} \ge -H_2(t), \quad t \in [0,a). \qquad (11.117)$$

It follows that

$$\begin{cases} x(t) \le x_0 + \displaystyle\int_0^t H_1(t)\,dt \equiv \alpha(t), \\ y(t) \ge y_0 - \displaystyle\int_0^t H_2(t)\,dt \equiv \beta(t), \end{cases} \quad t \in [0,a). \qquad (11.118)$$

Since $H_1(t)$ and $H_2(t)$ are continuous on the closed interval $[0,a]$, the above-defined functions $\alpha(t)$ and $\beta(t)$ are bounded on the interval $[0,a)$. It follows that

$$x(t) \le C_1, \quad y(t) \ge C_2, \quad \text{for } t \in [0,a), \qquad (11.119)$$

where C_1 and C_2 are constants. Using (11.114) again, we have

$$\frac{dy}{dt} \le H_4(t) e^{x(t)} \le e^{C_1} H_4(t), \quad \text{for } t \in [0,a).$$

It follows that there is a constant C_3, such that

$$y(t) \le C_3, \quad \text{for } t \in [0,a). \qquad (11.120)$$

Similarly, we have a constant C_4, such that

$$x(t) \ge C_4, \quad \text{for } t \in [0,a). \qquad (11.121)$$

It follows from (11.119), (11.120) and (11.121) that both $x(t)$ and $y(t)$ are bounded functions in $[0,a)$. However, this conclusion is in conflict with the limit (11.116).

Lemma 11.16 is thus proved by the method of contradiction. □

Lemma 11.17 *There is a constant $R_0 > 0$, such that for any solution*

$$x(t) = r(t)\cos\theta(t), \quad y(t) = r(t)\sin\theta(t) \qquad (11.122)$$

of (11.114), we have

$$\theta'(t) > 0, \quad \text{whenever } r(t) \ge R_0. \qquad (11.123)$$

Proof. It follows from (11.114) that

$$r\frac{d\theta}{dt} = [H_4(t) e^{r\cos\theta} - H_2(t)]\cos\theta$$

$$+[H_3(t)e^{r\sin\theta} - H_1(t)]\sin\theta. \tag{11.124}$$

Since $H_i(t)$ $(i = 1, \cdots, 4)$ are continuous periodic functions, we have the constants A and B, such that

$$A \leq H_i(t) \leq B \ (i = 1, \cdots, 4), \qquad \text{for } t \in \mathbb{R}^1.$$

Now, let $d_1 > 0$ and σ_1 $(0 < \sigma_1 < \frac{1}{4}\pi)$ be two constants to be determined later. Assume

$$r \geq d_1, \qquad \pi - \sigma_1 \leq \theta \leq \pi.$$

Using (11.124) yields

$$r\frac{d\theta}{dt} \geq A\cos\sigma_1 - B\sin\sigma_1 - Be^{-d_1\cos\sigma_1}$$

$$= e^{-d_1\cos\sigma_1}[(A\cos\sigma_1 - B\sin\sigma_1)e^{d_1\cos\sigma_1} - B]. \tag{11.125}$$

Then, choosing a constant σ_1, satisfying

$$0 < \sigma_1 < \min\left(\frac{1}{4}\pi, \arctan\frac{A}{B}\right),$$

which implies

$$(A\cos\sigma_1 - B\sin\sigma_1) > 0.$$

Then, take the positive constant

$$d_1 > \frac{1}{\cos\sigma_1}\log\frac{B}{A\cos\sigma_1 - B\sin\sigma_1},$$

which together with (11.125) yields

$$\theta'(t) > 0, \qquad \text{if } r(t) \geq d_1, \quad \pi - \sigma_1 \leq \theta(t) \leq \pi. \tag{11.126}$$

Similarly, we have constants $d_2 > 0$ and σ_2 $(0 < \sigma_2 < \frac{\pi}{4})$, such that

$$\theta'(t) > 0, \qquad \text{if } r(t) \geq d_2, \quad -\frac{\pi}{2} \leq \theta(t) \leq -\frac{\pi}{2} + \sigma_2. \tag{11.127}$$

Finally, it is easy to find a constant $d_3 > 0$, such that

$$\theta'(t) > 0, \tag{11.128}$$

when $r(t) \geq d_3$ and $\theta(t) \in [-\frac{\pi}{2} + \sigma_2, \pi - \sigma_1] \cup [\pi, \frac{3}{2}\pi]$. Letting

$$R_0 = \max\{d_1, d_2, d_3\},$$

then we get the desired conclusion (11.123) from (11.126), (11.127) and (11.128). □

It is clear that for any positive integer n, there is a constant $a_n > 0$, such that the solution (11.115) satisfies

$$r(t) \geq R_0, \quad \text{for } 0 \leq t \leq nT \quad \text{whenever } r(0) \geq a_n,$$

which yields the following conclusion.

Corollary 11.2 *Given a positive integer n, there is a constant $a_n > 0$, such that the solution (11.115) satisfies*

$$\theta'(t) > 0, \quad \text{for } 0 \leq t \leq nT \quad \text{whenever } r(0) \geq a_n. \quad (11.129)$$

Moreover, we have the following result.

Lemma 11.18 *Given a positive integer n, there is a constant $d_n > 0$, such that the solution (11.115) satisfies*

$$0 < \theta'(nT) - \theta'(0) < 2\pi, \quad \text{if } r(0) \geq d_n. \quad (11.130)$$

Proof. Let $a_n > 0$ be a large constant satisfying (11.129) and assume the solution (11.115) satisfies a sufficiently large initial condition $r(0) \geq d_n$, such that

$$r(t) \geq a_n, \quad \text{for } 0 \leq t \leq nT. \quad (11.131)$$

It follows from (11.129) that the solution (11.115) rotates in counter-clockwise direction around the origin during $0 \leq t \leq nT$. If it does not make any turn around the origin, we get the desired inequality (11.130). Otherwise, we have at least a time interval $[t_1, t_2]$ in $[0, nT]$, such that the motion stays in the third quadrant during $[t_1, t_2]$ satisfying

$$\theta(t_2) - \theta(t_1) = \frac{\pi}{2};$$

or we have two time intervals $[t_1, t_2]$ and $[t_3, t_4]$, such that

$$[\theta(t_2) - \theta(t_1)] + [\theta(t_4) - \theta(t_3)] = \frac{\pi}{2}.$$

It follows from (11.124) that

$$0 < a_n \theta'(t) \leq r(t)\theta'(t) \leq H_2(t) + H_1(t) \leq 2B,$$

for $t \in [t_1, t_2]$; or for $t \in [t_1, t_2] \cup [t_3, t_4]$. Using integration yields

$$\frac{\pi}{2} a_n \leq 2nTB.$$

It follows that if we take

$$a_n > \frac{4nTB}{\pi},$$

the above solution (11.115) with sufficiently large initial condition $r(0) \geq d_n$ can not rotate a turn around the origin during $[0, nT]$.

We have thus proved Lemma 11.18. □

Note that the Poincaré map f of the system (11.114) is defined by

$$(x_0, y_0) \quad \mapsto \quad (x(T, x_0, y_0), y(T, x_0, y_0)),$$

where $(x(T, x_0, y_0), y(T, x_0, y_0))$ is given by the solution (11.115) when $t = T$. It follows from Lemma (11.18) with $n = 1$ that the map f satisfies the boundary condition of Poincaré-Bohl fixed-point theorem on the disk

$$\mathfrak{D}_1: \quad x_0^2 + y_0^2 \leq d_1^2.$$

Hence, f has at least a fixed-point (ξ, η) in $\mathfrak{D}_1$. It follows from (11.115) that

$$x = x(t, \xi, \eta), \quad y = y(t, \xi, \eta),$$

is a T-periodic solution of (11.114). Hence,

$$p = e^{x(t,\xi,\eta)}, \quad q = e^{y(t,\xi,\eta)},$$

is a T-periodic (or harmonic) solution of the system (11.108).

Theorem 11.10 is thus proved. □

11.7.3 Existence of Subharmonic Solutions

Proof of Theorem 11.11. Let

$$x = u + \tilde{x}(t), \quad y = v + \tilde{y}(t),$$

where $(\tilde{x}(t), \tilde{y}(t))$ is a T-periodic solution of (11.108), which is given by the Theorem 11.10 as proved above. It follows from (11.108) that

$$\frac{du}{dt} = G_1(t)(1 - e^v), \qquad \frac{dv}{dt} = G_2(t)(e^u - 1), \qquad (11.132)$$

where the coefficient functions

$$G_1(t) = H_3(t)e^{\tilde{y}(t)} > 0, \qquad G_2(t) = H_4(t)e^{\tilde{x}(t)} > 0,$$

are T-periodic continuous in $t \in \mathbb{R}^1$. Further results of the system (11.132) can be proved as follows.

Lemma 11.19 *Let*

$$u(t) = \rho(t)\cos\phi(t), \qquad v(t) = \rho(t)\sin\phi(t), \qquad (11.133)$$

be any non-trivial solution of (11.132). Then

$$\phi'(t) > 0, \qquad \text{for } t \in \mathbb{R}^1. \qquad (11.134)$$

Proof. It follows from (11.132) that

$$\rho^2(t)\phi'(t) = G_2(t)(e^u - 1)u + G_1(t)(e^v - 1)v.$$

It can be easily proved that

$$\begin{cases} (e^u - 1)u > 0 & (u \neq 0); \\ (e^v - 1)v > 0 & (v \neq 0), \end{cases}$$

which together with $G_1(t) > 0$, $G_2(t) > 0$ implies

$$\phi'(t) > 0, \qquad \text{whenever } \rho(t) > 0.$$

Lemma 11.19 is thus proved. □

Lemma 11.20 *For any non-trivial solution (11.133), we have*

$$\lim_{t\to\infty} \phi(t) = \infty. \qquad (11.135)$$

Proof. Assume the contrary. Then we have a nontrivial solution given by (11.133), such that

$$\lim_{t\to\infty} \phi(t) = \alpha, \qquad (11.136)$$

with a constant α. Hence, given an arbitrarily small constant $\sigma > 0$, there is a constant $t_\sigma > 0$, such that the solution will stay in the fan-region

$$\Delta_\sigma: \qquad \rho > 0, \quad \alpha - \sigma < \phi < \alpha,$$

whenever $t > t_\sigma$. Since the constant $\sigma > 0$ is arbitrarily small, we can assume Δ_σ belongs to some quadrant. For definiteness, assume Δ_σ is in the first quadrant without destroying the generality since in other cases the proof is in a similar manner.

Now, consider an auxiliary autonomous system

$$\frac{du}{dt} = K_1(1 - e^v), \qquad \frac{dv}{dt} = K_2(e^u - 1), \tag{11.137}$$

where the coefficients $K_1 > 0$ and $K_2 > 0$ are constants. It is evident that system (11.137) has an integral

$$E(u, v) := K_2(e^u - u) + K_1(e^v - v) - (K_1 + K_2),$$

which defines a closed orbit

$$\Gamma_c: \qquad E(u, v) = c,$$

for any constant $c > 0$, and the degenerate orbit Γ_0 is a center of (11.137). Then, consider the derivative of $E(u, v)$ with respect to (11.132); that is,

$$\left.\frac{dE}{dt}\right|_{(11.132)} = \frac{\partial E}{\partial u}\frac{du}{dt} + \frac{\partial E}{\partial v}\frac{dv}{dt}$$

$$= (e^u - 1)(e^v - 1)[K_1 G_2(t) - K_2 G_1(t)]. \tag{11.138}$$

Since Δ_σ is in the first quadrant, we have

$$(e^u - 1)(e^v - 1) > 0, \qquad \text{for } (u, v) \in \Delta_\sigma.$$

It is obvious that there are two positive constants satisfying

$$a < G_i(t) < b \ (i = 1, 2), \qquad \text{for } t \in \mathbb{R}^1.$$

Hence, we have

$$K_1 G_2(t) - K_2 G_1(t) < K_1 b - K_2 a.$$

Now, set

$$K_1 = a, \qquad K_2 = b.$$

It follows from (11.138) that

$$\left.\frac{dE}{dt}\right|_{(11.132)} < 0, \qquad \text{for } (u, v) \in \Delta_\sigma.$$

Hence the motion of (11.132) in Δ_σ comes strictly into the above auxiliary closed orbits Γ_c and approaches directly to the origin $\mathfrak{o}$. Hence, we have

$$\lim_{t\to\infty} \rho(t) = 0,$$

which yields

$$\phi' = G_2(t)\cos^2\phi + G_1(t)\sin^2\phi + o(\rho), \qquad \text{as } \rho \to 0.$$

Then along the motion we have

$$\phi'(t) > a + o(\rho), \qquad \text{as } \rho \to 0.$$

It follows that

$$\lim_{t\to\infty} \phi(t) = \infty.$$

It contradicts the fact that the motion stays in Δ_σ forever as $t \to \infty$.
Lemma 11.20 is thus proved by the contradiction. $\square$

Since Lemma 11.18 holds for the system (11.114), it holds for the system (11.132). Therefore, we have the following result.

Lemma 11.21 *Given any integer $n > 0$, there is a constant $d_n > 0$, such that the solution (11.133) satisfies*

$$0 < \phi(nT) - \phi(0) < 2\pi, \qquad \text{if } \rho(0) \geq d_n. \tag{11.139}$$

Finally, it is evident that Theorem 11.11 is equivalent to the following theorem.

Theorem 11.12 *There is an integer $N > 1$ such that for any integer $n \geq N$, the periodic system (11.132) possesses at least $2S_n$ subharmonic solutions of order n, with the property that*

$$1 \leq S_N \leq \cdots \leq S_n \leq S_{n+1} \leq \cdots \quad (\to \infty). \tag{11.140}$$

Proof. Let C_1 be a circle centered at $\mathfrak{o}$ with radius $d_0 = 1$. Using Lemma 11.20 yields an integer $N > 1$, such that for any integer $n \geq N$, the solution (11.133) satisfies

$$\phi(nT) - \phi(0) > 4\pi, \qquad \text{when } \rho(0) = d_0 = 1.$$

Moreover, there are S_N prime integers p_i ($i = 1, \cdots, S_N$), such that

$$\phi(nT) - \phi(0) > 4p_i\pi \ (i = 1, \cdots, S_N), \qquad \text{when } \rho(0) = d_0 = 1. \tag{11.141}$$

Notice that $p_1 = 2$ and the sequence S_n has the property (11.140).

On the other hand, using Lemma 11.21, for any integer $n > 0$ we have a constant $d_n > 0$ ($d_n \to \infty$ as $n \to \infty$), such that (11.139) holds.

Consider the annulus

$$\mathcal{A}_n = \{\langle r, \theta \rangle \in \mathbb{R}^2 : \quad 1 \leq r \leq d_n, \quad \theta \in \mathbb{R}^1 \}.$$

It follows from (11.139) and (11.141) that the n-th iteration g^n is a twist map on the annulus $\mathcal{A}_n$, where g is the Poincaré map of (11.132). It is clear that g^n is analytic and area-preserving. Using the generalized Poincaré-Birkhoff twist theorem, we conclude that g^n has at least two distinct fixed points

$$\langle r_{ni}, \theta_{ni} \rangle \in \mathcal{A}_n, \qquad (i = 1, 2).$$

Correspondingly, the solution of (11.132) passing through the initial point $\langle r_{ni}, \theta_{ni} \rangle$ is nT-periodic. In a similar manner as in the proof of subharmonic solutions of Duffing equation, we can prove that the period nT of the above nT-periodic solution is minimal. Therefore, these nT-periodic solutions are subharmonic of order n.

The proof of Theorem 11.12 is thus completed. $\square$

Bibliography

[1] E. Akin, J. Auslander, K. Berg (1996), *When is a Transitive Map Chaotic?* "Conference in Ergodic Theory and Probability " (V. Berggeson, K. March, J. Rosenblatt, eds.), 25–40, Watter de Gruyter, Berlin.
[2] V. Arnold (1993), *On A. N. Kolmogorov, Golden Years of Moscow Mathematics* (History of Mathematics), Vol.**6**, 129-153, (AMS LMS).
[3] S. Aranson et al (1996), *Introduction to the Qualitative Theory of Dynamical Systems on Surfaces*, Translations of Mathematical Monographs, Vol.153.
[4] A. Bahri and H. Berestycki (1984), *Existence of Forced Oscillations for Some Nonlinear Differential Equations*, Comm. Pure Appl. Math., Vol.**37**, 403–442.
[5] J. Banks, J. Brooks, G. Cairns, G. Davis, and P. Stacey (1996), *On Devaney's Definition of Chaos*, Amer. Math. Monthly, **99**, 332–333.
[6] I. Bendixson (1901), *Sur les Courbes Défines par des Equations Differentielles*, Acta Mathematica, Vol.**24**, 1–88.
[7] S. Bernfeld and J. Haddock (1977), *A Variation of Razumikhin's Method for Retarded Functional Differntial Equations*, Non-Linear Systems and Applications (An International Conference, Ed. V. Lakshmikantham). 561–566.
[8] G. D. Birkhoff (1912), *Quelques theoremes sur les mouvements des systems dynamiques*, Bulletin de la Societe Mathematique de France, **40**, 305–323.
[9] G. D. Birkhoff (1913), *Proof of Poincaré's Geometric Theorem*, Trans. A. M. S. **14**, 14–22.
[10] G. D. Birkhoff (1925), *An Extension of Poincaré Last Geometric Theorem*, Acta Math. **47**, 297–311.
[11] G. D. Birkhoff (1927), *Dynamical Systems*, AMS Colloq. Publications. **9**, reprinted in 1966. Amer. Math. Soc., Providence.
[12] G. Birkhoff and P. Smith (1928), *Structure Analysis of Surfaces Transformations*, J. Math. Pure Appl., Vol.**7**, 9:345–379.
[13] S. Bochner and T. Martin (1948), *Several Complex Variables*, Princeton University Press.
[14] H. Bohr (1947), *Almost Periodic Functions*, Chelsea, New York.
[15] M. Brown (1984), *A New Proof of Brouwer's Lemma on Translation Arcs*, Houston Journal of Math., **10**, 35–41.

[16] M. Brown and W. Neumann (1977), *Proof of the Poincaré-Birkhoff Fixed Point Theorem*, Michigan Math. J., Vol.24, 21–31.

[17] A. Capietto, J. Mawhin and F. Zanolin (1990), *A Continuation Approach to Superlinear Periodic Boundary Value Problems*, J. Differential Equations, Vol.**88**, 347–395.

[18] M. Cartwright and J. Littlewood (1947), *On Nonlinear Differential Equations of the Second Order*, Annals of Math. (2) **48**, 472–494.

[19] L. Cesari (1947), *Nonlinear Analysis*, 'A Collection of Papers in Honor of E. H. Rothe' (L. Cesari et al, ed.).

[20] L. Cesari (1978), *Nonlinear Problems across a Point of Resonance for Non-Self-Adjoint Systems*, Nonlinear Analysis (Eds., L. Cesari, R. Kannan and H. Weinberger).

[21] E. Coddington and N. Levinson (1955), *Theory of Ordinary Differential Equations*, McGraw-Hill, New York.

[22] K. Cooke and J. Yorke (1973), *Some Equations Modelling Growth Process and Gonorrhea Epidimies*, Math. Biosci., Vol.**16**, 75–101.

[23] R. Devaney (1986), *An Introduction to Chaotic Dynamical Systems*, Benjaming/Cummings; Menlo Park, CA.

[24] T. Ding (1957), *An Oscillation Theorem for a Differential System of Fourth Order*, Acta Sci. Nat. Univ. Pekin., No.**3**, 317–320.

[25] T. Ding (1963), *An Example resolving the Reeb's Problem* (in Chinese), Acta Scientiarum Naturalium: Universitatis Pekinensis, no. 3, 139–142.

[26] T. Ding (1965), *A Boundary Value Problem for the Periodic Brillouin Focusing System*, Ibid, No.**1**, 31–38.

[27] T. Ding (1980), *A Necessary and Sufficient Condition for Convergence of Gear's Method for Numerical Initial Value Problems*, Acta Appl. Math. Sinica, Vol.**3**, 293–300.

[28] T. Ding (1980), *A Necessary and Suffificient Condition for Convergence of Solutions of the Gear's Difference Method* (in Chinese), Acta of Applied Mathematics, Vol.**3**, 293–300.

[29] T. Ding (1981), *Fundamentals of Oridinary Differential Equations*, Shanghai Press of Science and Technology.

[30] T. Ding (1981), *An Answer to the Birkhoff's Conjecture*, Acta Math. Sinica, Vol.**24**, 64–68.

[31] T. Ding (1981), *Some Fixed-Point Theorems and Periodically Perturbed Non-dissipative Systems*, Chin. Ann. of Math., Vol.**2**, 281–300.

[32] T. Ding (1981), *Existence of Forced Periodic Solutions of High Frequency with Small or Large Amplitude*, Chin. Ann. of Math., Vol.**2**, 93–104, (English Issue).

[33] T. Ding (1981), *Asymptotic Behavior of Solutions of Some Retarded Differential Equations*, Scientia Sinica (Series A), Vol.**25**, 263-371.

[34] T. Ding (1982), *Nonlinear Oscillations at a Point of Resonance*, Scientia Sinica (Series A), Vol.**25**, 918–931.

[35] T. Ding (1982), *An Infinite Class of Periodic Solutions of Periodically Perturbed Duffing Equation at Resonance*, Proc. Amer. Math. Soc., Vol.**86**, 47–54.

[36] T. Ding (1983), *Unbounded Perturbations of Forced Harmonic Oscillations at Resonance*, Ibid, Vol.**88**, 59–66.
[37] T. Ding (1984), *Unbounded Solutions of Conservative Oscillations under Roughly Periodic Perturbations*, Chin. Ann. of Math. (Series B), Vol.**5**, 687–694.
[38] T. Ding (1986), *Boundedness of Solutions of Duffing's Equation*, Journal of Differential Equations, Vol.**61**, 178–207.
[39] T. Ding (1987), *Some Problems in Nonlinear Oscillations*, Special Issue in Mathematics, Nanjing University, 1–7.
[40] T. Ding (1988), *A Response to Birkhoff's Conjecture on Recurrent Motions of Analytic Dynamical Systems* (English Translation of [30]), Atti del Seminaro Matematico e Fisco dell' Univ. di Modena, Vol.**36**, 273–280.
[41] T. Ding (1989), *An Extension of the Massera Theorem*, Acta Math. Sinica (New Series), Vol.**5**, 159–164.
[42] T. Ding (1989), *On Subharmonic Solutions of a Special Class of Sublinear Duffing's Equations*, Acta Applied Math. Sinica, Vol.**12**, 449–455.
[43] T. Ding (1991), *On the Morse Conjecture of Metric Transitivity*, Scientia Sinica (Series A), Vol.**34**, 138–146.
[44] T. Ding (1991), *Topological Transitivity and Metric Transitivity on* $\mathbb{T}^2$, Advanced Series in Dynamical Systems (Ed. Shiraiwa), World Scientific, Vol.**9**, 65–71.
[45] T. Ding (1991), *Topological Transitivity and Metric Transitivity on* $\mathbb{T}^2$, Advanced Series in Dynamical Systems, (Ed. K. Shiraiwa), World Scientific, Vol.**9**, 65–71.
[46] T. Ding (1992), *An Existence Theorem for Harmonic Solutions of Periodically Perturbed Systems of Duffing's Type*, Acta Sci. Nat. Uni. Pekin., Vol.**28**, 71–78.
[47] T. Ding (1994), *Non-existence Theorem of Homoclinic Solutions for Some Dissipative Duffing's Equations*, Proceedings of International Conference on *Dynamical Systems and Chaos*, (Ed. N. Aoki, K. Shiraiwa and Y. Takahashi), Vol.**1**, 28–34.
[48] T. Ding (1999), *An Ergodic Theorem for Flows on Closed Surface*, Nonlinear Analysis (TMA), Vol.**35**, 669–676.
[49] T. Ding (2002), *On the Poincaré-Hopf's Index Formula of Singular Points*, Advances of Mathematics (in China), Vol.**31**, 543–548.
[50] T. Ding (2003), *An Example of a Minimal Torus of Weak Mixing for Nearly Integrable Hamiltonian Systems*, Nonlinearity, Vol.**16**, 507-519.
[51] T. Ding (2004), *Qualitative Method of Ordinary Differential Equations and Its Applications*, Press of Higher Education, Beijing.
[52] T. Ding (2006), *Non-Existence of Strong Mixing Flow on the Minimal Torus*, Preprint.
[53] T. Ding, H. Huang and F. Zanolin (1995), *A Priori Bounds and Periodic Solutions for a Class of Planar Systems with Applications to Lotka-Volterra Equations*, Discrete and Continuous Dynamical Systems, Vol.**1**, 103–117.
[54] T. Ding, R. Iannacci and F. Zanolin (1991), *On Periodic Solutions of Sublinear Duffing Equations*, J. Math. Anal. Appl., Vol.**158**, 316–332.

[55] T. Ding, R. Iannacci and F. Zanolin (1993), *Existence and Multiplicity Results for Periodic Solutions of Semi-Linear Duffing Equations*, J. of Differential Equations, Vol.**105**, 364–409.

[56] T. Ding and C. Li (1991), *A Course of Ordinary Differential Equations*, Beijing: Press of Higher Education; (Taiwa: Fan-Yi Press, 1995).

[57] T. Ding and B. Liu (1996), *Periodic and Quasi-Periodic Solutions of Second Order Duffing's Equations*, Lecture Notes in Pure and Applied Math: Differential Equations and Control Theory (Ed. Deng and etc.), 35–44, (1996).

[58] T. Ding and W. Ding (1985), *Periodic Solutions of Duffing's Equations at Resonance*, Chin. Ann. of Math. (Series B), Vol.**6**, 427–432.

[59] T. Ding and F. Zanolin (1991), *Time-Maps for the Solvability of Periodically Perturbed Nonlinear Duffing Equations*, Nonlinear Analysis, (TMA), Vol.**17**, 635–653.

[60] T. Ding and F. Zanolin (1992), *Periodic Solutions of Duffing's Equations with Super-Quadratic Potential*, J. of Differential Equations, Vol.**97**, 326–378.

[61] T. Ding and F. Zanolin (1993), *Harmonic Solutions and Subharmonic Solutions for Periodic Lotka-Volterra Systems*, In: Proceedings of the Special Program at Nankai Institute of Mathematics: *Dynamical Systems* (Ed. S. Liao, Y. Ye and T. Ding), pp.55–65, World Scientific.

[62] T. Ding and F. Zanolin (1993), *Subharmonic Solutions of Second Order Nonlinear Equations: A Time- Map Approach*, Nonlinear Analysis (TMA), Vol.**20**, 509–532.

[63] W. Ding (1982), *Fixed-Points of Twist Mapping and Periodic Solutions of Ordinary Differential Equation*, Acta Mathematical Sinica, Vol.**25**, 227–235.

[64] W. Ding (1983), *A Generalization of the Poincaré-Birkhoff Theorem*, Proc. Amer. Soc., Vol.**88**, 341–346.

[65] J. Franks (1988), *Generalization of the Poincaré-Birkhoff Theorem*, Ann. of Math., Vol.**128**, 139–151.

[66] B. Füller (1952), *Note on Trajectories in a Solid Torus*, Ann. of Math., Vol.**56**, 438–439.

[67] S. Fučik and V. Lovicar (1975), *Periodic Solution of the Equation:* $\ddot{x} + g(x) = p(t)$, Casopis Pest. Mat., Vol.**100**, 160-175.

[68] C. Gardner (1976), *Another Elementary Proof of Peano's Existence Theorem*, Amer. Math. Monthly, Vol.**83**, 556–559.

[69] W. Gao (1989), *Global Semi-Stability of a Planar System of Third-Degree Differential Equations*, Acta of Mathematica, Vol.**32**, 35–41.

[70] W. Gear (1971), *Numerical initial-value problems in Ordinary Differential Equations*, Englewood Cliffs. N. J. Prentice-Hall.

[71] A. Harvey (1963), *Periodic Solutions of the Differential Equation:* $\ddot{x} + g(x) = p(t)$, Contributions to Differential Equations, Vol.**1**, 425–451, New York: Wiley.

[72] P. Hartman (1982), *Ordinary Differential Equations*. Second Edition. Boston-Basel-Stuttgart: Birkhäuser.

Bibliography

[73] A. Hausrath and R. Manasevich (1991), *Periodic Solutions of a Periodically Perturbed Lotka-Voltera Equation Using the Poincaré-Birkhoff Theorem*, Journal of Math. Anal. and Appl. Vol.**157**, 1-9.

[74] M. Hirsch (1976), *Differential Topology*, Springer-Verlag.

[75] M. Hirsch, S. Smale, R. Devaney (2004), *Differential Equations, Differential Systems & An Introduction to Chaos* (Second Edition), Elsevier (USA).

[76] L. Graves (1956), *The Theory of Functions of Real Variables*, McGraw-Hill Book Company, INC. New York.

[77] J. Hale (1963), *Oscillations in Nonlinear Oscillations*, McGraw-Hill Book Company, INC. New York.

[78] E. Ince (1956), *Ordinary Differential Equations*, New York: Dover.

[79] R. Ingraham (1992), *A Survey of Nonlinear Dynamics ("Chaos Theory")*, World Scientific, Singapore.

[80] A. Kochergin (1972), *The Absence of Mixing in Special Flows over a Rotation of the Circle and in Flows on a Two-Dimensional Torus*. English Translation in Soviet Math. Dokl. **13** (1972), 949–952.

[81] A. Kochergin (1975), *On Mixing in Special Flows over a Shifting of Segments and in Smooth Flows on Surfaces*. English Translation in Math. USSR-Sb. **25** (1976).

[82] A. Kochergin (2004), *Hölder Time Change and the Mixing Rate in a Flow on a Two-Dimensional Torus*, (Russian). Translation in Proc. Steklov Inst. Math. no. 1 (**244**), 201–232.

[83] G. Kuperberg and K. Kuperberg (1996), *Generalized Counterexample to the Seifert Conjecture*, Annals of Mathematics, Vol. **144**, 239-268.

[84] A. Lavrentief (1925), *Sur une Equation Differentielle du Premier Order*, Math. Zeitschr, Vol.**23**, 197–209.

[85] A. Lazer and D. Leach (1969), *Bounded Perturbations of Forced Harmonic Oscillations at Resonance*, Ann. Mat. Pura Appl. Vol.**82**, 49–68.

[86] S. Lefschetz (1957), *Differential Equations: Geometrical Theory*, Interscience Publishers, INC., New York.

[87] D. Leach (1970), *On Poincaré's Perturbations Theorem and a Theorem of W. S. Loud*, J. Differential Equations, Vol.**7**, 34–53.

[88] N. Levinson (1943), *On the Existence of Periodic Solutions for Second Order Differential Equations with a Forcing Term*, Journal of Math. and Phys., Vol.**22**, 41–48.

[89] W. Li (1986), *Some Results on the Periodic Solutions and Boundary-Value Problems of Duffing Equations* (in chinese), Ma D - Thesis, Peking Univ.

[90] W. Li (1990), *A Necessary and Sufficient Condition on the Existence and Uniqueness of 2π-Periodic Solution of Duffing Equation*, Chin. Ann. of Math., Vol.**11**, 342–345.

[91] J. Littlewood (1968), *Some Problems in Real and Complex Analysis*, Health, Lexington, Mass.

[92] W. Loud (1967), *Periodic Solutions of Nonlinear Differential Equations of Duffing Type*, in "Differential and Functional Equations" (W. Harris and Y. Sibuya, Eds.), pp. 199–244, Benjamin, New York.

[93] J. Massera (1950), *The Existence of Periodic Solutions of Systems of Differential Equations*, Duke Math. J., Vol.**17**, 457–475.
[94] J. Mahwin and M. Willem (1989), *Critical Point Theory and Hamiltonian Systems*, New York: Springer-Verlag.
[95] M. Morse (1946), *George David Birkhoff and His Mathematical Works*, Bulletin Amer. Math. Soc., Vol. **52**, 5:357–391.
[96] M. Morse (1973), *A Short Biography of George David Birkhoff*, Dictionary of Scientific Biography, **II**, 143–146, Charles Scribner's Sons, New York.
[97] J. Moser (1973), *Stable and Random Motions in Dynamical Systems*, Princeton University Press, Princeton and Oxford.
[98] J. Moser and E. Zehnder (2005), *Notes on Dynamical Systems*, Courant Lectures Notes in Mathematics 12, AMS.
[99] V. Nemytskii (1949), *Topological Problems of the Theory of Dynamical Systems*, Uspehi Mat. Nauk. (N.S.), Vol. **4**, no. 6: 91-153.
[100] V. Nemytskii and V, Stepanov (1989), *Qualitative Theory of Differential Equations*, Dover, New York; (Press of Science, Beijing, in Chinese, 1959).
[101] Z. Opial (1960), *Sur les Solutions Periodique del Equation Differentielle: $\ddot{x} + g(x) = p(t)$*, Bull. Acad. Pol. Sci. Math. Astr. Phy., Vol.**8**, 151–156.
[102] L. Piccinini et al. (1984), *Ordinary Differential Equations in $\mathbb{R}^n$*, Applied Math. Sci.(Vol. **39**), Springer-Verlag.
[103] H. Poincaré (1912), *Sur un théorème de géométrie*, Rend. Cir. Mat. Palermo **33**, 375-407.
[104] Y. Qin (1984), *Integral Curves defined by Differential Equations*, Vols **1** and **2** (in Chinese). Science Publisher.
[105] C. Rebelo and F. Zanolin (1996), *Twist Conditions and Periodic Solutions of Differential Equations*, Proceedings of Dynamical Systems and Applications, Vol.**2**, 469–476.
[106] R. Reissig (1975), *Contraction Mappings and Periodically Perturbed Nonconservative Systems*, Atti Accad. Naz. Lincei Rend. Cl. Sci. Fis. Mat. Natur., Vol.**58**, 696–702.
[107] W. Robbin (1968), *On the Existence Theorem for Differential Equations*, Proc. Amer. Math. Soc., Vol.**19**, 1005–1106.
[108] C. Robinson (1999), *Dynamical Systems: Stability, Symbolic Dynamics, and Chaos*, CRC Press, London.
[109] G. Sansone (1949), *Ordinary Differential Equations* (second edition translated in Russian in 1954), University of Bologna.
[110] P. Schweitzer (1974), *Counterexamples to the Seifert Conjecture and Opening Closed Leaves of Foliationd*, Ann. of Math., **100**, 386-400.
[111] L. Shekhter (1986), *A Boundary Value Problem of Periodic Type for a Nonlinear Second-Order Ordinary Differential Equation*, (Translation of Differential'nye Uranvneniya), Vol.**22**, 1080–1084.
[112] M. Struwe (1980), *Multiple Solutions of Anticoercive Boundary Value Problems for a Class of Ordinary Differential Equation of Second Order*, Journal of Diff. Eqs., Vol.**37**, 285–295.
[113] P. Schweitzer (1974), *Counterexamples to the Seifert Conjecture and Opening Closed Leaves of Foliationd*, Ann. of Math., **100**, 386-400.

[114] F. Verhult (1996), *Nonlinear Differential Equations and Dynamical Systems* (second, revised and expanded edition), Springer-Verlag Berlin Heidelberg (Printed in Germany).
[115] J. de Vries (1993), *Elements of Topological Dynamics*, Mathematics and Its Applications, Vol.**257**, Dorchrecht: Kluwer.
[116] J. Walvogel (1986), *The Period in the Lotka-Volterra System Is Monotone*, J. Math. Anal. Appl., Vol.**114**, 178–184.
[117] D. Wang (1984), *On the Existence of 2π-Periodic Solution of $\ddot{x} + g(x) = p(t)$*, Chin. Ann. of Math., Vol.**5**(Series A), 61–72.
[118] S. Wiggins (1990), *Introduction to Applied Nonlinear Dynamical Systems and Chaos*, Springer-Verlag Berlin Heidelberg (Printed in the United States of America).
[119] Y. Ye (1984), *Theory of Limit Cycles* (in Chinese), Shanghai Science and Technology Publisher.
[120] Ye and Wang (1978), *A Nonlinear Differential Equation in the Electronic Focusing Theory* (in Chinese), Acta of Applied Mathematics, Vol.**1**, 13-41.
[121] T. Yi and L. Huang (2007), *A Generalization of the Bernfeld-Haddock Conjecture and Its Proof*, Math. Sinica (in Chinese), Vol.**50**, 261-270.
[122] M. Zhang (1996), *Periodic Solutions of Liénard Equations with Singular Forces of Repulsive Type*, J. Math. Anal. and Appl., Vol.**203**, 254-269.
[123] Z. Zhang et al (1997), *Qualitative Theory of Differential Equations* (in Chinese: the first edition was published in 1985), Beijing: Press of Science; (Translations of Mathematical Monographs, Volume 101, American Mathematical Society, Providence, 1992).